Studies in Computational Intelligence

Volume 657

Series editor

Janusz Kacprzyk, Polish Academy of Sciences, Warsaw, Poland
e-mail: kacprzyk@ibspan.waw.pl

About this Series

The series "Studies in Computational Intelligence" (SCI) publishes new developments and advances in the various areas of computational intelligence—quickly and with a high quality. The intent is to cover the theory, applications, and design methods of computational intelligence, as embedded in the fields of engineering, computer science, physics and life sciences, as well as the methodologies behind them. The series contains monographs, lecture notes and edited volumes in computational intelligence spanning the areas of neural networks, connectionist systems, genetic algorithms, evolutionary computation, artificial intelligence, cellular automata, self-organizing systems, soft computing, fuzzy systems, and hybrid intelligent systems. Of particular value to both the contributors and the readership are the short publication timeframe and the worldwide distribution, which enable both wide and rapid dissemination of research output.

More information about this series at http://www.springer.com/series/7092

Vassil Sgurev · Ronald R. Yager
Janusz Kacprzyk · Krassimir T. Atanassov
Editors

Recent Contributions in Intelligent Systems

Springer

Editors
Vassil Sgurev
Institute of Information and Communication Technologies
Bulgarian Academy of Sciences
Sofia
Bulgaria

Ronald R. Yager
Machine Intelligence Institute, Hagan School of Business
Iona College
New Rochelle, NY
USA

Janusz Kacprzyk
Systems Research Institute
Polish Academy of Sciences
Warsaw
Poland

Krassimir T. Atanassov
Department of Bioinformatics and Mathematical Modelling, Institute of Biophysics and Biomedical Engineering
Bulgarian Academy of Sciences
Sofia
Bulgaria

ISSN 1860-949X ISSN 1860-9503 (electronic)
Studies in Computational Intelligence
ISBN 978-3-319-82354-6 ISBN 978-3-319-41438-6 (eBook)
DOI 10.1007/978-3-319-41438-6

Softcover reprint of the hardcover 1st edition 2016

Printed on acid-free paper

This Springer imprint is published by Springer Nature
The registered company is Springer International Publishing AG Switzerland

Preface

This volume can be viewed from different perspectives. First of all, it is a result of a special project initiated by the editors, and then implemented thanks to an enthusiastic response of the contributors to the invitation to present their new ideas and solutions. The idea of the volume has been a result of, first of all, discussions of the editors between themselves and with the participants at the well-known IEEE Intelligent Systems, IEEE IS, which since 2002 have been a popular venue for the international research community interested in broadly perceived intelligent systems theory and applications. These idea of these conferences was born in Bulgaria in the beginning of the new century and the first IEEE IS conferences were held in 2002, 2004 and 2008 in Varna, Bulgaria, where both the scientific level and a very attractive venue at the Black Sea coast attracted many participants from all over the world. Due to a growing importance of the IEEE ISs, and a growing interest from the international research community, the third and fifth conferences, in 2006 and 2010, were organized in London, UK and the sixth conference, in 2012 was held in Sofia, Bulgaria.

Following the tradition that had existed since the very beginning, that is, since IEEE IS'2002, during the IEEE IS'2012 the International Program Committee again decided to choose the best papers, both from the point of view of their novelty of ideas and tools, and technical content, to be included in a special volume meant as some sort of a summary of the state of the art and new trends in broadly perceived intelligent systems.

This volume has resulted from that decision and careful analyses of both the theoretical and applied contents of the papers and interests of the participants and the entire research community, with an emphasis on, on the one hand, what has been presented in the best papers at the conference, and on the other hand, with some emphasis of what has been proposed by leading Bulgarian scientists and scholars who have inspired many people around the world with new ideas and solutions.

In this short preface, we will briefly summarize the content of the consecutive papers included in the volume, emphasizing novel concepts, and ideas.

Samuel Delepoulle, André Bigand Christophe Renaud and Olivier Colot (Chapter "Low-Level Image Processing Based on Interval-Valued Fuzzy Sets and Scale-Space Smoothing") present a new approach for image analysis and restoration based on interval-valued fuzzy sets and scale-space smoothing. To show the effectiveness and efficiency of their solution, two specific and significant image processing applications are considered: no-reference quality evaluation of computer-generated images and speckle noise filtering.

In his paper Dimitar G. Dimitrov (Chapter "Generalized Net Representation of Dataflow Process Networks") presents translation rules for mapping from a given dataflow process network to a generalized net which is a novel, highly effective and efficient model of, among others, discrete event processes and systems.

Stefka Fidanova, Miroslav Shindarov and Pencho Marinov (Chapter "Wireless Sensor Positioning Using ACO Algorithm") deal with spatially distributed sensors which communicate wirelessly and form a wireless sensor network. The minimization of the number of sensors and energy consumption by the network is then performed using an Ant Colony Optimization (ACO) algorithm.

In the paper by Petia Georgieva, Luis Alberto Paz Suárez and Sebastião Feyo de Azevedo (Chapter "Time Accounting Artificial Neural Networks for Biochemical Process Models") the problem of developing more efficient computational schemes for the modeling of biochemical processes is discussed. A theoretical framework for the estimation of process kinetic rates based on different temporal (time accounting) Artificial Neural Network architectures is introduced.

Tomohiro Hara, Tielong Shen, Yasuhiko Mutoh and Yinhua Liu (Chapter "Periodic Time-Varying Observer-Based Learning Control of A/F Ratio in Multi-cylinder IC Engines") present an air-fuel ratio control scheme via individual fuel injection for multi-cylinder internal combustion (IC) engines. Their concern is to improve the air-fuel ratio precision by a real-time compensation of the unknown off-set in the fuel path of the individual cylinder, which represents the effect of the cylinder-to-cylinder imbalance caused by the perturbations in each injector gain or disturbances in the dynamics of fuel injection path.

Tatjana Kolemishevska-Gugulovska, Mile Stankovski, Imre J. Rudas, Nan Jiang and Juanwei Jing (Chapter "Fuzzy T–S Model-Based Design of Min–Max Control for Uncertain Nonlinear Systems") present an approach to robust control synthesis for uncertain nonlinear systems through the use of the Takagi–Sugeno fuzzy model and fuzzy state observer. The existence conditions the output feedback min-max control in the sense of Lyapunov asymptotic stability are derived, and a convex optimization algorithm is used to obtain the minimum upper bound on the performance and the optimum parameters of mini-max controller. An example of an inverted pendulum is shown and the results are promising.

A novel application of a generalized net is described in the paper by Maciej Krawczak, Sotir Sotirov and Evdokia Sotirova (Chapter "Modeling Parallel Optimization of the Early Stopping Method of Multilayer Perceptron") for the parallel optimization of the multilayer perception (MLP) based on an early stopping algorithm.

Paper Jinming Luo and Georgi M. Dimirovski (Chapter "Intelligent Controls for Switched Fuzzy Systems: Synthesis via Nonstandard Lyapunov Functions") investigate the synthesis of intelligent control algorithms for switched fuzzy systems by employing non-standard Lyapunov functions and some combined, hybrid techniques. The control plants are assumed to be nonlinear and to be represented by some specific Takagi–Sugeno fuzzy models.

The latest advances in the field of switching adaptive control based on hybrid multiple Takagi–Sugeno (T–S) models are presented in paper by Nikolaos A. Sofianos and Yiannis S. Boutalis (Chapter "A New Architecture for an Adaptive Switching Controller Based on Hybrid Multiple T-S Models").

Ketty Peeva (Chapter "Optimization of Linear Objective Function Under *min* – Probabilistic Sum Fuzzy Linear Equations Constraint") presents a method for the solution of a linear optimization problem when the cost function is subject to the constraints given as fuzzy linear systems of equations.

Tania Pencheva and Maria Angelova (Chapter "Intuitionistic Fuzzy Logic Implementation to Assess Purposeful Model Parameters Genesis") are concerned with the derivation of intuitionistic fuzzy estimations of model parameters of the process of yeast fed-batch cultivation. Two kinds of simple genetic algorithms with the operator sequence selection-crossover-mutation and mutation-crossover-selection are considered, and both applied for the purposes of parameter identification of *S. cerevisiae* fed-batch cultivation.

Patrick Person, Thierry Galinho, Hadhoum Boukachour, Florence Lecroq and Jean Grieu (Chapter "Dynamic Representation and Interpretation in a Multiagent 3D Tutoring System") present an intelligent tutoring system aimed at decreasing the students' dropout rate by offering a possibility of a personalized follow up. An architecture of an intelligent tutoring system is described and the experimental results of the decision support system used as the core of the intelligent tutor are given.

The dynamics of the upper extremity is modeled in Simeon Ribagin, Vihren Chakarov and Krassimir Atanassov (Chapter "Generalized Net Model of the Scapulohumeral Rhythm") as the motion of an open kinematic chain of rigid links, attached relatively loosely to the trunk.

A method for the interpretation of propositional binary logic functions that allows the logical concepts 'true' and 'false' to be treated as stochastic variables is described in Vassil Sgurev and Vladimir Jotsov (Chapter "Method for Interpretation of Functions of Propositional Logic by Specific Binary Markov Processes"). Examples are presented and a numerical realization is done by using some functions of propositional logic by binary Markov processes.

Shannon, A.G., B. Riecan, E. Sotirova, K. Atanassov, M. Krawczak, P. Melo-Pinto, R. Parvathi and T. Kim (Chapter "Generalized Net Models of Academic Promotion and Doctoral Candidature") propose a new generalized net based model for the analysis of the process of academic promotion through the hierarchy in higher education and the preparation of PhD candidates.

The generalized net model, described in paper "Maria Stefanova-Pavlova, Velin Andonov, Todor Stoyanov, Maia Angelova, Glenda Cook, Barbara Klein, Peter

Vassilev and Elissaveta Stefanova's paper (Chapter "Modeling Telehealth Services with Generalized Nets"), presents the processes related to the tracking of changes in health status (diabetes) of adult patients. The progress in telecommunications and navigation technologies allow this model to be extended to the case of active and mobile patient.

Yancho Todorov, Margarita Terziyska and Michail Petrov (Chapter "State-Space Fuzzy-Neural Predictive Control") give a novel view of potentials of the state–space predictive control methodology based on a fuzzy-neural modeling technique and different optimization procedures for process control. The proposed controller methodologies are based on the Fuzzy-Neural State-Space Hammerstein model and variants of Quadratic Programming optimization algorithms.

Vesela Vasileva and Kalin Penev (Chapter "Free Search and Particle Swarm Optimisation Applied to Global Optimisation Numerical Tests from Two to Hundred Dimensions) investigate two methods of global optimization, Free Search (FS) and Particle Swarm Optimisation (PSO), and show results of some numerical tests on difficult examples. The objective is to identify how to facilitate the evaluation of effectiveness and efficiency of heuristic, evolutionary, adaptive, and other optimisation and search algorithms.

Peter Vassilev (Chapter "Intuitionistic Fuzzy Sets Generated by Archimedean Metrics and Ultrametrics") investigates a general metric approach for the generation of intuitionistic fuzzy sets, notably the cases when the generation is done by a norm on R^2 and a field norm on Q^2.

Boriana Vatchova and Alexander Gegov (Chapter "Production Rule and Network Structure Models for Knowledge Extraction from Complex Processes Under Uncertainty") consider processes with many inputs, some of which are measurable, and many outputs from different application areas, and in which uncertainty plays a key role.

We wish to thank all the contributors to this volume. We hope that their papers, which constitute a synergistic combination of foundational and application oriented works, including relevant real world implementations, will be interesting and useful for a large audience interested in broadly perceived intelligent systems.

We also wish to thank Dr. Tom Ditzinger, Dr. Leontina di Cecco, and Mr. Holger Schaepe from Springer for their dedication and help to implement and finish this publication project on time maintaining the highest publication standards.

Sofia, Bulgaria — Vassil Sgurev
New Rochelle, USA — Ronald R. Yager
Warsaw, Poland — Janusz Kacprzyk
Sofia, Bulgaria — Krassimir T. Atanassov
March 2015

Contents

Low-Level Image Processing Based on Interval-Valued Fuzzy Sets and Scale-Space Smoothing

Samuel Delepoulle, André Bigand, Christophe Renaud and Olivier Colot

Abstract In this paper, a new technique based on interval-valued fuzzy sets and scale-space smoothing is proposed for image analysis and restoration. Interval-valued fuzzy sets (IVFS) are associated with type-2 semantic uncertainty that makes it possible to take into account usually ignored (or difficult to manage) stochastic errors during image acquisition. Indeed, the length of the interval (of IVFS) provides a new tool to define a particular resolution scale for scale-space smoothing. This resolution scale is constructed from two smoothed image histograms and is associated with interval-valued fuzzy entropy (IVF entropy). Then, IVF entropy is used for analyzing the image histogram to find the noisy pixels of images and to define an efficient image quality metric. To show the effectiveness of this new technique, we investigate two specific and significant image processing applications: no-reference quality evaluation of computer-generated images and speckle noise filtering.

1 Introduction

Low-level image processing is very important to provide good quality images to further stages of digital image processing. Low-level image processing is based on *image acquisition*, which is viewed as a composition of blurring, ideal sampling and added noise [1]. In this model, (ideal) intensity distribution is first affected by aberrations in the optics of the real camera (blurring). This process attenuates high-

O. Colot
LAGIS-UMR CNRS 8219, Université Lille 1, 59655 Villeneuve d'Ascq Cedex, France
e-mail: Olivier.Colot@univ-lille1.fr

S. Delepoulle (✉) · A. Bigand · C. Renaud
LISIC-ULCO, 50 rue Ferdinand Buisson - BP 699, 62228 Calais Cedex, France
e-mail: delepoulle@lisic.univ-littoral.fr

A. Bigand
e-mail: bigand@lisic.univ-littoral.fr

C. Renaud
e-mail: renaud@lisic.univ-littoral.fr

V. Sgurev et al. (eds.), *Recent Contributions in Intelligent Systems*,
Studies in Computational Intelligence 657,
DOI 10.1007/978-3-319-41438-6_1

frequency components in the image. Then in the case of blurred images issued from video cameras (CCD array images), the image acquisition process is classically modeled using two important hypotheses.

First, each pixel intensity is considered as the weighted mean of the intensity distribution in a window around the ideal pixel position (ideal sampling, that is to say the image intensity distribution is mapped onto a discrete intensity distribution, or histogram h in the sequel). The second classic hypothesis is about added noise that is classically assumed to be an additive stationary random field (quantization effects are ignored). These assumptions are often set as prior knowledge in many vision models. In two special cases, we will try to estimate the blurred, distorted intensity distribution h using a new technique.

The first hypothesis is not verified in the case of **image synthesis** using global illumination methods. The main goal of global illumination methods is to produce *computer-generated images with photo-realistic quality*. For this purpose, photon propagation and light interaction with matter have to be accurately simulated. Stochastic methods were proposed for more than 20 years in order to reach these goals. They are generally based on the Path Tracing method proposed by Kajiya [2] where stochastic paths are generated from the camera point of view towards the $3D$ scene. Because paths are randomly chosen, the light gathering can greatly change from one path to another generating high-frequency intensity variations through the image [3]. The Monte Carlo theory however ensures that this process will converge to the correct image when the number of samples (the paths) grows. But no information is available about the number of samples that are really required for the image being considered as *visually* satisfactory. The human visual system (HVS) is endowed with powerful performances, and most HVS models provide interesting results but are complex and still incomplete due to the internal system complexity and its partial knowledge. So, objective metrics have been developed: full-reference metrics (using a reference image and PSNR or SSIM quality index [4]) and no-reference metrics (with no-reference image available). Image quality is governed by a variety of factors such as sharpness, naturalness, contrast, noise, etc. To develop a no-reference objective image quality metric by incorporating all attributes of images without referring to the original ones is a difficult task. Hence, we shall concentrate on the work of the no-reference image sharpness metric (or blurriness metric which is inversely related to sharpness metric). In [5], a general review of no-reference objective image sharpness/blurriness metrics is given. They generally require relatively long computation times and are often difficult to use and to compute. So, in this paper a novel no-reference image content metric is proposed, based on type-2 fuzzy sets entropy (to model uncertainty brought by blur and noise affecting the image computer generation). This metric does not require any prior knowledge about the test image or noise. Its value increases monotonically either when image becomes blurred or noisy. So it may be used to detect both blur and noise.

Speckle is the term used for granular patterns that appear on some types of images (in particular ultrasonic images), and it can be considered as a kind of multiplicative noise. Speckle degrades the quality of images and hence it reduces the ability of human observer to discriminate fine details. Ordinary filters, such as mean or median filters are not very effective for edge preserving smoothing of images corrupted with speckle noise. So particular filter, like Frost filter, was developed [6]. This filter assumes multiplicative noise and stationary noise statistics. So, in a more general case, we propose to use type-2 fuzzy sets in he same way than for computer-generated images. We want to demonstrate the general ability of type-2 fuzzy sets entropy to quantify noise level for each stage of image generation (or acquisition) that is to say blurring, stochastic errors at the pixel level, and added multiplicative noise.

Type-1 fuzzy sets (or FS in the sequel) are now currently used in image processing [7–9], since greyscale images and fuzzy sets are modeled in the same way, [10, 11]. The major concern of these techniques is that spatial ambiguity among pixels (imprecision about ideal pixel position) has inherent vagueness rather than randomness. However, some sources of uncertainties are not managed using FS [12]: the meaning of the words that are used, measurements may be noisy, the data used to tune the parameters of FS may also be noisy. Imprecision and uncertainty are naturally present in image processing [13], and particularly these three kinds of uncertainty. The concept of type-2 fuzzy set was introduced first by Zadeh [14] as an extension of the concept of FS. Mendel [12, 15, 16] has shown that type-2 fuzzy sets (or FS2) may be applied to take into account these three kinds of uncertainty. Type-2 fuzzy sets have membership degrees that are themselves fuzzy. Hence, the membership function of a type-2 fuzzy set is three dimensional, and it is the new third dimension that provides new design degrees of freedom for handling uncertainties.

In this paper, we consider the special case of interval type-2 fuzzy set (or interval-valued fuzzy sets) for image pre-processing. An interval valued-fuzzy set [14] is defined by interval-valued membership functions; each element of an IVFS (related to an universe of discourse *X*) is associated with not just a membership degree but also the length of its membership interval that makes it possible IVFSs to deal with other dimensions of uncertainty. One way to model uncertainty in the pixel values of a greyscale image is to use fuzzy mathematical morphology (FMM) [10]. FMM has also been extended to interval-valued FMM ([11]) and makes it possible IVFSs generation. Melange et al. [11] investigated with success the construction of IVFS for image processing using these tools. Sussner et al. presented an approach towards edge detection based on an interval-valued morphological gradient [17], and obtained good results applying this method to medical images processing. The interval length of IVFS can also be used to take into account the dispersion of the probability of occurrences associated to each pixel grey level of an image and due to noise. In a previous paper [18], we have shown that interval-valued fuzzy entropy is a particularly adapted tool to detect relevant information in noisy images, and then impulse noise removal from images becomes possible. We have also shown [19] that this method is efficient to remove speckle noise and to define a no-reference computer-generated images quality metric [20] (and its application to denoising). According to these interesting results, it seems important to link **IVFSs and stochastic analysis of image acquisition (or image generation)** using a scale-space smoothing frame, to provide a generic low-level image pre-processing technique. Indeed, a classic way to model uncertainty in the pixel values of a greyscale image is to use information brought by image histograms, that allows the construction of IVFSs from extension of **scale-space theory**. IVFS is constructed from the association of two non-additive Gaussian kernels (the upper and lower bounds of membership interval are obtained from two smoothed histograms). In this paper we focus on IVFSs construction for the special case of stochastic image analysis and filtering. The paper is organized as follows: Sect. 2 describes some preliminary definitions about scale-space smoothing and IVFSs; Sect. 3 introduces the design of the IVFS image quality evaluation; Sect. 4 presents the design of the IVFS speckle noise detection and filtering. Conclusion and potential future work are considered in Sect. 5.

2 IVFS and Scale-Space Smoothing

2.1 Scale-Space Smoothing

Scale-space smoothing is an interesting framework for stochastic image analysis [21]. Smoothed histograms are classically generated using kernel convolution, and we propose to extend this technique to interval-valued fuzzy membership function generation for image processing.

A kernel is a [0, 1]-valued function K defined on a domain X verifying the summative normalization property

$$\int_{x \in X} K(x)dx = 1. \tag{1}$$

Since Babaud et al. [22] proved that Gaussian kernel is the only linear filter that gives a consistent scale-space theory, we only consider this kind of kernel. As well as classic kernels K (normal, Epanechnikov, ... kernels), Gaussian (FS) fuzzy numbers may be considered as non-additive kernels. Like an additive kernel, a non-additive kernel (or maxitive kernel as defined by Rico et al. [23]) is a [0, 1]-valued function π defined on a domain X verifying the maxitive normalization property

$$Sup_{x \in X} \pi(x) = 1. \tag{2}$$

where $\pi(x)$ can be seen as a possibility distribution or as the membership function of a normalized FS of X. Jacquey et al. [24] successfully used this technique to define *maxitive filters* applied on noisy images, as well as Strauss et al. [23] for histogram smoothing. Then we consider a (Gaussian) membership function $\mu(x; g, \sigma)$ seen as non-additive kernel for the sequel, ($\mu(x; g, \sigma)$ is represented in Fig. 1), centered on g and defined using a free constant parameter σ, considered as the scale value in the following ($x \in [0, G-1]$):

$$\mu(x; g, \sigma) = exp\left[-\frac{1}{2}\left(\frac{x-g}{\sigma}\right)^2\right]. \tag{3}$$

Let "G" measured data points be $(g, h(g))$, with g the image pixels grey level, $(g = 0, \ldots, G-1)$ and *h(g)* the image histogram value at the gth location. Averaging or scale-space smoothing is used to reduce the effects of noise. The main idea of this approach is to map the discrete histogram h to a continuous one $H(x)$ (interpolation method). $H(x)$ is called the scale-space representation of h, at scale σ and is given by the following equation:

$$H(x) = \sum_{g=0}^{G-1} \left[K(x-g).h(g)\right] \tag{4}$$

with $x \in X, x = 0, \ldots, G-1$. The scale-space theory has several advantages, and particularly it works for all scales. The effect of additive noise can easily be estimated. It makes it possible to compare the real intensity distribution with the interpolated distribution. A classic and well-known application of this theory is sub-pixel edge detection [25], so this framework is a good candidate to take into account the stochastic errors involved in computer-generated images and for multiplicative noise. Nevertheless, there remains a huge drawback about the choice of scales (and particularly for small-scale parameters). So we propose a new strategy to remedy this drawback. IVFSs, applied to digital images, can be viewed as a special kind of resolution scale constructed from differencing two smoothed image histograms.

2.2 *Interval-Valued Fuzzy Sets*

Ordinary fuzzy sets (precise fuzzy sets) used in image processing are often fuzzy numbers. A fuzzy set defines the meaning representation of the vagueness associated with a linguistic variable A in a natural language. However, it is not possible to say which membership function is the best one. The major motivation of this work is to remove the uncertainty of membership values by using interval-valued fuzzy sets (and so incorporate uncertainty carried by the grey values of pixels of an image).

Let X be the universe of discourse. An ordinary fuzzy set A of a set X is classically defined by its membership function $\mu_A(x)$ (with $x \in X$) written as

$$\mu_A \quad : X \rightarrow [0, 1] \tag{5}$$

Let $S([0, 1])$ denote the set of all closed subintervals of the interval $[0, 1]$. An interval-valued fuzzy set (IVFS) A in a non-empty and crisp universe of discourse X is a set such that [14, 26]

$$A = \{\left(x, M_A(x) = \left[\mu_{AL}(x), \mu_{AU}(x)\right]\right) \mid x \in X\}. \tag{6}$$

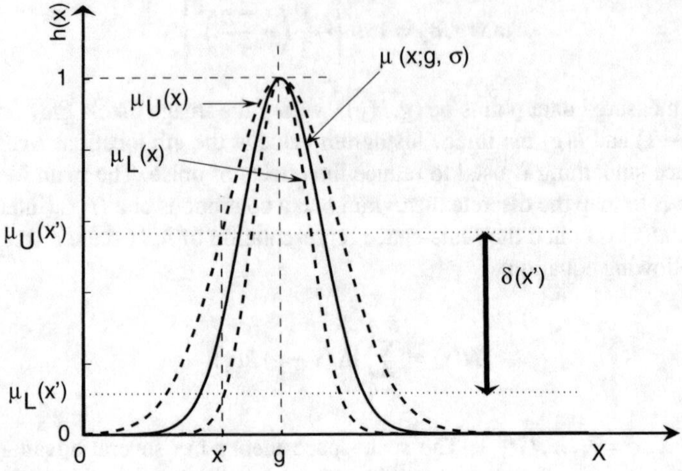

Fig. 1 Interval-valued fuzzy set

The membership function (MF) M_A defines the degree of membership of an element x to A as follows:

$$M_A : X \longrightarrow S([0,1]) \tag{7}$$

For each IVFS A, we denote by $\delta_A(x)$ the amplitude of the considered interval ($\delta_A(x) = \mu_{AU}(x) - \mu_{AL}(x)$). So non-specific evidence (an interval of membership values) for x belonging to a linguistic value A is identified by IVFS. Figure 1 presents (Gaussian) membership functions of an IVFS in X, and particularly illustrates interval $\delta(x')$ associated with an element $x' \in X$. For each element $x \in X$ of the IVFS, the imprecision of the FS is defined by closed intervals delimited by the upper membership function $\mu_U(x)$ and the lower membership function $\mu_L(x)$. These membership functions $\mu_L(x)$ and $\mu_U(x)$ are two FS membership functions, which fulfill the following condition:

$$0 \leq \mu_L(x) \leq \mu_U(x) \leq 1. \tag{8}$$

2.3 *Multiscale Image Analysis Using IVFSs*

According to multiscale technique, a signal can be decomposed into components of different scales. In that way, many 2D wavelet transforms algorithms [27] have been introduced. All these methods have advantages and drawbacks. Following the content of the data, and the nature of the noise, each of these models can be considered as optimal. So, we propose to use IVFS to build a generic model using a resolution scale interval instead of a multiresolution decomposition. A resolution scale interval is constructed from differencing two smoothed image histograms, one smoothed image histogram at scale σ_1, the other smoothed image histogram at scale σ_2 (σ_1 and σ_2 are defined using σ and α in the following membership functions $\mu_U(x)$ and $\mu_L(x)$). An important case of partial information about a random variable *h(g)* is when we know that *h(g)* is within a given interval $h_U(g) - h_L(g)$ (with probability equal to 1), but we have no information about the probability distribution within this interval. Thus, IVFS is defined with the following membership functions $\mu_U(x)$ and $\mu_L(x)$:

- upper limit: $\mu_U(x) : \mu_U(x) = [\mu(x; g, \sigma)]^{1/\alpha}$, (with α to be determined),
- lower limit: $\mu_L(x) : \mu_L(x) = [\mu(x; g, \sigma)]^{\alpha}$

So, it is natural to use IVFSs to take into account uncertainty regarding the measured grey levels. In this model, a pixel in the image domain is not longer mapped onto one specific occurrence $h(g)$ (associated to grey value g), but onto an interval of occurrences to which the uncertain value $h(g)$ is expected to belong. The previous smoothing technique is refined to construct IVFSs from histogram smoothing as we present now.

First, we generate two smoothed histograms H_U and H_L by performing Gaussian convolution with scales σ_1 and σ_2 on the initial histogram, where σ_1 is smaller than σ_2. The Gaussian kernel was previously presented and H_U and H_L are computed using the functions $\mu_U(x)$ and $\mu_L(x)$

$$
\begin{aligned}
H_U &= \sum_{g=0}^{G-1} \left[h(g).\left(\mu_U(x)\right)\right] = \sum_{g=0}^{G-1} \left[h(g).\left([\mu(x; g, \sigma)]^{\alpha}\right)\right] \\
H_L &= \sum_{g=0}^{G-1} \left[h(g).\left(\mu_L(x)\right)\right] = \sum_{g=0}^{G-1} \left[h(g).\left([\mu(x; g, \sigma)]^{1/\alpha}\right)\right]
\end{aligned}
\tag{9}
$$

2.4 Interval-Valued Fuzzy Entropy

As mentioned before, one way to model uncertainty in the pixel values of a greyscale image is to use information brought by image histograms. Image histograms are probability distributions, and information is typically extracted using information measures, and particularly entropy. The terms *fuzziness degree* [28] and *entropy* [29] provide the measurement of fuzziness in a set and are used to define the vagueness degree of the process. These well-known concepts have been developed in a previous paper [18]. Particularly, the linear index of fuzziness proposed by Pal [30] reflects the average amount of ambiguity present in an image A. So, for a $M \times N$ image subset $A \subseteq X$ with G grey levels $g \in [0, G-1]$, the histogram $h(g)$, and for each previous smoothed image histogram, we can define the linear indices of fuzziness

$$\gamma_U(A) = \frac{1}{M \times N} \sum_{g=0}^{G-1} \left[h(g).\left(\mu_U(g)\right)\right]$$
$$\gamma_L(A) = \frac{1}{M \times N} \sum_{g=0}^{G-1} \left[h(g).\left(\mu_L(g)\right)\right] \tag{10}$$

Nevertheless, the total amount of uncertainty is difficult to calculate in the case of fuzzy sets (FS), and particularly when images (represented using a FS) are corrupted with noise, so we introduced the IVFS imprecision degree (imprecision of approximation of a fuzzy set; Many authors named it interval-valued fuzzy entropy for historical reasons). Burillo [31] presented an interesting study relative to the entropy of an IVFS A in X. The defined entropy measures how interval-valued a set is with respect to another fuzzy set.

Tizhoosh [9], first **intuitively** showed that it is very easy to extend the previous concepts of FS (linear index of fuzziness proposed by Pal [30]) for IVFS, and to define (linear) index of ultrafuzziness. The construction of IVFS using histogram smoothing we propose leads to an IVF entropy $\Gamma(x)$ that fulfills the conditions required by Burillo [31] for IVFS entropy. So, it is very easy to extend the previous concepts of linear ultrafuzziness index $\gamma(x)$ proposed by Tizhoosh [9], and to define the IVF entropy $\Gamma(x)$ as follows:

$$\begin{aligned}
\Gamma(x) &= \frac{1}{M \times N} \sum_{g=0}^{G-1} \left[h(g).\left(\mu_U(x) - \mu_L(x)\right)\right] \\
&= \frac{1}{M \times N} \sum_{g=0}^{G-1} \left[h(g).\left([\mu(x; g, \sigma)]^{1/\alpha} - [\mu(x; g, \sigma)]^{\alpha}\right)\right] \\
&= \frac{1}{M \times N} \sum_{g=0}^{G-1} \left[\Delta_g(x)\right]
\end{aligned} \tag{11}$$

Tizhoosh used linear ultrafuzziness index as a tool to threshold greyscale images with success. Bustince et al. [32] proposed a very similar method to generalize Tizhoosh's work using IVFS. They introduced the concept of *ignorance function* and proposed a thresholding method based on this function (they look for the best threshold to define background and objects of an image). Using the same definition of the linear index of ultrafuzziness, we apply this performing index to greyscale images multithresholding [18]. We want to show that IVF entropy, built using scale-space smoothing, is a generic tool to characterize stochastic errors during image generation. This IVF entropy is a good candidate to define a no-reference quality metrics for images synthetic and to handle images distorted with multiplicative noise, that we will present in the next sections.

2.5 Scale Parameters Tuning

Now, let us consider the different parameters involved in the characterization of an IVFS. σ_1 and σ_2 are defined using the parameter σ (of the previous Gaussian kernel defined by $\mu(x; g, \sigma)$ and the exponent α). Tizhoosh has applied IVFS to image thresholding, where IVFS was defined with $\alpha = 2$. It is almost impossible to find these optimum values depending on the input images, and this is why an iterative scheme is adopted. In our case, from a large quantity of tests, we have observed that filtering effects were better for $\alpha > 2$. In this application, the value of α was kept constant to $\alpha = 3$, but values from 2 to 5 yield good results as well.

The method for the automatic tuning of σ is data dependent, and, consequently, sensitive to the noise present in the input image. The proposed technique adopts a multipass procedure that takes into account IVF entropy Γ. In fact, IVFS construction and image processing are carried out both together and this action is performing until IVF entropy Γ becomes maximum to conduct the best image processing (and IVFS construction). This procedure operates as follows:

- An image I corrupted by stochastic errors is assumed as input data.
- By varying the value σ of IVFS from a minimum value ($\sigma = 10$) to a maximum one ($\sigma = 125$), a collection of scale-space representations with their IVF entropy Γ is evaluated. Let $\sigma(p_1)$ the value that corresponds to the global maximum of Γ, Max(Γ) (p_i corresponds to the number of iteration).
- The resulting scale-space representation is assumed as input data in order to perform a second scale-space representation.
- Again, by varying the value of parameter σ, a collection of scale-space representations is obtained. Let $\sigma(p_2)$ the value that correspond to Max(Γ). If MAX($\Gamma(p_2)$) > MAX($\Gamma(p_1)$) proceed to the next step, otherwise stop the procedure and consider the previous resulting scale-space representation as the data output.
- The result represents the resolution scale corresponding to image I.

2.6 Noise Filtering

To complete the pre-processing scheme it is possible to filter the considered corrupted image. Let denote $I(n, m)$ the grey level value of the pixel (n, m) of a $M \times N$ noisy image. The noise detection process results in dividing the histogram into different zones, according to the maxima of IVF entropy Γ_{max}. So pixels are classified into two groups: noise-free pixels (i.e. belonging to one of the K classes), and noisy pixels. Each of the K pixels class is associated with a region R_k in the restored image. Then, median filter ($med(n, m)$) is applied to all pixels $I(n, m)$ identified as corrupted while leaving the rest of the pixels identified as noise-free. Let $med(n, m)$ be the value of pixels in a local window of size 3×3 in the surrounding of $I(n, m)$ (i.e. a windowing $(3, 3)$ around the pixel (n, m)). This method reduces the distortion generated in the restored image, introducing the necessary spatial (local) information to obtain spatially connected regions. So, the restored image $J(n, m)$ appears as follows:

$$J(n,m) = \begin{cases} I(n,m) & \text{if } I(n,m) isnoise-free \\ med(n,m) & \text{if } I(n,m) isnoisy. \end{cases} \tag{12}$$

3 The Design of the IVFS Image Quality Evaluation

We propose now to use IVF entropy to extract a new blur (and noise) level index, and thus to introduce a new no-reference image quality metrics.

3.1 *Proposed Scheme*

The blur and noise level measure scheme is divided into two steps. In the first one, we perform histogram analysis using the IVF entropy $\Gamma(x)$ applied to a block of the processed image I (I is divided into *Ki* patches (or blocks)). We use the efficient peak detection method (dominating peaks in the histogram analysis represent homogeneous regions of *I*) presented by Cheng [33] to obtain the maximum IVF entropy for each block. In the second step, we applied the denoising treatment on that block. The implementation of image quality evaluation based on IVF entropy Γ is illustrated by the following experiment. Consider an edged image patch (chessboard example, see Fig. 2) added with white Gaussian noise with different variances σ_n^2. The average value of Γ is plotted in Fig. 2 (100 simulations are carried out with independent noise realizations). Then the test patch was blurred first and added by white Gaussian noise with $\sigma_n = 0.1$, the average Γ is also plotted Fig. 2 (the edged patch is blurred by applying a Gaussian smoothing filter with a growing standard deviation σ). From this set of experiments we can see that for edged patches, the value of the metric Γ drops monotonically as the image content becomes more and more blurred and/or noisy. In other words, we experimentally demonstrate that Γ is an indicator of local signal to noise ratio.

3.2 *Algorithm*

The implementation of image quality evaluation based on IVFSs and IVF entropy Γ is given by the following algorithm (Algorithm 1).

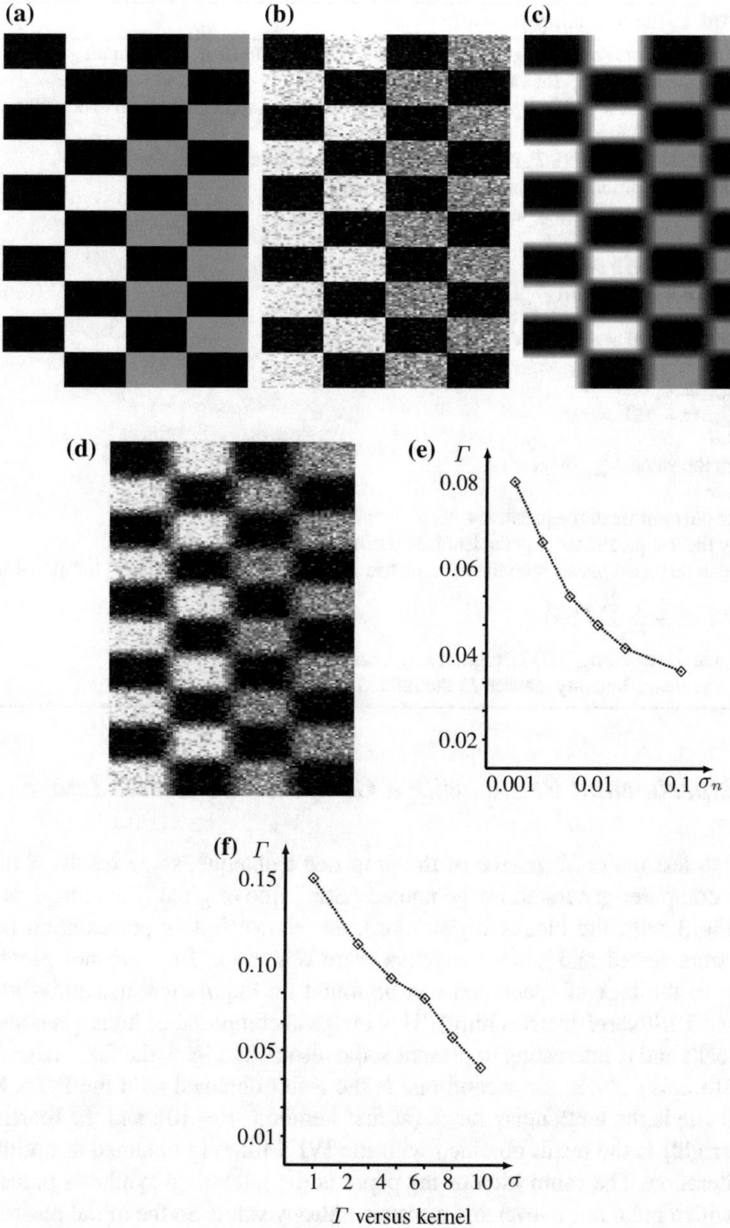

Fig. 2 Simulations using both random noise and blur for an edged patch. **a** Clean patch. **b** White Gaussian noise added to the clean patch, $\sigma_n = 0.02$. **c** Blurred clean patch. **d** Blurred clean patch and added with white Gaussian noise, $\sigma_n = 0.1$. **e** Γ versus noise. **f** Γ versus kernel

Algorithm 1 Image quality measure

Require: an input noisy $M \times N$ grey-level image I, divided into Ki non-overlapping patches and calculate the entropy Γ^k for each patch k $(0 < k < Ki)$,
for example a 512×512 image is divided into 16 non-overlapping blocks of size 128×128

Compute the k-patch image histogram $h^k(g)$
Select the shape of MF (MF, membership function, with σ approximation)
Initialize the position of the membership function
Shift the MF along the grey-level range
$\Gamma^k_{max} \leftarrow 0$
for each position g **do**
  Compute $\mu_U(g)$ and $\mu_L(g)$
  Compute $\Gamma^k(g) = \frac{1}{M \times N} \sum_{g=0}^{L-1} h^k(g) \times \left[\mu_U(g) - \mu_L(g)\right]$
  if $\Gamma^k_{max} \leq \Gamma^k(g)$ **then**
    $\Gamma^k_{max} \leftarrow \Gamma^k(g)$
  end if
  Keep the value Γ^k_{max} for patch k
end for
Iterate the number of the patch: $k + 1$
Apply the test parameter to get a denoised patch
For each denoised patch, keep the local metric $\Gamma^k(g)$. Compute the value Γ for global image I with $\Gamma(g) = \frac{1}{K_i} \sum_{k=1}^{K_i} \Gamma^k(g)$
Compute the new image (J) histogram $h^J(g)$, normalized to 1

Ensure: The image I quality metrics Γ, the filtered image J

3.3 Experimental Results with a Computer-Generated Image

In order to test the performance of the proposed technique, some results obtained with the computer-generated image named "Bar", (the original (Bar) image is presented Fig. 3 with the blocks registration), are shown is this presentation (other images were tested and same behaviors were observed. They are not presented here due to the lack of space and can be found on http://www-lisic.univ-littoral.fr/~bigand/IJIS/Noref-metrics.html). This image is composed of homogeneous and noisy blocks and is interesting to present some results. In Fig. 4, the first image (left) is the fifth noisy block, the second one is the result obtained with the IVFS filter, the third one is the tenth noisy block (at first iteration, $\sigma = 10$), and the fourth one (bottom right) is the result obtained with the IVFS filtering obtained at optimized (tenth) iteration. The **main idea** of the paper is the following: synthesis process is started with a great noise level and a certain entropy value. So the initial position of the synthesis process is unknown but the observed behavior measured at each iteration of the image synthesis process brings us information. The average information quantity gained at each iteration is entropy. The measured entropy using IVFS seems to be an interesting measure of noise level and supplies a no-reference quality evaluation used as denoising scheme in the proposed image synthesis process. It proves the advantage of this approach qualitatively (more uncertainty is taken into account using IVFSs, as previously suggested).

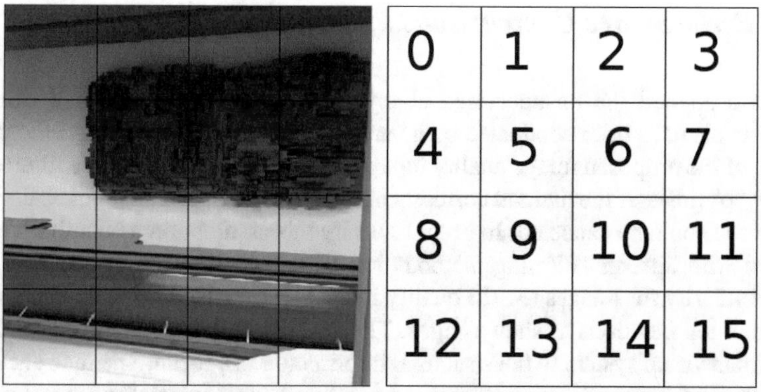

Fig. 3 Reference image "Bar"

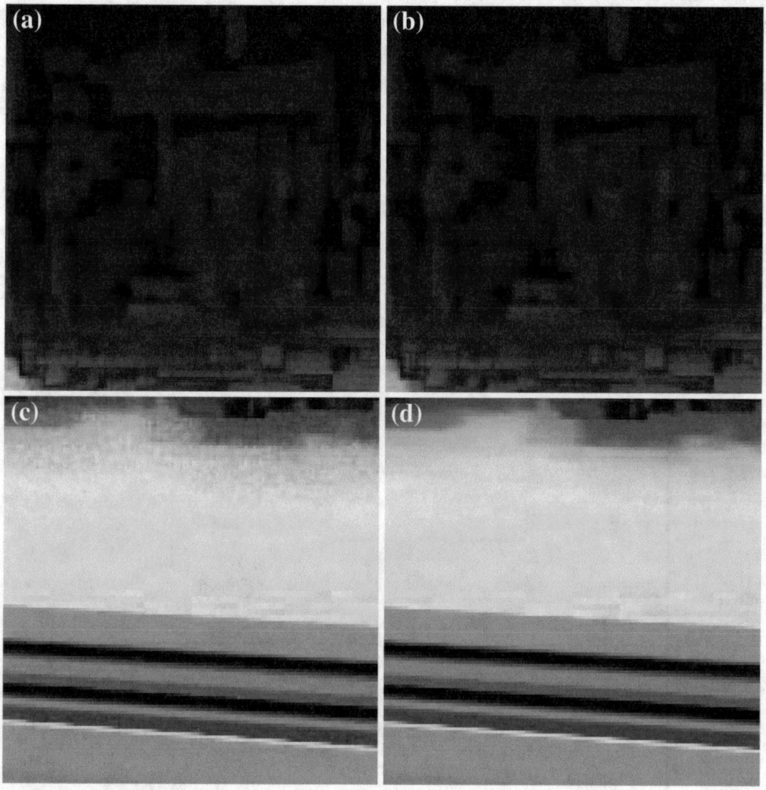

Fig. 4 Blocks denoising using Γ. **a** Noisy image, block 5. **b** Denoised image, block 5. **c** Noisy image, block 10. **d** Denoised image, block 10

3.4 Performance Comparisons of Image Quality Measures

We have applied the unsupervised algorithm we propose further, and extensive experiments have been conducted on a variety of test images to evaluate the performance of the proposed image quality index. The previous image will be filtered iteratively "p" times. If the method correctly operates, the image should not be degraded but noise should be canceled. In order to verify this assumption a quantitative comparison with different filtering methods has been made. So the measure of structural similarity for images (SSIM quality index [4]) has also been calculated on the course of ten iterations as shown Fig. 5. This measure is based on the adaptation of the human visual system to the structural information in a scene. The index accounts for three different similarity measures, namely luminance, contrast and structure. The closer the index to unity, the better the result. It is easy to see that Wiener and averaging filters have a stable behavior for recursive filtering (but small SSIM values) while IVFS filtering presents the best results in term of SSIM. We would like to highlight the advantages of the proposed measure: this measure is simple; it is parameter free and avoids additional procedures and training data for parameter determination.

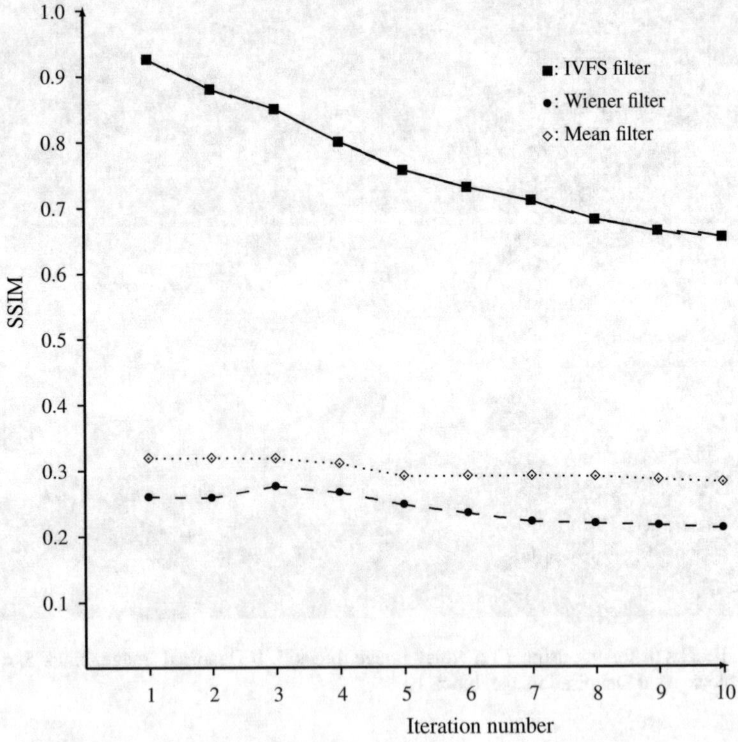

Fig. 5 SSIM index versus iteration number

4 The Design of the IVFS Speckle Noise Detection and Filtering

After the proposition of a new no-reference image quality metrics, speckle noise detection is now considered.

4.1 Proposed Scheme

The restoration scheme of image I corrupted with speckle noise is divided into two steps. In the first one, we perform histogram analysis of image I using the IVF entropy Γ_{max} that is used as a tool to find all major homogeneous regions of I as previously. In the second step, median filter ($med(n, m)$) is applied to all pixels $I(n, m)$ identified as corrupted while leaving the rest of the pixels identified as noise-free. Let $med(n, m)$ be the value of pixels in a local window of size 3×3 in the surrounding of $I(n, m)$ (i.e. a windowing $(3, 3)$ around the pixel (n, m)). This method reduces the distortion generated in the restored image, introducing the necessary spatial (local) information to obtain spatially connected regions.

4.2 Algorithm

The implementation of image restoration based on IVFS and IVF entropy Γ is made according to the algorithm presented in [19] for impulse noise filtering.

4.3 Experimental Results with a Synthetic Image

In order to test the performance of the proposed technique, a classic synthetic image (named "Savoyse", Fig. 6), and composed of five areas on an uniform background, added with a multiplicative speckle noise (Variance = 0.2) is first tested. We present Fig. 6 the original and noisy images respectively. *How does it work?* It is now well-known that images and fuzzy sets can be defined in the same way. To each pixel x_i corresponds a membership value $\mu(x_i)$, when using FS. Using IVFS, to each pixel x_i corresponds a membership value $\mu_L(x_i)$ and $\mu_U(x_i)$, respectively. Let us consider one pixel and its neighbors. If these pixels are uncorrupted by noise, they have about the same grey level, and consequently the same membership values. So we obtain about the same results with FS and IVFS. On the other hand, if some pixels in the neighboring areas are noisy, the FS-entropy consider only $\mu(x_i)$, while IVFS-entropy consider the difference $\mu_L(x_i) - \mu_U(x_i)$. The maximum difference represents the uncertainty about the grey-level value of the pixel considered and consequently the noise-free

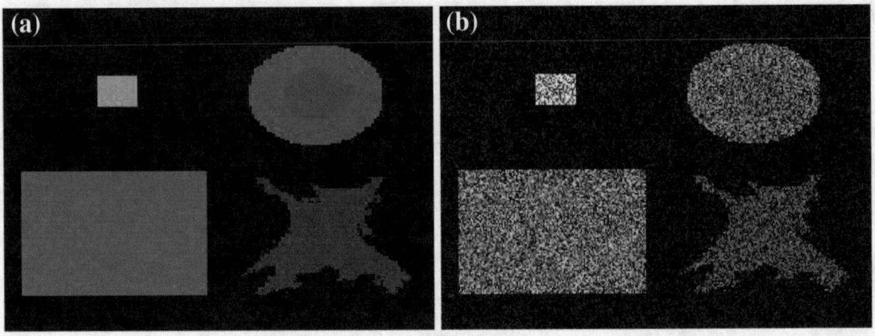

Fig. 6 The original and noisy images. **a** Original image Savoyse. **b** Noisy image Savyse, Speckle noise (Variance = 0.2)

accepted range of that pixel. So, noise is taken into account (the greater the difference $\mu_L(x_i) - \mu_U(x_i)$ is, the bigger the entropy is and the better the filtering effects should be).

In order to verify these assumptions, we have compared fuzzy sets entropy approach to its counterpart with IVFSs. In particular, we can notice that the peak values of the entropy IVF entropy are more important than their counterparts using type-1 fuzzy sets (FS entropy, Fig. 7, where the first two modes of the histogram have disappeared when using FS), so corresponding regions will be easier to extract in a noisy image, and this proves the advantage of this approach qualitatively (more uncertainty is taken into account using IVFSs, as it is previously suggested). This is well illustrated by the results obtained with the synthetic image. It is clear that IVFSs lead to better results. IVFSs are able to model imprecision and uncertainty which FS cannot afford to handle efficiently. Local entropy in information theory represents the variance of local region and catches the natural properties of transition region. So IVFS being able to deal with a greater amount of uncertainty than FS, transi-

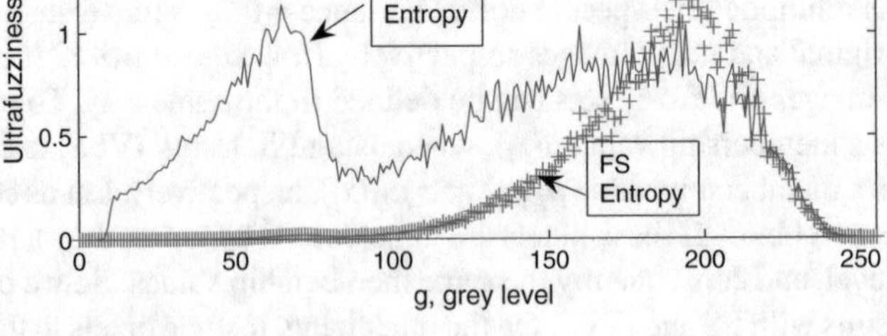

Fig. 7 Ultrafuzziness curves

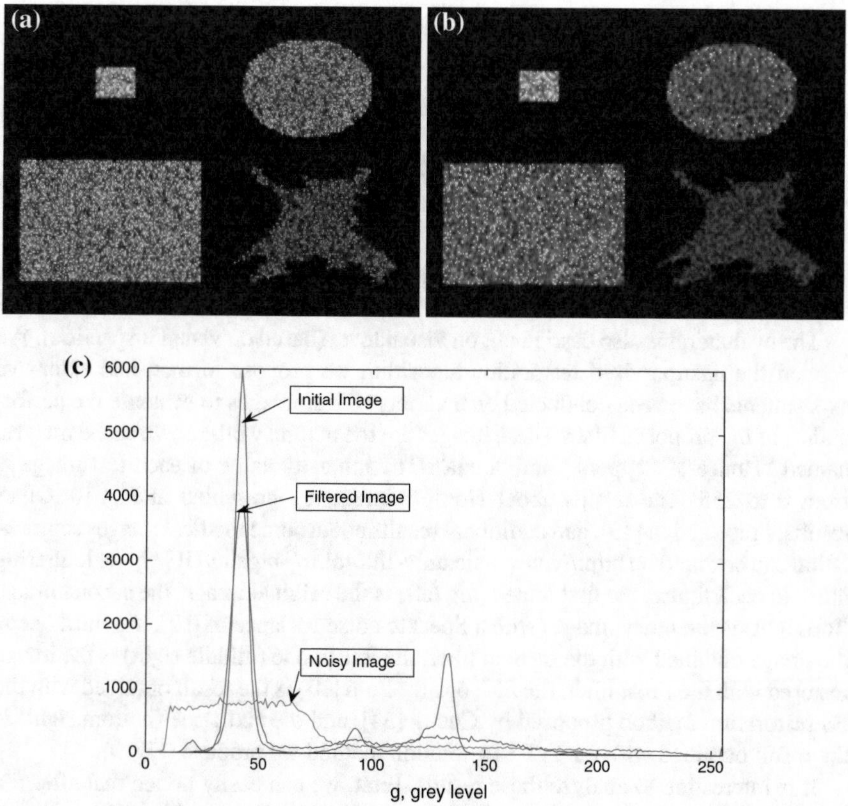

Fig. 8 Restored images and comparison of histograms, image 'Savoyse'. **a** Restored image using FS. **b** Restored image using IVFS. **c** Filtering results: comparison of histograms (number of pixels versus *grey* level)

tion regions are more acute and homogeneous regions are better drawn, as we can see with the inside circle of the Savoyse image. Finally, we present the comparative results we obtain for fuzziness (FS, Fig. 8a) and IVF entropy (IVFS, Fig. 8b). The histograms shown in Fig. 8c confirm the efficiency of the proposed method to reduce speckle noise in images.

4.4 Experimental Results

Anyway, running time is less than one second, and in the same order of time than the other tested filters. The peak signal-to-noise ratio (PSNR) is used to quantitatively evaluate the restoration performance, which is defined as

$$PSNR = 10\log_{10}\frac{255^2}{MSE} \tag{13}$$

where mean square error (MSE) is defined as:

$$MSE = \frac{1}{M.N}\sum_{m=1}^{M}\sum_{n=1}^{N}(n_i - n_f)^2 \tag{14}$$

where n_i and n_f denote the original image pixels and the restored image pixels, respectively, and $M \times N$ is the size of the image.

The evaluation is also carried out on visual level (based on visual inspection). We applied the unsupervised restoration algorithm we propose further, and extensive experiments have been conducted on a variety of test images to evaluate the performance of the proposed filter. These images are the natural well-known scene images, named "House", "Peppers," and "Lena". The intensity value of each test image is from 0 to 255. The results about House's image are presented in Fig. 10. Other results, presented in [19], and additional results about some medical images segmentation can be found on http://www-lisic.univ-littoral.fr/~bigand/IJIS/Specklefiltering.html. In each figure, the first image (top left) is the original image, the second image (top right) is the noisy image (with a Speckle noise variance of 0.2), the third one is the image obtained with the median filter, the fourth one (middle right) is the image restored with the Frost filter, the fifth one (bottom left) is the result obtained with the FS restoration method proposed by Cheng [34], and the sixth one (bottom right) is the result obtained with the IVFS restoration method we propose (Fig. 9).

It is interesting to analyze these results. First, we can easily notice that after the restoration process, filtering using IVF entropy is very effective. Mendel has shown that the amount of uncertainty associated to IVFS is characterized by its lower and upper membership functions, and these results confirm this assertion. So we are intuitively able to explain these results (compared with a FS for example, Fig. 10a and b). On the synthetic image (Savoyse), the two regions corresponding to the two concentric discs are correctly extracted. This result shows that the method is able to handle unequiprobable and overlapping classes of pixels. The restoration of the other (natural) images is challenging because of the presence of shadows and highlight effects.

Comparisons have been carried out between the IVFS filter and other filters (classical filters like mean, median and Frost filters, and the Cheng's FS filter) in terms of capability of noise removal. Table 1 shows that the proposed filter performed well, providing an improvement in PSNR on the other methods. It is well-known that PSNR depends on the type of images, so the average value of the PSNR for these four images is calculated and presented in Table 1. The following figure (Fig. 9) displays the results from processing a classic "Lena" image corrupted with different values of noise level, demonstrating the superior results obtained by our proposed method when compared to the FS, Frost and the median filters. The displayed values for the median filter and the FS filter follow behaviors presented in the literature. Relatively

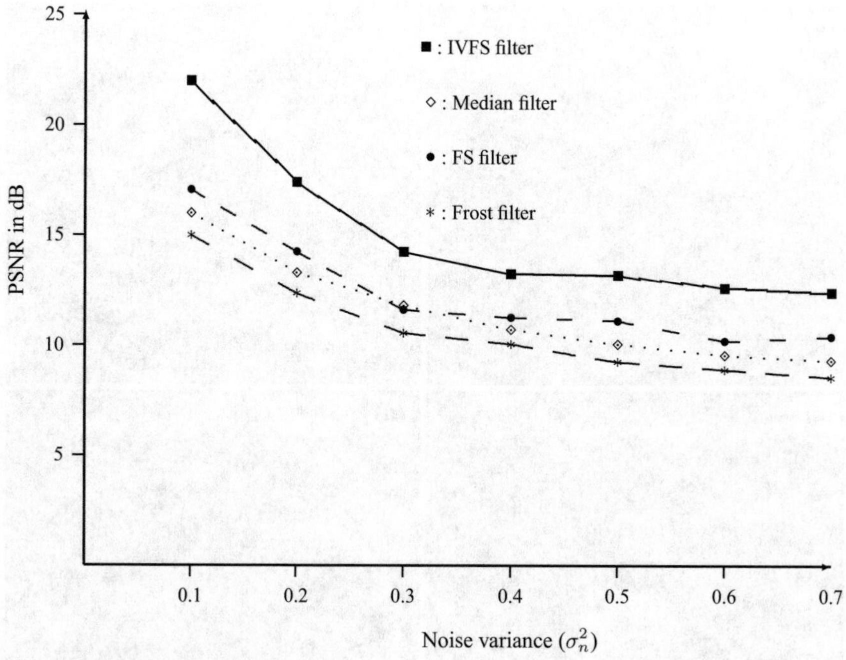

Fig. 9 PSNR versu noise level

good results obtained with FS filter confirm the results presented by Nachtegael et al. [35] for impulse and Gaussian noises filtering. These results also confirm the significant improvement that our proposed method successfully achieves.

5 Conclusion

The central idea of this paper was to introduce the application of IVFSs, to take into account the total amount of uncertainty present at the image acquisition stage. IVF entropy is defined from discrete (h, image histogram) and continuous (H, smoothed histogram at scale σ) scale-space representations and is consistent with scale-space theory. The stochastic and deterministic properties of H can be studied separately. So in our case, we used the deterministic part of H to define IVF entropy and to propose a new no-reference image quality metrics. Like other techniques based on image thresholding, this technique is simple and computationally efficient. It makes it possible to characterize computer-generated images efficiently and to restore images corrupted with multiplicative noise. Particularly, it assumes no "a-priori" knowledge of a specific input image, no numerous tunable parameters, yet it has superior performance compared to other existing fuzzy and non-fuzzy filters for the full range

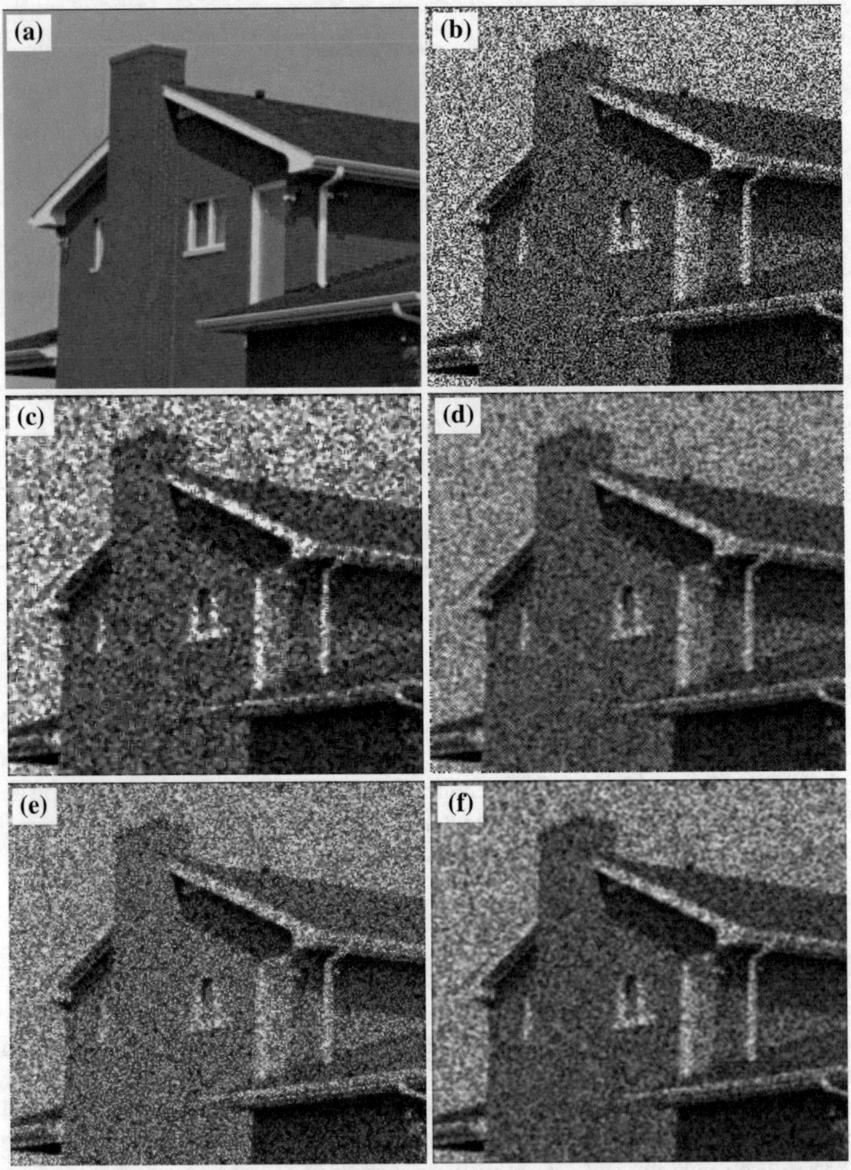

Fig. 10 Comparison of different method of noise reduction, speckle variance of 0.2 **a** Original image House. **b** Noisy image House, Speckle noise (Variance = 0.2). **c** Restored image House using median filter. **d** Restored image House using Frost filter. **e** Restored image House using FS filter. **f** Restored image House using IVFS

Table 1 Comparative restoration results in PSNR (dB) for Speckle noise ($\sigma_n^2 = 0.2$)

Image name	Size (in pixels)	Filters				
		Mean	Median	Frost	FS	IVFS
Savoyse	320 × 200	9.42	17.16	16.12	17.99	20.82
House	256 × 256	4.62	12.85	11.99	13.73	16.69
Peppers	512 × 512	5.50	13.64	12.62	14.65	18.45
Lena	512 × 512	5.22	13.30	12.35	14.25	17.61
Average		6.19	14.23	13.27	15.15	18.39

of Speckle noise level. In a previous paper, we showed that this new technique can also deal with impulsive and Gaussian noise [18]. Uncertainty is correctly treated, and interplay between scale-space theory and IVFS seems very fruitful for stochastic analysis of image generation (stochastic detection of errors and image filtering). Nevertheless there remains some open questions to effectively establish a link between the characteristics of noise affecting the image (noise level) and the choice of the IVFS. So the stochastic properties of the errors fields present during image generation have to be studied now with this new framework. In particular, more extensive investigations on other measures of entropy and the effect of parameters influencing the width (length) of IVFS are under investigation, and could lead to establish a general noise model in images using a link between IVFS and level and type of noise (particularly with color images).

References

1. Pratt, B.: Digital Image Processing. Wiley-Interscience (1978)
2. Kajiya, J.T.: The rendering equation. ACM Comput. Graph. **20**(4), 143–150 (1986)
3. Shirley, P., Wang, C.Y., Zimmerman, K.: Monte Carlo techniques for direct lighting calculations. ACM Trans. Graph. **15**(1), 1–36 (1996)
4. Wang, Z., Bovik, A.C., Sheikh, H.R., Simoncelli, E.P.: Image quality assessment: from error visibility to structural similarity. IEEE Trans. Image Process. **13**(4) (2004)
5. Ferzli, R., Karam, L.: No-reference objective wavelet based noise immune image sharpness metric. Int. Conf. Image Process. (2005)
6. Lopes, A., Nezri, E., Touzi, R., Laur, H.: Maximum a posteriori speckle filtering and first order texture models in SAR images. IGARSS (1990)
7. Bigand, A., Bouwmans, T., Dubus, J.P.: Extraction of line segments from fuzzy images. Pattern Recogn. Lett. **22**, 1405–1418 (2001)
8. Cheng, H.D., Chen, C.H., Chiu, H.H., Xu, H.J.: Fuzzy homogeneity approach to multilevel thresholding. IEEE Trans. Image Process. **7**(7), 1084–1088 (1998)
9. Tizhoosh, H.R.: Image thresholding using type 2 fuzzy sets. Pattern Recogn. **38**, 2363–2372 (2005)
10. Bloch, I.: Lattices of fuzzy sets and bipolar fuzzy sets, and mathematical morphology. Inf. Sci. **181**, 2002–2015 (2011)

11. Nachtegael, M., Sussner, P., Melange, T., Kerre, E.E.: On the role of complete lattices in mathematical morphology: from tool to uncertainty model. Inf. Sci. **181**, 1971–1988 (2011)
12. Mendel, J.M., Bob John, R.I.: Type-2 fuzzy sets made simple. IEEE Trans. Fuzzy Syst. **10**(2), 117–127 (2002)
13. Bloch, I.: Information combination operators for data fusion: a comparative review with classification. IEEE Trans. SMC—Part B **26**, 52–67 (1996)
14. Zadeh, L.A.: The concept of a linguistic variable and its application to approximate reasoning. Inf. Sci. **8**, 199–249 (1975)
15. Wu, H., Mendel, J.M.: Uncertainty bounds and their use in the design of interval type-2 fuzzy logic systems. IEEE Trans. Fuzzy Syst. **10**(5), 622–639 (2002)
16. Liang, Q., Karnish, N.N., Mendel, J.M. Connection admission control in ATM networks using survey-based type-2 fuzzy logic systems. IEEE Trans. Syst. Man Cyber—Part B **30**(3), 329–339 (2000)
17. Sussner, P., Nachtegael, M., Esmi, E.: An approach towards edge detection and watershed segmentation based on an interval-valued morphological gradient. ICPV'11 (2011)
18. Bigand, A., Colot, O.: Fuzzy filter based on interval-valued fuzzy sets for image filtering. Fuzzy Sets Syst. **161**, 96–117 (2010)
19. Bigand, A., Colot, O.: Speckle noise reduction using an interval type-2 fuzzy sets filter. In: Intelligent Systems IS'12 IEEE Congress (Sofia, Bulgaria) (2012)
20. Delepoulle, S., Bigand, A., Renaud, C.: an interval type-2 fuzzy sets no-reference computer-generated images quality metric and its application to denoising. In: Intelligent Systems IS'12 IEEE Congress (Sofia, Bulgaria) (2012)
21. Astrom, K., Heyden, A.: Stochastic analysis of scale-space smoothing. ICPR (1996a)
22. Babaud, J., Witkin, A.P., Baudin, M., Duda, R.O.: Uniqueness of the Gaussian kernel for scape-space filtering. IEEE Trans. PAMI-8 **8**, 26–33 (1986)
23. Strauss, O.: Quasi-continuous histograms. Fuzzy Sets Syst. **160**, 2442–2465 (2009)
24. Jacquey, F., Loquin, K., Comby, F., Strauss, O.: Non-additive approach for gradient-based edge detection. ICIP (2007)
25. Astrom, K., Heyden, A.: Stochastic analysis of sub-pixel edge detection. ICPR (1996b)
26. Bustince, H., Barrenechea, E., Pergola, M., Fernandez, J.: Interval-valued fuzzy sets constructed from matrices: application to edge detection. Fuzzy Sets Syst. **160**(13), 1819–1840 (2009)
27. Starck, J.L., Murtagh, F., Bijaoui, A.: Image processing and data analysis: the multiscale approach. Cambridge University Press (1998)
28. Kaufmann, A.: Introduction to the Theory of Fuzzy Set—Fundamental Theorical Elements, vol. 28. Academic Press, New York (1975)
29. Deluca, A., Termini, S.: A definition of a nonprobabilistic entropy in the setting of fuzzy set theory. Inf. Control **20**(4), 301–312 (1972)
30. Pal, N.R., Bezdek, J.C.: Measures of Fuzziness: A Review and Several Classes. Van Nostrand Reinhold, New York (1994)
31. Burillo, P., Bustince, H.: Entropy on intuitionistic fuzzy sets and on interval-valued fuzzy sets. Fuzzy Sets Syst. **78**, 305–316 (1996)
32. Bustince, H., Barrenechea, E., Pergola, M., Fernandez, J., Sanz, J.: Comments on: image thresholding using type 2 fuzzy sets. Importance of this method. Pattern Recogn. **43**, 3188–3192 (2010)
33. Cheng, H.D., Sun, Y.: A hierarchical approach to color image segmentation using homogeneity. IEEE Trans. Image Process. **9**(12), 2071–2081 (2000)
34. Cheng, H., Jiang, X., Wang, J.: Color image segmentation based on homogram thresholding and region merging. Pattern Recogn. **35**(2), 373–393 (2002)
35. Nachtegael, M., Schulte, S., Weken, D.V., Witte, V.D., Kerre, E.E.: Fuzzy filters for noise reduction: the case of Gaussian noise. Int. Conf. Fuzzy Syst. (2005)

Generalized Net Representation of Dataflow Process Networks

Dimitar G. Dimitrov

Abstract This paper presents translation rules for mapping from a given dataflow process network (DPN) to a generalized net (GN). The so obtained GN has the same behaviour as the corresponding DPN. A reduced GN that represents the functioning and the results of the work of an arbitrary DPN is also defined.

Keywords Generalized nets · Dataflow process networks · Kahn process networks

1 Introduction

Generalized Nets (GNs) are defined as extensions of ordinary Petri nets, as well as of other Petri nets modifications [1]. The additional components in GN definition give more and greater modeling possibilities and determine the place of GNs among the separate types of Petri nets, similar to the place of the Turing machine among finite automata. GNs can describe wide variety of modeling tools such as Petri nets and their extensions [1, 2], Unified modeling language (UML) [7], Kahn Process Networks (KPN) [4], etc. In this paper we shall show how GNs can adequately represent dataflow process networks (DPN).

Dataflow process networks are a model of computation (MoC) used in digital signal processing software environments and in other contexts. DPN are a special case of Kahn Process Networks [5, 6].

The structure of this paper is as follows. In next section a brief introduction of DPN is presented. In Sect. 3 a procedure for translating a concrete DPN to a GN is given. In the next section, a universal GN that represents the functioning and the

This work is supported by the National Science Fund Grant DID 02/29 "Modelling of Processes with Fixed Development Rules (ModProFix)".

D.G. Dimitrov (✉)
Faculty of Mathematics and Informatics, Sofia University,
5 James Bourchier Blvd., Sofia, Bulgaria
e-mail: dgdimitrov@fmi.uni-sofia.bg

V. Sgurev et al. (eds.), *Recent Contributions in Intelligent Systems*,
Studies in Computational Intelligence 657,
DOI 10.1007/978-3-319-41438-6_2

results of the work, an arbitrary DPN is formally defined. After that the advantages of using GN instead of DPN are discussed. Finally, future work on this subject is proposed, as well as software implementation details are given.

2 Dataflow Process Networks

DPNs consist of a set of data processing nodes named actors which communicate through unidirectional unbounded FIFO buffers. Each actor is associated with a set of firing rules that specify what tokens must be available in its inputs for the actor to fire. When an actor fires, it consumes tokens from its input channels and produces tokens in its output channels [8, 9]. Figure 1 shows a DPN with four actors.

A *(dataflow) actor* with m inputs and n outputs is a pair $\langle R,f \rangle$, where

$$R = \{R_1, R_2, \ldots, R_k\}$$

is a set of *firing rules*. Each firing rule constitutes a set of patterns, one for each input:

$$R_i = \{R_{i,1}, R_{i,2}, \ldots, R_{i,m}\}$$

A pattern $R_{i,j}$ is a finite sequence of data elements from j-th channel's alphabet. A firing rule R_i is satisfied, iff for each $j \in \{1, 2, \ldots, m\}$ $R_{i,j}$ forms a prefix of the sequence of unconsumed tokens at j-th input. An actor with $k = 0$ firing rules is always enabled. $R_{i,j} = \perp$ denotes that any available sequence of tokens is acceptable from input j. "*" denotes a token wildcard (i.e., any value is acceptable).

$f : S^m \rightarrow S^n$ is a function that calculates the token sequences that are to be output by the actor, where S^m is the set of all m-tuples of token sequences.

DPNs are an untimed model. Firing rules are not bounded to a specific time moment.

DPNs do not over specify an execution algorithm. There are many different execution models with different strengths and weaknesses.

To avoid confusion with tokens in GNs, in this paper we shall denote channel tokens in DPN as *data elements*.

With $pr_i A$ we shall denote the i-th projection of the n-dimensional set A where $n \in \mathbb{N}, 1 \leq i \leq n$.

Fig. 1 Example DPN

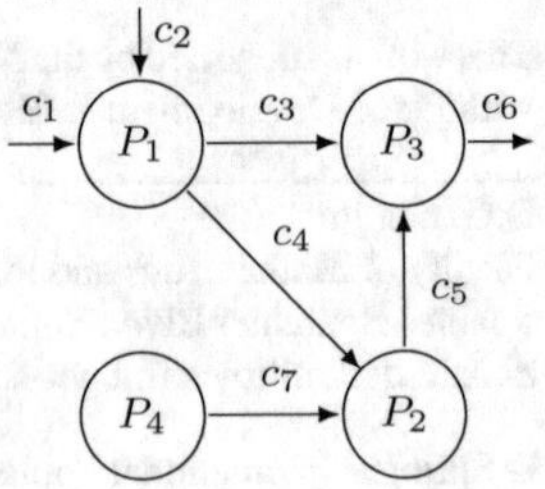

3 Translating a DPN to a GN

In generalized nets, as in Petri nets, transitions represent discrete events, and places represent conditions (either preconditions or postconditions, depending on arc directions). Dataflow actors consume tokens from its input channels and produce tokens to its output channels. Similarly when a GN transition is activated, it transfers tokens from its input places to its output places. Dataflow actors fire when tokens in the inputs satisfy given conditions (firing rules). The same way token transfer in GN occurs when predicates associated with transitions are evaluated as true. Thus actors in DPN can obviously be mapped to GN transitions and firing rules can be seen as a special case of GN transition predicates. GN transitions' concept of firing rules—transition types—will not be used, because they ensure only the presence of a token in a given input, not its contents.

In GN, places have characteristic functions. They calculate the new characteristics of every token that enters an output place. This can be used as analog of DPN actors' firing function that calculates output sequences based on actors' input data. Channels in DPN which connect dataflow actors can be translated to GN places. Similar approach of mapping from dataflow actors to Petri net transitions and channels to places is also used in [10]. DPN tokens can be directly mapped to GN tokens but in order to preserve their ordering, whole data sequence can be represented as a single GN token.

Table 1 summarizes the translation rules from DPN to GN.

Figure 2 shows a GN representation of the example DPN in Fig. 1.

Token splitting and merging must be enabled for the GN, e.g., operator $D_{2,2}$ must be defined for it [2].

In this paper we shall use the following function to check whether an actor can fire:

$$c(R, I) = (\exists i \in \{1, \dots, |R|\} : \forall R_{i,j} \in R_i : R_{i,j} \sqsubseteq I_j)$$

where R is a set of firing rules, I is a list of data element sequences, one for each input channel, and $p \sqsubseteq q$ is true iff the sequence p is a prefix of q.

Each dataflow actor $\langle R, f \rangle$ can be translated to a transition in the following form:

$$Z = \langle \{l'_1, \dots, l'_m, l, l^*\}, \{l''_1, \dots, l''_n, l, l^*\}, *, *, r, *, * \rangle$$

Table 1 Mapping from dataflow process networks to generalized nets

	DPN	GN
Actor	Node	Transition
Channel	Arc	Place
Firing rules	(First component of an actor)	Transition predicate
Firing function	(Second component of an actor)	Characteristic function
Data	Tokens with single data elements	Token with one or more data elements

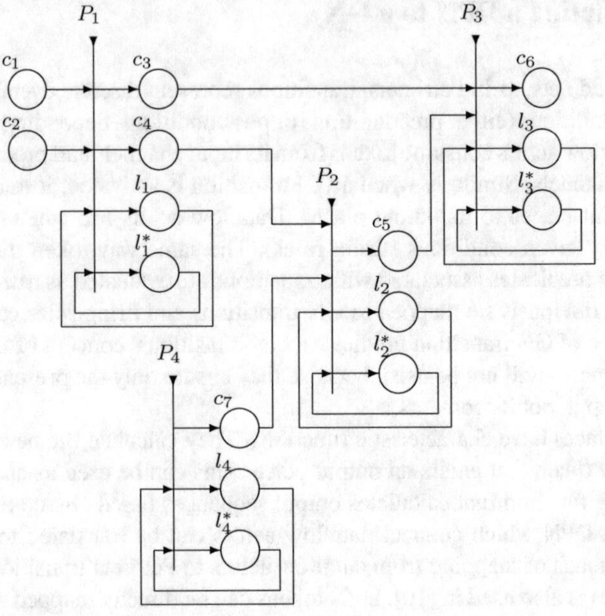

Fig. 2 GN representation of the example GN

where

- $l'_1, \ldots, l'_m$ are places that correspond to the actor's inputs;
- $l''_1, \ldots, l''_n$ are places that correspond to the actor's outputs;
- l is a place that collects all input data consumed by one firing of the actor and l's characteristic function calculates the actor's output sequences;
- l^* is a place that keeps all input data before it can be consumed;
- tokens in l^* are merged into token δ with the following characteristic:

  $$x^{\delta}_{s+1} = \{\langle l_i, get(x^{\delta}_s, l_i).x^{\delta_{l_i}} \rangle | l_i \in \{l'_1, \ldots, l'_m\}\}$$

  where δ_{l_i} is the token (if such available) in the input place l_i, x^{δ}_s is the previous characteristic of token δ which loops in l^* (if available, empty set otherwise), "." is sequence concatenation and *get* is a function that gets the content of a given channel stored as characteristic in δ;
- r is the following index matrix (IM):

$$r = \begin{array}{c|cccc} & l''_1 \; \ldots \; l''_n & l^* & l \\ \hline l'_1 & & & \\ \vdots & false & true & false \\ l'_m & & & \\ l^* & false & true & W_{l^*,l} \\ l & true & false & false \end{array}$$

where predicate $W_{l^*,l}$ checks whether exists a firing rule from R that is satisfied by current input:

$$W_{l^*,l} = c(R, pr_2 x^\delta)$$

- the characteristic function Φ is defined for l as follows:

$$\Phi_l = \{\langle l''_j, pr_j f(pr_1 x^\delta)\rangle | 1 \le j \le n\}$$

where f is the function associated with the dataflow actor corresponding to Z. As a side effect Φ_l removes consumed data from the characteristic of the token in l^* (which may result in empty characteristic of this token, if there is no unconsumed data);
- Φ for outputs $l''_1, \ldots, l''_n$ retains only the data sequence which corresponds to the given output place (previously the token contains information for all output channels):

$$\Phi_{l''_j} = get(x^\delta, l''_j), 1 \le j \le n$$

A special case are dataflow actors without inputs. Such actors are always enabled [9]. The so defined GN transition Z is capable of representing such actors. An empty token should always be present in l^* meaning that there is no input data. Predicate $W_{l^*,l}$ is always evaluated as true.

4 Universal GN for Dataflow Process Networks

Below is a formal definition of a reduced GN E that represents any DPN N. The graphical structure of E is shown in Fig. 3. Token splitting and merging must be enabled for the net.

$$E = \langle\langle\{Z_1, Z_2\}, *, *, *, *, *, *\rangle, \langle\{\alpha\}, *, *\rangle, \langle *, *, *\rangle, \langle\{x_0^\alpha\}, *, *\rangle, \Phi, *\rangle$$

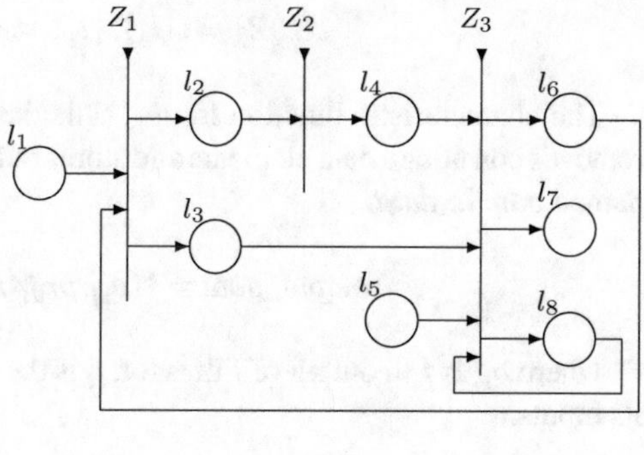

Fig. 3 Graphical structure of the universal GN representing dataflow process networks

The first transition is responsible for dividing dataflow actors into two sets—ones that can be fired, according to their firing rules, and ones, that cannot:

$$Z_1 = \langle \{l_1, l_6\}, \{l_2, l_3\}, *, *, r_1, *, * \rangle$$

Token α enters place l_1 containing as a characteristic information about the DPN in the following form:

$$x^\alpha = \{\langle\langle R, f\rangle, \{\langle i_1, s_{i_1}\rangle, ..., \langle i_{m_k}, s_{i_{m_k}}\rangle\}\rangle\}$$

where $\langle R, f\rangle$ denotes a dataflow actor and s_{i_j} is a sequence of initial data elements in j-th input of the actor.

A token may be available in either l_1 or l_6. There is a token in l_1 only in the first step of the functioning of E.

Token α is transferred to l_2 if it contains at least one actor that can be fired. If there is at least one actor whose firing rules are not satisfied by input data, the token splits and goes to l_3 too. The predicates of Z_1 are the following:

$$r_1 = \begin{array}{c|cc} & l_2 & l_3 \\ \hline l_1 & W_1 & W_2 \\ l_6 & W_1 & W_2 \end{array}$$

where $W_1 = (\exists a \in x^\alpha : c(pr_1 pr_1 a, pr_2 pr_2 a)$ and $W_2 = (\exists a \in x^\alpha : \neg c(pr_1 pr_1 a, pr_2 pr_2 a)$

The characteristic function for l_2 and l_3 retains only firable and non-firable actors, respectively:

$$\Phi_{l_2} = \{\langle\langle R, f\rangle, \{\langle i_1, s_{i_1}\rangle, ..., \langle i_{m_k}, s_{i_{m_k}}\rangle\}\rangle | c(R, \{i_1, \dots, i_{m_k}\})\}$$

$$\Phi_{l_3} = \{\langle\langle R, f\rangle, \{\langle i_1, s_{i_1}\rangle, ..., \langle i_{m_k}, s_{i_{m_k}}\rangle\}\rangle | \neg c(R, \{i_1, \dots, i_{m_k}\})\}$$

After passing the second transition, actors are fired and their firing function f is calculated. Data elements are read from inputs and output data is written to outputs.

$$Z_2 = \langle \{l_2\}, \{l_4\}, *, *, *, *, * \rangle$$

The characteristic function for Φ_{l_4} calls the firing functions of each actor in α_1, removes consumed data elements and stores output sequences in a new characteristic named *output_data*:

$$output\_data = \{\langle o_{i,j}, pr_j f_i(I_i)\rangle | 1 \leq i \leq |x^\alpha|\rangle\}$$

where $o_{i,j}$ is j-th output of i-th actor, f_i is the firing function of i-th actor and I_i are its inputs.

The last transition Z_3 merges actors in a single token α again. It also collects input data from input channels (which is not generated by an actor from the DPN).

$$Z_3 = \langle\{l_3, l_4, l_5\}, \{l_6, l_7, l_8\}, *, *, r_3, *, \square_2\rangle$$

Tokens carrying data elements corresponding to input channels enter in l_5. Characteristics have the following form:

$$x^\delta = \{\langle c, s\rangle\}$$

where c is a channel and s is a sequence of data elements.

Z_3 has the following predicate matrix:

$$r_3 = \begin{array}{c|ccc} & l_6 & l_7 & l_8 \\ \hline l_3 & W & false & false \\ l_4 & W & true & false \\ l_5 & false & false & true \\ l_8 & W & false & \neg W \end{array}$$

Predicate W checks whether tokens exist in both l_3 and l_4. If there is incoming data but no actor tokens, input data is collected in l_8 and waits.

In order for input data and actor information to be merged easily, all actors must be available:

$$\square_2 = \vee(\wedge(l_3, l_4), l_5, l_8)$$

Tokens in place l_6 are merged and the new characteristic is the union of the characteristics of α_1 and α_2. The characteristic of the token from l_8 (if such is available) is merged with the characteristic named *output_data* (the two sets do not intersect). Merging is executed before the calculation of Φ_{l_6}.

In l_8 tokens are merged. The new characteristic is calculated in similar way as in place l^* from Sect. 3:

$$x^\delta_{s+1} = \{\langle i, get(x^\delta, i).get(x^{\delta'}, i)\rangle | i \in pr_1 x^\delta \cup pr_1 x^{\delta'}\}$$

where δ denotes the token from l_8 and δ' the new token coming from l_5.

In l_6 after tokens are merged, the characteristic function writes actors' output data into the corresponding channels. After that it removes the *output_data* characteristic from α:

$$\Phi_{l_6} = \{\langle\langle R, f\rangle, \{\langle i_j, s_{i_j}.get(output\_data, i_j)\rangle | 1 \leq j \leq m_k\}\rangle\}$$

Data written to output channels (which do not act as input channels for other actors) leaves the net through place l_7. The new characteristic that tokens receive in l_7 consists of only output data written to such channels (previously tokens contains output data for all channels).

$$\Phi_{l_7} = \{s \in output\_data | pr_1 s \notin \bigcup pr_1 pr_2 x^{\alpha}\}$$

5 Conclusion

The so constructed in this paper GNs are reduced ones, i.e., this work shows that even simple class of GNs is capable of describing DPNs. If one uses a GN to model a real process, usually modeled by a DPN, he will receive many advantages, since GNs are more detailed modeling tool. First, token characteristics in GNs have history, so in the universal GN all data elements that pass through a given channel are remembered. Complex dataflow actors' mapping functions by default are translated to characteristic functions but they can also be represented as GNs. If a real process that runs in a given time is represented by a GN instead of a DPN, the modeler can use the global time component of the GN, so process time can be mapped to the time scale. Unlike DPNs and KPNs, GNs support different types of time. As in Kahn process networks' universal GN [4], a scheduler that manages the execution order of actors and channel capacities can easily be integrated.

As a future work on the topic we can define GNs for other models of computation, as well as for some special cases of DPN such as synchronous DPN.

GN IDE [3], the software environment for modeling and simulation with GNs, can be extended to support different modeling instruments such as KPNs, DPNs, as well as Petri nets, and their various extensions. The results of current research imply that the above functionality can easily be implemented without modifying GNTicker—the software interpreter for GNs, used by GN IDE.

Another potential direction is to introduce fuzzyness in DPN. Several fuzzy GN extensions are defined [2] and can be used to model such fuzzy DPNs.

References

1. Atanassov, K.: Generalized Nets. World Scientific, Singapore, New Jersey, London (1991)
2. Atanassov, K.: On Generalized Nets Theory. Prof. Marin Drinov Academic Publishing House, Sofia (2007)
3. Dimitrov, D.G.: GN IDE—a software tool for simulation with generalized nets. In: Proceedings of Tenth International Workshop on Generalized Nets, pp. 70–75. Sofia, 5 Dec 2009
4. Dimitrov, D.G., Marinov, M.: On the representation of Kahn process networks by a generalized net. In: 6th IEEE International Conference on Intelligent Systems (IS'12), pp. 168–172. Sofia, Bulgaria, 6–8 Sept 2012
5. Geilen, M., Basten, T.: Requirements on the execution of kahn process networks, In: Degano, P. (eds.) Programming Languages and Systems. In: 12th European Symposium on Programming, ESOP 2003, Proceedings, pp. 319–334. Warsaw, Poland, 7–11, LNCS 2618. Springer, Berlin, Germany (2003)
6. Kahn, G.: The semantics of a simple language for parallel programming. In: Rosenfeld, J.L. (ed.) Information Processing 74. In: Proceedings of IFIP Congress 74, North-Holland, Stockholm, Sweden, 5–10 Aug 1974

7. Koycheva E., Trifonov, T., Aladjov, H.: Modelling of UML sequence diagrams with generalized nets. In: International IEEE Symposium on Intelligent Systems, (79 84), Varna, IEEE (2002)
8. Lee, E., Matsikoudis, E.: The semantics of dataflow with firing. In: Huet, G., Plotkin, G., Lvy, J.-J., Bertot, Y. (eds.) Chapter in From Semantics to Computer Science: Essays in Memory of Gilles Kahn, Preprint Version, 07 March 2008, Copyright (c) Cambridge University Press
9. Lee, E., Parks, T.: Dataflow process networks. Readings in Hardware/Software Co-design. Kluwer Academic Publishers Norwell, pp. 59–85 (2002). ISBN:1-55860-702-1
10. Rocha, J.-I., Gomes, L., Dias, O.P.: Dataflow model property verification using Petri net translation techniques. INDIN 2011, pp. 783-788, 26–29 July 2011

Wireless Sensor Positioning Using ACO Algorithm

Stefka Fidanova, Miroslav Shindarov and Pencho Marinov

Abstract Spatially distributed sensors, which communicate wirelessly form a wireless sensor network (WSN). This network monitors physical or environmental conditions. A central gateway, called high energy communication node, collects data from all sensors and sends them to the central computer where they are processed. We need to minimize the number of sensors and energy consumption of the network, when the terrain is fully covered. We convert the problem from multi-objective to mono-objective. The new objective function is a linear combination between the number of sensors and network energy. We propose ant colony optimization (ACO) algorithm to solve the problem. We compare our results with the state of the art in the literature.

Keywords Wireless sensor network · Ant colony optimization · Metaheuristics

1 Introduction

The development of new technologies during the last decades gives a possibility for wireless data transmission. Thus a new types of networks, called wireless networks, was created.

Wireless sensor networks (WSN) allow the monitoring of large areas without the intervention of a human operator. Their working is based on the exchange of local information between nodes in order to achieve a global goal. Cooperation between the sensor nodes is an important feature when solving complex tasks.

S. Fidanova (✉) · M. Shindarov · P. Marinov
Institute of Information and Communication Technologies–BAS,
Acad. G. Bonchev Str. Block 25A, 1113 Sofia, Bulgaria
e-mail: stefka@parallel.bag

M. Shindarov
e-mail: miroslavberberov@abv.bg

P. Marinov
e-mail: pencho@parallel.bas.bg

V. Sgurev et al. (eds.), *Recent Contributions in Intelligent Systems*,
Studies in Computational Intelligence 657,
DOI 10.1007/978-3-319-41438-6_3

The WSN can be used in areas where traditional networks fail or are inadequate. They find applications in a variety of areas, such as climate monitoring, military use, industry, and sensing information from inhospitable locations. Unlike other networks, sensor networks depend on deployment of sensors over a physical location to fulfill a desired task. Sometimes deployments imply the use of hundreds or thousands of sensor nodes in small areas to ensure effective coverage range in a geographical field.

For a lot of applications wireless sensors offer a lower cost method for collecting system health data to reduce energy usage and better manage resources. Wireless sensors are used to effectively monitor highways, bridges, and tunnels. Other applications are continually monitor office buildings, hospitals, airports, factories, power plants, or production facilities. The sensors can sense temperature, voltage, or chemical substances. A WSN allows automatically monitoring of almost any phenomenon. The WSN gives a lot of possibilities and offers a great amount of new problems to be solved. WSN have been used in military activities, such as reconnaissance, surveillance, [5], environmental activities, such as forest fire prevention, geophysical activities, such as volcano eruptions study [16], health data monitoring [19] or civil engineering [12].

A WSN node contains several components including the radio, battery, microcontroller, analog circuit, and sensor interface. In battery-powered systems, higher data rates, and more frequent radio use consumes more power. There are several open issues for sensor networks, such as signal processing [13], deployment [18], operating cost, localization, and location estimation.

The wireless sensors, have two fundamental functions: sensing and communicating. The sensing can be of different types (seismic, acoustic, chemical, optical, etc.). However, the sensors which are far from the high energy communication node (HECN) cannot communicate with him directly. Therefore, the sensors transmit their data to this node, either directly or via hops, using nearby sensors as communication relays.

When deploying a WSN, the positioning of the sensor nodes becomes one of the major concerns. The coverage obtained with the network and the economic cost of the network depend directly of it. Since many WSN can have large numbers of nodes, the task of selecting the geographical positions of the nodes for an optimally designed network can be very complex. Therefore, metaheuristics seem an interesting option to solve this problem.

In this paper, we propose an algorithm which solves the WSN layout problem using ACO. We focus on both minimizing the energy depletion of the nodes in the network and minimizing the number of the nodes, while the full coverage of the network and connectivity are considered as constraints. The problem is multi-objective. We convert it to mono-objective. The new objective function is a combination of the two objective functions of the original problem. We learn the algorithm performance and influence of the number of ants on achieved solutions.

Jourdan [8] solved an instance of WSN layout using a multi-objective genetic algorithm. In there formulation a fixed number of sensors had to be placed in order to maximize the coverage. In some applications most important is the network energy.

In [7] is proposed ACO algorithm and in [17] is proposed evolutionary algorithm for this variant of the problem. In [6] is proposed ACO algorithm taking in to account only the number of the sensors. In [10] are proposed several evolutionary algorithms to solve the problem. In [9] is proposed genetic algorithm which achieves similar solutions as the algorithms in [10], but it is tested on small test problems.

The number of sensor nodes should be kept low for economical reasons and the network needs to be connected. Finally, the energy of the network is a key issue that has to be taken into account, because the life time of the network depends on it.

The rest of the paper is organized as follows. In Sect. 2 the WSN is described and the layout problem is formulated. Section 3 presents the ACO algorithm. The existing state of the art is briefly reviewed in Sect. 4. In Sect. 4 the experimental results obtained are shown. Finally, several conclusions are drown in Sect. 5.

2 Wireless Sensor Network Layout Problem

A WSN consists of spatially distributed sensors which monitor the conditions of sensing area, such as temperature, sound, vibration, pressure, motion, or pollutants [1, 14]. The development of WSN was motivated by military applications, such as surveillance and are now used in many industrial and civilian application areas, including industrial process monitoring and control, machine health monitoring, environment and habitat monitoring, healthcare applications, home automation, and traffic control [14].

A sensor node might vary in size from that of a box to the size of a grain of dust [14]. The cost of sensor nodes is similarly variable, ranging from hundreds of dollars to a few cents, depending on the complexity required of individual sensor nodes [14]. Size and cost constraints on sensor nodes result in corresponding constraints on resources such as energy, memory, computational speed and bandwidth [14].

A WSN consists of sensor nodes. Each sensor node sense an area around itself called its sensing area. The sensing radius determines the sensitivity range of the sensor node and thus the sensing area. The communication radius determines how far the node can send his data. A special node in the WSN called High Energy Communication Node is responsible for external access to the network. Therefore, every sensor node in the network must have communication with the HECN. Since the communication radius is often much smaller than the network size, direct links are not possible for peripheral nodes. A multi-hop communication path is then established for those nodes that do not have the HECN within their communication range. They transmit their date by other nodes which are closer to the HECN.

The WSN layout problem aims to decide the geographical position of the sensor nodes that form a WSN. In our formulation, sensor nodes has to be placed in a terrain providing full sensitivity coverage with a minimal number of sensors and minimizing the energy spent in communications by any single node, while keeps the connectivity of the network. Minimal number of sensors means cheapest network for constructing. Minimal energy means cheapest network for exploitation. The energy of the network defines the lifetime of the network, how frequently the batteries need to be replaced. These are opposed objectives since the more nodes there are the lesser share of retransmissions they bear and in opposite, when we try to decrease the energy consumption normally we include nodes. Thus, we look for a good balance between number of sensors and energy consumption.

In order to determine the energy spent by communications, the number of transmissions every node performs is calculated. The WSN operates by rounds: In a round, every node collects the data and sends it to the HECN. Every node transmits the information packets to the neighbor that is closest to the HECN, or the HECN itself if it is within the communication range. When several neighbors are tied for the shortest distance from the HECN, the traffic is distributed among them. That is, if a node has n neighbors tied for shortest distance from HECN, each one receives 1/n of its traffic load. Therefore, every node has a traffic load equal to 1 (corresponding to its own sent data) plus the sum of all traffic loads received from neighbors that are farther from the HECN. On one hand the sensing area need to be fully covered. On the other hand, the number of sensor nodes must be kept as low as possible, since using many nodes represents a high cost of the network, possibly influences of the environment. The objectives of this problem is to minimize network energy and the number of sensors deployed while the area is fully covered and connected.

3 ACO for WSN Layout Problem

The WSN Layout problem is a hard combinatorial optimization problem which needs an exponential amount of computational resources (NP-hard). Exact methods and traditional numerical methods are unpractical for this kind of problems. Therefore normally the NP problems are solved with some metaheuristic method. Many of the existing solutions of WSN Layout problem come from the field of evolutionary computation [2, 10]. After analyzing them, we noticed that the ACO is based on solution construction, most of other metaheuristics are based on improving current solution. Therefore ACO is appropriate for problems with restrictive constraints. The ACO is one of the most successful metaheuristics which outperforms others for a lot of classes of problems. The idea for ACO comes from real ant behavior.

Real ants foraging for food lay down quantities of pheromone (chemical cues) marking the path that they follow. An isolated ant moves essentially guided by previously laid pheromone. After the repetition of the above mechanism if more ants follow a trail, the more attractive that trail becomes. Thus ants can find the shortest path between the nest and sours of the food.

The ACO algorithm uses a colony of artificial ants that behave as cooperative agents in a mathematic space were they are allowed to search and reinforce pathways (solutions) in order to find the optimal ones.

On the Fig. 1 is shown a pseudocode of ACO algorithm.

The problem is represented by graph and the ants walk on the graph to construct solutions. The solution is represented by a path in the graph. After initialization of the

Fig. 1 Pseudocode for ACO

```
Ant Colony Optimization

Initialize number of ants;
Initialize the ACO parameters;
while not end-condition do
    for k=0 to number of ants
        ant k starts from a random node;
        while solution is not constructed do
            ant k selects higher probability node;
        end while
    end for
    Local search procedure;
    Update-pheromone-trails;
end while
```

pheromone trails, ants construct feasible solutions, starting from random nodes, then the pheromone trails are updated. At each step ants compute a set of feasible moves and select the best one (according to some probabilistic rules based on a heuristic function) to carry out the rest of the tour. The structure of ACO algorithm is shown in Fig. 1. The transition probability p_{ij}, to chose the node j when the current node is i, is based on the heuristic information η_{ij} and on the pheromone trail level τ_{ij} of the move, where $i, j = 1, \ldots, n$.

$$p_{ij} = \frac{\tau_{ij}^{\alpha}\eta_{ij}^{\beta}}{\sum\limits_{k \in allowed} \tau_{ik}^{\alpha}\eta_{ik}^{\beta}} \tag{1}$$

The higher value of the pheromone and the heuristic information, the more probable is to select this move. The pheromone corresponds to the global memory of the ants, their experience in problem solving from previous iterations. The heuristic information is an a priori knowledge for the problem which is used to manage the search process to improve the solution. In the beginning, the initial pheromone level is set to a small positive constant value τ_0 and then ants update this value after completing the construction stage [3]. ACO algorithms adopt different criteria to update the pheromone level.

In our implementation we use MAX–MIN Ant System (MMAS) [15], which is one of the more popular ant approaches. The main feature of MMAS is using a fixed upper bound τ_{max} and a lower bound τ_{min} of the pheromone trails. Thus, the accumulation of big amounts of pheromone by part of the possible movements and repetition of same solutions is partially prevented on one side, and the amount of the pheromone to decrease a lot of and to become close to zero and the element to be unused is partially prevented in another side.

The pheromone trail update rule is given by:

$$\tau_{ij} \leftarrow \rho\tau_{ij} + \Delta\tau_{ij}, \tag{2}$$

$$\Delta\tau_{ij} = \begin{cases} 1/C(V_{best}) & \text{if } (i,j) \in \text{best solution} \\ 0 & \text{otherwise} \end{cases},$$

where V_{best} is the iteration best solution and $i, j = 1, \ldots, n$, $\rho \in [0, 1]$ models evaporation in the nature. By parameter ρ we decrease the influence of the old information and we keep it in some level. $C(V)$ is the objective function.

The WSN layout problem is a NP-hard multi-objective problem. We simplify it converting it to mono-objective. Normally converting from multi- to mono- objective make worse the achieved solutions. Therefore it is very important how the new objective function will be constructed. It is one of our contributions in this work. The new objective function is a combination of the number of sensors and the energy of the network and we search for solution which minimizes it. The new objective function is as follows:

$$C(V_k) = \frac{f_1(V_k)}{\max_i f_1(V_i)} + \frac{f_2(V_k)}{\max_i f_2(V_i)} \tag{3}$$

where V_k is the solution constructed by the ant k and $f_1(V_k)$ and $f_2(V_k)$ are the number of sensors and energy corresponding to the solution V_k. Dividing f_1 by $\max f_1$ and f_2 by $\max f_2$ is a sort to normalize the values of f_1 and f_2, respectively, the maximum is over the solutions from the first iteration.

Thus, when the energy and/or number of sensors decreases the value of the objective function will decrease and the two components have equal influence.

To avoid stagnation of the search, the range of possible pheromone values on each movement is limited to an interval $[\tau_{min}, \tau_{max}]$. τ_{max} is an asymptotic maximum of τ_{ij} and $\tau_{max} = 1/(1-\rho)C(V^*)$, while $\tau_{min} = 0.087\tau_{max}$. Where V^* is the optimal solution, but it is unknown, therefore we use V_{best}, the current best value of the objective function, instead of V^*.

One of the crucial point of the ACO algorithms is construction of the graph of the problem. We need to chose which elements of the problem will correspond to the nodes and the meaning of the arcs, where is more appropriate to deposit the pheromone—on the nodes or on the arcs. In our implementation the WSN layout problem is represented by two graph, it is one of our contributions. The terrain is modeled by grid $G = \{g_{ij}\}_{N \times M}$, where M and N are the size of the sensing region. By the graph G we calculate the coverage of the terrain. We use another graph $G1_{N1 \times M1}$, on which nodes we map the sensors, where $N1 \leq N$ and $M1 \leq M$. The parameters $M1$ and $N1$ depend of the sensing and communication radius. Thus, we decrease the number of calculations the algorithm performs, respectively the running time. The pheromone is related with location sites $Ph = \{ph_{ij}\}_{N1 \times M1}$, the initial pheromone can be a small value, for example, $1/n_{ants}$. The central point, where the HECN is located, is included in the solutions like first point (zero point).

Every ant starts to create the rest of the solution from a random node which communicates with central one, thus the different start of every ant in every iteration is guaranteed. The ant chooses the next position by the ACO probabilistic rule (Eq. 1). It chooses the point having the highest probability. If there are more than one point with same probability, the ant chooses one of them randomly.

The construction of heuristic information is another crucial points of the ACO algorithm. The heuristic information needs to be constructed thus, to manage the ants to look for better solutions and to avoid unfeasible solutions. For some kinds of problems it is not obvious how to prepare it. One needs to combine different elements of the problem to most appropriate way.

Other contribution of this paper is proposition of appropriate heuristic information, which will guarantee construction of feasible solutions. Our heuristic information is a product of three parameters as follows:

$$\eta_{ij}(t) = s_{ij} l_{ij} (1 - b_{ij}), \tag{4}$$

where s_{ij} is the number of points which the new sensor will cover, and

$$l_{ij} = \begin{cases} 1 \text{ if communication exists} \\ 0 \text{ if there is not communication} \end{cases} \tag{5}$$

b is the solution matrix and the matrix element $b_{ij} = 1$ when there is sensor on the node (i,j) of the graph $G1$, otherwise $b_{ij} = 0$. With s_{ij} we try to locally increase the covered points, more new covered points leads eventually to less number of sensors. With l_{ij} we guarantee that all sensors will be connected; with rule $(1 - b_{ij})$ we guarantee that the position is not chosen yet and no more than one sensor will be mapped on the same node of the graph $G1$. When $p_{ij} = 0$ for all values of i and j the search stops. Thus, the construction of the solution stops if no more free positions, or all points are covered or new communication is impossible.

4 Experimental Results

We will contribute with this work to improve the state of the art of the use of metaheuristics for solving the WSN layout problem. Our aim is to provide an efficient solving method by comparing a set of state-of-the-art metaheuristic techniques applied in the same scenario. We want to solve a new flexible instance in which, for the first time (to the best of our knowledge), both the number and positions of the sensors can be freely chosen, with full coverage of the sensor field guaranteed, and treating the energy efficiency and the overall cost of the network. Besides this, our interest is to tackle complex instances in which the WSN size is in the same order of magnitude as real WSN, with several hundred nodes.

With our algorithm we can solve WSN layout problem on any rectangular area and the HECN can be fixed on any point on the area. The reported results are for an WSN problem instance where a terrain of 500×500 points has to be covered using sensors with coverage and communication radii covered 30 points (see Table 1). We

Table 1 Problem parameters

Terrain	500 × 500
Coverage radius	30
Communication radius	30

Table 2 Algorithm parameters

Ants	1–10
Iterations	60
ρ	0.5
α	1
β	1

choose this example because other authors use the same and to can compare achieved results. The terrain has an area of 250,000 points in total, and each sensor covers 2,827 points, meaning that in ideal conditions only 89 would be necessary. Now, these ideal conditions do not exist since they would imply that no overlap exists between any two nodes sensing areas, which is impossible due to their geometrical shape (circle). Therefore, the expected minimum number of nodes for full coverage is higher than 89. Thus, the graph G consists of 500×500 nodes. When we apply our ACO algorithm on this example the graph $G1$ consists of 50×50 nodes which are 100 times less than the graph G. The number of nodes of the graph $G1$ is proportional to the number of the nodes of the graph G and is proportional to the number of points covered by coverage and communication radius. Thus, the nodes of the graph $G1$ are mapped on nodes of the graph G and coverage and communication radii cover 3 points from the graph $G1$. In our example, the HECN is fixed in the center of the area.

An example of solution that achieves full coverage of the region is a square grid formed by the sensors separated by 30 points. Starting at the HECN, 250 points have to be covered to each side of the terrain, requiring eight sensors. Therefore, the grid has 17 (8 + 8 + 1) rows and 17 columns, thus 289 sensors including the HECN. In this symmetrical configuration there are four nodes directly connected to the HECN, so the complete traffic of the network 288 messages per round is evenly divided among them. This results in the most loaded nodes having a load of 72 messages. So this candidate solution obtains (288, 72).

After several runs of the algorithm we specify the most appropriate values of its parameters (see Table 2). We apply MAX–MIN ant algorithm with the following parameters: $\alpha = \beta = 1, \rho = 0.5$. When the number of ants is double the running time is doubled. The number of used ants is from 1 to 10 and the maximum number of iterations is 60 (about 3 h per ant).

In Table 3 are reported best found results (with respect to the sensors and with respect to the energy) achieved by several metaheuristic methods. We compare our

Table 3 Comparison with other metaheuristics

Algorithm	Min sensors	Min energy
Symmetric	(288, 72)	(288, 72)
MOEA	(260, 123)	(291, 36)
NSGA-II	(262, 83)	(277, 41)
$IBEA_{HD}$	(265, 83)	(275, 41)
ACO	(227, 61)	(239, 50)

ACO algorithm results with results obtained by the evolutionary algorithms in [10] and the symmetric solution. These evolutionary algorithms performs like multi-objective and reports non-dominated solutions. A solution dominates other if all components of the first are better than the second. A solution is non-dominated if and only if no other solution dominates it.

We perform 30 independent runs of the ACO algorithm, because for statistical point of view the number of runs need to be minimum 30. The achieved numbers of sensors are in the interval [227, 247]. Thus, the worst number of sensors achieved by ACO algorithm is less than the best number of sensors achieved by other mentioned algorithms.

Let compare achieved solutions with minimal number of sensors. The solutions achieved by mentioned evolutionary algorithms have very high energy more then symmetric solution. Thus they are not good solutions. The ACO solution with minimal number of sensors dominates other solutions with minimal number of sensors and symmetric solution. Thus it is a good solution. Let compare solutions with minimal energy achieved by mentioned algorithms. MOEA algorithm achieves solution with very small value for energy, but too many sensors, more than symmetric solution, thus it is not good solution. Other two evolutionary algorithms achieve solutions with a less energy than symmetric and a little bit less number of sensors. Thus they are not bed solutions. The ACO solution dominates symmetric one. Its energy is a little bit more than the evolutionary algorithms, but the number of sensors is much less. We can conclude that our ACO algorithm achieves very encouraging solutions.

We learn the influence of the number of ants on algorithm performance and quality of the achieved solutions. Our aim is to find optimal number of ants, first according the solutions and second—according the computational resources.

We run our ACO algorithm 30 times. We prepare Pareto front for over 30 runs for every ant number extracting the achieved non-dominated solutions. On the Table 4 we show the achieved non-dominated solutions (Pareto fronts). In the left column are the number of sensors and in other columns is the energy corresponding to this number of sensors and number of ants. We observe that the front achieved by eight ants dominates the fronts achieved by 1, 2, 3, 4, 5, 6, and 7 ants. The fronts achieved by 8, 9, and 10 ants do not dominate each other. To estimate them we prepare Pareto front by the solutions (Pareto fronts) achieved by 8, 9, and 10 ants and we call it common Pareto front. In our case the common Pareto front

Table 4 Pareto fronts

Sensors	Ants									
	1	2	3	4	5	6	7	8	9	10
239	55			50						
238										
237	56									
236										
235		52	52	52					52	
234					51	51	51	48		51
233									54	54
232								51	55	55
231	58	58			56	56	56		51	
230		59	57	57	57	57	57	54	57	
229		60		60	60	60	58	56		
228	60									
227			60	61				57	58	58
226								58		

Table 5 Distances from extended front

Ants	8	9	10
Distance	3	4	3

is $\{(226,58),(230,57),(231,56),(232,55),(233,54),(234,51)\}$. We introduce the concept for "extended front." If for some number of sensors there are not corresponding energy in the common Pareto front, we put the energy to be equal to the point of the front with less number of sensors. We can do this because if we take some solution and if we include a sensor close to the HECN it will not increase the value of the energy and will increase with 1 only the number of the sensors. Thus, there is corresponding energy to any number of nodes. In our case the extended front is $\{(226,58),(227,58),(228,58),(229,58),(230,57),(231,56),$ $(232,55),(233,54),(234,51),(235,51)\}$.

We include additional criteria to estimate Pareto front. We calculate the distance between a Pareto front and the Extended front. To calculate the distance we extend every of the Pareto fronts in a similar way as the extended front. The distance between any Pareto front and the Extended front is the sum of the distances between the points with a same number of sensors, or it is the difference between their energy. These distances are positive because the Extended front dominates the Pareto fronts. Thus, the best Pareto front is the closest to the Extended front.

Regarding Table 5 we observe that the distance of the fronts achieved by 8 and 10 ants is 3 and the distance of the front achieved by 9 ants is 4. Thus, the fronts achieved by 8 and 10 ants are with a equal quality and they are better than the front achieved by 9 ants. We consider that the algorithm performs better with 8 ants because the solution (226, 58) dominates solution (227, 58) and when the number of ants is less the used computational resources are less, running time is less and the used memory is less.

5 Conclusion

We have defined WSNs layout problem with its connectivity constraint. A very large instance consisting of 500×500 points area has to be covered, using sensors whose sensing and communication radii are 30 points in a way that minimizes both the number of sensors and the traffic load in the most loaded sensor node. We simplify the problem converting it from multi-objective to mono-objective. The new objective function is a linear combination of the number of sensors and energy of the network. Thus, the both "previous" objective functions have equal influence in the "new" objective function. We propose ACO algorithm to solve this problem and we compare it with existing state-of-the-art algorithms. We decrease the number of computations describing the problem by two graphs (square grids) with different size. We learn the algorithm performance with different number of ants and the same number of iterations. We conclude that our algorithm achieves best results when the number of ants is 8. We compare our algorithm with state of the art in the literature and can conclude that our solutions dominate most of the solutions achieved by other methods. The results of the experiments indicate very encouraging performance of ACO algorithm.

Acknowledgments This work has been partially supported by the Bulgarian National Scientific Fund under the grants Modeling Processes with fixed development rules—DID 02/29 and Effective Monte Carlo Methods for large-scale scientific problems—DTK 02/44.

References

1. Akuildiz, I.F., Su, W., Sankarasubramaniam, Y., Cayrci, E.: Wireless sensor networks: a survey. Comput. Netw. **38**(4), 393–422 (2001). Elsevier
2. Alba, E., Molina, G.: Optimal Wireless Sensor Layout with Metaheuristics: Solving a Large Scale Instance, Large-Scale Scientific Computing, LNMCS 4818, pp. 527–535. Springer (2008)
3. Bonabeau, E., Dorigo, M., Theraulaz, G.: Swarm Intelligence: From Natural to Artificial Systems. Oxford University Press (1999)
4. Cahon, S., Melab, N., Talbi, El.-G.: Paradiseo: a framework for the reusable design of parallel and distributed metaheuristics. J. Heuristics **10**(3), 357–380 (2004)
5. Deb, K., Pratap, A., Agrawal, S., Meyarivan, T.: A fast and elitist multiobjective genetic algorithm: NSGA-II. IEEE Trans. Evol. Comput. **6**(2), 182–197 (2002)
6. Fidanova, S., Marinov, P., Alba, E.: ACO for optimal sensor layout. In: Filipe, J., Kacprzyk, J. (eds.) Proceedings of International Conference on Evolutionary Computing, Valencia, Spain,

pp. 5–9. SciTePress-Science and Technology Publications Portugal (2010). ISBN 978-989-8425-31-7
7. Hernandez, H., Blum, C.: Minimum energy broadcasting in wireless sensor networks: an ant colony optimization approach for a realistic antenna model. J. Appl. Soft Comput. **11**(8), 5684–5694 (2011)
8. Jourdan, D.B.: Wireless sensor network planing with application to UWB localization in gps-denied environments. Massachusetts Institute of Technology, Ph.D. thesis (2000)
9. Konstantinidis, A., Yang, K., Zhang, Q., Zainalipour-Yazti, D.: A multi-objective evolutionary algorithm for the deployment and power assignment problem in wireless sensor networks. J. Comput. Netw. **54**(6), 960–976 (2010)
10. Molina, G., Alba, E., Talbi, El.-G.: Optimal sensor network layout using multi-objective metaheuristics. Univ. Comput. Sci. **14**(15), 2549–2565 (2008)
11. Nemeroff, J., Garcia, L., Hampel, D., DiPierro, S.: Application of Sensor Network Communications. In: IEEE Military Communication Conference, pp. 336–341 (2011)
12. Paek, J., Kothari, N., Chintalapudi, K., Rangwala, S., Govindan, R.: The performance of a wireless sensor network for structural health monitoring. In: Proceedings of 2nd European Workshop on Wireless Sensor Networks, Istanbul, Turkey, Jan 31–Feb 2 (2005)
13. Pottie, G.J., Kaiser, W.J.: Embedding the internet: wireless integrated network sensors. Commun. ACM **43**(5), 51–58 (2000)
14. Romer, K., Mattern, F.: The design space of wireless sensor networks. IEEE Wirel. Commun. **11**(6), pp. 54–61 (2004). ISSN 1536-1284
15. Stutzle, T., Hoos, H.H.: MAX-MIN ant system. Future Gener. Comput. Syst. **16**, 889–914 (2000)
16. Werner-Allen, G., Lorinez, K., Welsh, M., Marcillo, O., Jonson, J., Ruiz, M., Lees, J.: Deploying a wireless sensor nnetwork on an active volcano. IEEE Internet Comput. **10**(2), 18–25 (2006)
17. Wolf, S., Mezz, P.: Evolutionary local search for the minimum energy broadcast problem. In: Cotta, C., van Hemezl, J. (eds.) VOCOP 2008. Lecture Notes in Computer Sciences, vol. 4972, pp. 61–72. Springer, Germany (2008)
18. Xu, Y., Heidemann, J., Estrin, D.: Geography informed energy conservation for Ad Hoc routing. In: Proceedings of the 7th ACM/IEEE Annual International Conference on Mobile Computing and Networking, Italy, pp. 70–84, 16–21 July 2001
19. Yuce, M.R., Ng, S.W., Myo, N.L., Khan, J.Y., Liu, W.: Wireless body sensor network using medical implant band. Med. Syst. **31**(6), 467–474 (2007)
20. Zitzler, E., Knzli, S.: Indicator-based selection in multiobjective search. PPSN'04, LNCS 3242, pp. 832–842. Springer (2004)

Time Accounting Artificial Neural Networks for Biochemical Process Models

Petia Georgieva, Luis Alberto Paz Suárez and Sebastião Feyo de Azevedo

Abstract This paper is focused on developing more efficient computational schemes for modeling in biochemical processes. A theoretical framework for estimation of process kinetic rates based on different temporal (time accounting) artificial neural network (ANN) architectures is introduced. Three ANNs that explicitly consider temporal aspects of modeling are exemplified: (i) Recurrent Neural Network (RNN) with global feedback (from the network output to the network input); (ii) time-lagged feedforward neural network (TLFN), and (iii) reservoir computing network (RCN). Crystallization growth rate estimation is the benchmark for testing the methodology. The proposed hybrid (dynamical ANN and analytical submodel) schemes are promising modeling framework when the process is strongly nonlinear and particularly when input–output data is the only information available.

1 Introduction

The dynamics of chemical and biochemical processes are usually described by mass and energy balance differential equations. These equations combine two elements, the phenomena of conversion of one reaction component into another (i.e., the reaction kinetics) and the transport dynamics of the components through the reactor. The identification of such mathematical models from experimental input/output data is still a challenging issue due to the inherent nonlinearity and complexity of

P. Georgieva (✉)
Signal Processing Lab, IEETA, DETI, University of Aveiro,
3810-193 Aveiro, Portugal
e-mail: petia@ua.pt

L.A.P. Suárez · S.F. de Azevedo
Faculty of Engineering, Department of Chemical Engineering,
University of Porto,
4200-465 Porto, Portugal
e-mail: sfeyo@fe.up.pt

V. Sgurev et al. (eds.), *Recent Contributions in Intelligent Systems*,
Studies in Computational Intelligence 657,
DOI 10.1007/978-3-319-41438-6_4

this class of processes (for example polymerization or fermentation reactors, distillation columns, biological waste water treatment, etc.) The most difficult problem is how to model the reaction kinetics and more particularly, the reaction rates. The traditional way is to estimate the reaction rates in the form of analytical expressions, Bastin and Dochain [2]. First, the parameterized structure of the reaction rate is determined based on data obtained by specially designed experiments. Then, the respective parameters of this structure are estimated. Reliable parameter estimation is only possible if the proposed model structure is correct and theoretically identifiable, Wolter and Pronzato [15]. Therefore, the reaction rate analytical structure is usually determined after a huge number of expensive laboratory experiments. It is further assumed that the initial values of the identified parameters are close to the real process parameters, Noykove et al. [11], which is typically satisfied only for well-known processes. The above considerations motivated a search for alternative estimation solutions based on computationally more attractive paradigms as are the ANNs. The interest in ANNs as dynamical system models is nowadays increasing due to their good nonlinear time-varying input–output mapping properties. The balanced network structure (parallel nodes in sequential layers) and the nonlinear transfer function associated with each hidden and output nodes allows ANNs to approximate highly nonlinear relationships without a priori assumption. Moreover, while other regression techniques assume a functional form, ANNs allow the data to define the functional form. Therefore, ANNs are generally believed to be more powerful than many other nonlinear modeling techniques.

The objective of this work is to define a computationally efficient framework to overcome difficulties related with poorly known kinetics mechanistic descriptors of biochemical processes. Our main contribution is the analytical formulation of a modeling procedure based on time accounting ANNs, for kinetic rates estimation. A hybrid (ANN and phenomenological) model and a procedure for ANN supervised training when target outputs are not available are proposed. The concept is illustrated on a sugar crystallization case study where the hybrid model outperforms the traditional empirical expression for the crystal growth rate.

The paper is organized as follows. In the next section, a hybrid model of a general chemical or biochemical process is introduced, where a time accounting ANN is assumed to model the process kinetic rates in the framework of a nonlinear state-space analytical process model. In Sect. 3 three temporal ANN structures are discussed.

In Sect. 4 a systematic ANN training procedure is formulated assuming that all kinetics coefficients are available but not all process states are measured. The proposed methodology is illustrated in Sect. 5 for crystallization growth rate estimation.

2 Knowledge-Based Hybrid Models

The generic class of reaction systems can be described by the following equations, Bastin and Dochain [2]

$$\frac{dX}{dt} = K\varphi(X,T) - DX + U_x \tag{1.1}$$

$$\frac{dT}{dt} = b\varphi(X,T) - d_0T + U_T \tag{1.2}$$

where, for $n, m \in \mathrm{N}$, the constants and variables denote

$X = (x_1(t), \ldots, x_n(t)) \in R^n$	Concentrations of total amounts of n process components
$K = [k_1, \ldots, k_m] \in R^{n \times m}$	kinetics coefficients (yield, stoichiometric, or other)
$\varphi = (\varphi_1, \ldots, \varphi_m)^T \in R^m$	Process kinetic rates
T	Temperature
$b \in R^m$	Energy-related parameters
q_{in}/V	Feeding flow/volume
D	Dilution rate
d_o	Heat transfer rate-related parameter

U_x and U_T are the inputs by which the process is controlled to follow a desired dynamical behavior. The nonlinear state-space model (1) proved to be the most suitable form of representing several industrial processes as crystallization and precipitation, polymerization reactors, distillation columns, biochemical fermentation, and biological systems. Vector (φ) defines the rate of mass consumption or production of components. It is usually time varying and dependent of the stage of the process. In the specific case of reaction process systems φ represents the reaction rate vector typical for chemical or biochemical reactions that take place in several processes, such as polymerization, fermentation, biological waste water treatment, etc. In nonreaction processes as, for example, crystallization and precipitation, φ represents the growth or decay rates of chemical species. In both cases (reaction or nonreaction systems) φ models the process kinetics and is the key factor for reliable description of the components concentrations. In this work, instead of an exhaustive search for the most appropriate parameterized reaction rate structure, three temporal (time accounting) ANN architectures are applied to estimate the vector of kinetic rates. The ANN submodel is incorporated in the general dynamical model (1) and the mixed structure is termed knowledge-based hybrid model (KBHM), see Fig. 1. A systematic procedure for ANN-based estimation of reaction rates is discussed in the next section.

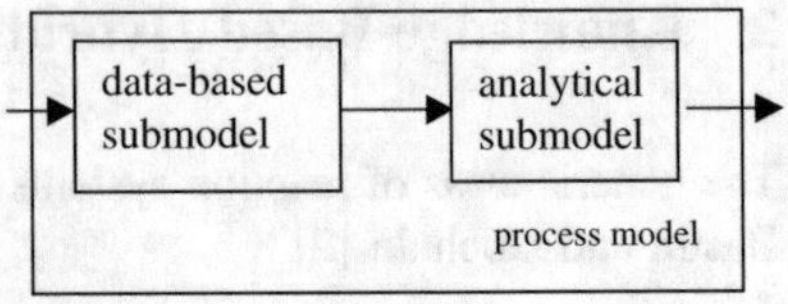

Fig. 1 Knowledge-based hybrid model (KBHM)

3 Time Accounting Artificial Neural Networks

The ANN is a computational structure inspired by the neurobiology. An ANN is characterized by its architecture (the network topology and pattern of connections between the nodes), the method of determining the connection weights, and the activation functions that employs. The multilayer perceptron (MLP), which constitute the most widely used network architecture, is composed of a hierarchy of processing units organized in a parallel-series sets of neurons and layers. The information flow in the network is restricted to only one direction from the input to the output, therefore a MLP is also called feedforward neural network (FNN). FNNs have been extensively used to solve static problems as classification, feature extraction, pattern recognition. In contrast to the FNN, the recurrent neural network (RNN) processes the information in both (feedforward and feedback) directions due to the recurrent relation between network outputs and inputs, Mandic and Chambers [10]. Thus, the RNN can encode and learn time dependent (past and current) information which is interpreted as memory. This paper specifically focuses on comparison of three different types of RNNs, namely, (i) RNN with global feedback (from the network output to the network input); (ii) Time lagged feedforward neural network (TLFN), and (iii) Reservoir Computing Network (RCN).

3.1 Recurrent Neural Network (RNN) with Global Feedback

An example of RNN architecture where past network outputs are fed back as inputs is depicted in Fig. 2. It is similar to Nonlinear Autoregressive Moving Average with eXogenios input (NARMAX) filters, Haykin [7]. The complete RNN input consists of two vectors formed by present and past network exogenous inputs (r) and past fed back network outputs (p), respectively.

The RNN model implemented in this work is the following

$$\mathbf{u}_{NN} = [\mathbf{r}, \mathbf{p}] \text{ (complete network input)} \tag{2.1}$$

$$\mathbf{r} = [r_1(k), \ldots r_1(k-l), \ldots r_c(k), \ldots r_c(k-l)] \text{ (network exogenous inputs)} \tag{2.2}$$

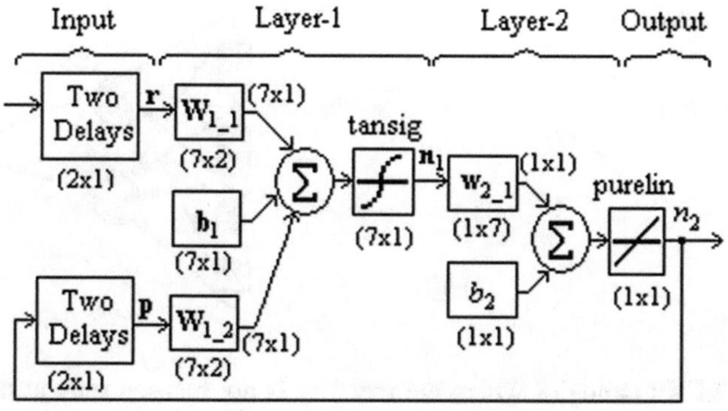

Fig. 2 RNN architecture

$$\mathbf{p} = [n_2(k-1), \ldots n_2(k-h)] \text{ (recurrent network inputs)} \tag{2.3}$$

$$\mathbf{x} = \mathbf{W}_{11} \cdot \mathbf{r} + \mathbf{W}_{12} \cdot \mathbf{p} + \mathbf{b}_1 \text{ (network states)} \tag{2.4}$$

$$\mathbf{n}_1 = (e^{\mathbf{x}} - e^{-\mathbf{x}})/(e^{\mathbf{x}} + e^{-\mathbf{x}}) \text{ (hidden layer output)} \tag{2.5}$$

$$n_2 = \mathbf{w}_{21} \cdot \mathbf{n}_1 + b_2 \text{ (network output),} \tag{2.6}$$

where $\mathbf{W}_{11} \in R^{m\times 2}$, $\mathbf{W}_{12} \in R^{m\times 2}$, $\mathbf{w}_{21} \in R^{1\times m}$, $\mathbf{b}_1 \in R^{m\times 1}$, $b_2 \in R$ are the network weights (in matrix form) to be adjusted during the ANN training, m is the number of nodes in the hidden layer. l is the number of past exogenous input samples and h is the number of past network output samples fed back to the input. The RNNs are a powerful technique for nonlinear dynamical system modeling, however, their main disadvantage is that they are difficult to train and stabilize. Due to the simultaneous spatial (network layers) and temporal (past values) aspects of the optimization, the static Backpropagation (BP) learning method has to be substituted by the Backpropagation through time (BPTT) learning. BPTT is a complex and costly training method, which does not guarantee convergence and often is very time consuming, Mandic and Chambers [10].

3.2 *Time Lagged Feedforward Neural Network (TLFN)*

TLFN is a dynamical system with a feedforward topology. The dynamic part is a linear memory, Principe et al. [13]. TLFN can be obtained by replacing the neurons in the input layer of an MLP with a memory structure, which is sometimes called a tap delay-line (see Fig. 3). The size of the memory layer (the tap delay) depends on the number of past samples that are needed to describe the input characteristics in time and it has to be determined on a case-by-case basis. When the memory is at the input the TLFN is also called focused time-delay neural network (TDNN). There

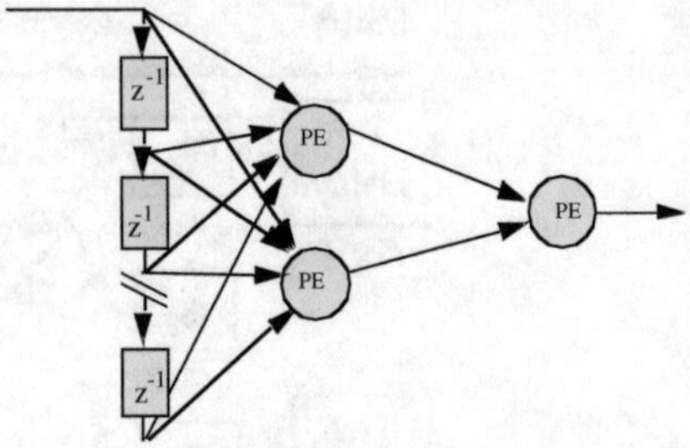

Fig. 3 RNN architecture

are other TLFN topologies where the memory is not focused only at the input but can be distributed over the next network layers. The main advantage of the TDNN is that it can be trained with the static BP method.

3.3 Reservoir Computing Network (RCN)

RCN is a concept in the field of machine learning that was introduced independently in three similar descriptions, namely, Echo State Networks [8], Liquid State Machines [9] and Backpropagation-Decorelation learning rule [14]. All three techniques are characterized by having a fixed hidden layer usually with randomly chosen weights that is used as a reservoir of rich dynamics and a linear output layer (termed also readout layer), which maps the reservoir states to the desired outputs (see Fig. 4). Only the output layer is trained on the response to input signals, while the reservoir is left unchanged (except when making a reservoir re-initialization). The concepts behind RCN are similar to ideas from both kernel methods and RNN theory. Much like a kernel, the reservoir projects the input signals into a higher dimensional space (in this case the state space of the reservoir), where a linear regression can be performed. On the other hand, due to the recurrent delayed connections inside the hidden layer, the reservoir has a form of a short-term memory, called the fading memory which allows temporal information to be stored

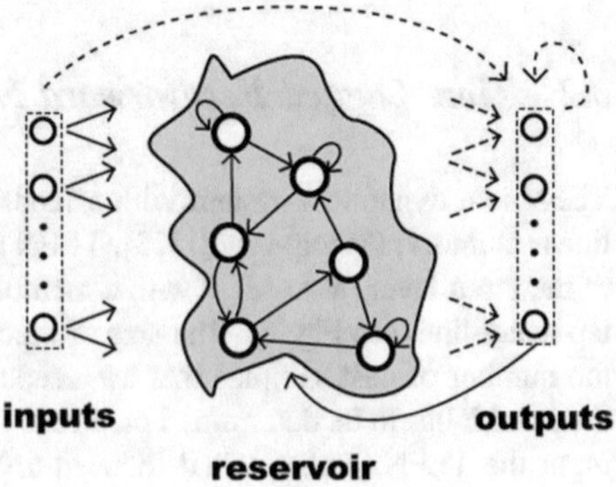

Fig. 4 Reservoir Computing (RC) network with fixed connections (*solid lines*) and adaptable connections (*dashed lines*)

in the reservoir. The general state update equation for the nodes in the reservoir and the readout output equation are as follows:

$$x(k+1) = f\left(W_{res}^{res}x(k) + W_{inp}^{res}u(k) + W_{out}^{res}y(k) + W_{bias}^{res}\right) \tag{3}$$

$$y(k+1) = W_{res}^{out}x(k+1) + W_{inp}^{out}u(k) + W_{out}^{outr}y(k) + W_{bias}^{out} \tag{4}$$

where: $u(k)$ denotes the input at time k; $x(k)$ represents the reservoir state; $y(k)$ is the output; and $f\,()$ is the activation function (with the hyperbolic tangent tan$h()$ as the most common type of activation function). The initial state is usually set to $x(0) = 0$. All weight matrices to the reservoir (denoted as W^{res}) are initialized randomly, while all connections to the output (denoted as W^{out}) are trained. In the general state update Eq. (3), it is assumed a feedback not only between the reservoir neurons expressed by the term $W_{res}^{res}x(k)$, but also a feedback from the output to the reservoir accounted by $W_{out}^{res}y(k)$. The first feedback is considered as the short-term memory, while the second one as a very long-term memory. In order to simplify the computations Following the idea of Antonelo et al. [1], for the present study the second feedback is discarded and a scaling factor α is introduced in the state update equation

$$x(k+1) = f\left((1-\alpha)x(k) + \alpha W_{res}^{res}x(k) + W_{inp}^{res}u(k) + W_{bias}^{res}\right) \tag{5}$$

Parameter α serves as a way to tune the dynamics of the reservoir and improve its performance. The value of α can be chosen empirically or by an optimization. The output calculations are also simplified [1] assuming no direct connections from input to output or connections from output to output

$$y(k+1) = W_{res}^{out}x(k+1) + W_{bias}^{out} \tag{6}$$

Each element of the connection matrix W_{res}^{res} is drawn from a normal distribution with mean 0 and variance 1. The randomly created matrix is rescaled so that the spectral radius $|\lambda_{max}|$ (the largest absolute eigenvalue) is smaller than 1. Standard settings of $|\lambda_{max}|$ lie in a range between 0.7 and 0.98. Once the reservoir topology is set and the weights are assigned, the reservoir is simulated and optimized on the training data set. It is usually done by linear regression (least-squares method) or ridge regression, Bishop [3]. Since the output layer is linear, regularization can be easily applied by adding a small amount of Gaussian noise to the RCN response.

The main advantage of RCN is that it overcomes many of the problems of traditional RNN training such as slow convergence, high computational requirements, and complexity. The computational efforts for training are related to computing the transpose of a matrix or matrix inversion. Once trained, the resulting RCN-based system can be used for real time operation on moderate hardware since the computations are very fast (only matrix multiplications of small matrices).

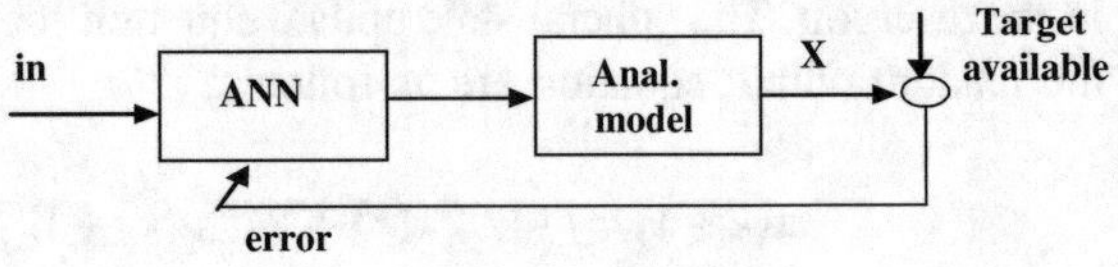

Fig. 5 Hybrid ANN training structure

4 Kinetic Rates Estimation by Time Accounting ANN

The ANNs are a data-based modeling technique where during an optimization procedure (termed also learning) the network parameters (the weights) are updated based on error correction principle. At each iteration, the error between the network output and the corresponding reference has to be computed and the weights are changed as a function of this error. This principle is also known as supervised learning. However, the process kinetic rates are usually not measured variables, therefore, targets (references) are not available and the application of any data-based modeling technique is questionable. A procedure is proposed in the present work to solve this problem. The idea is to propagate the ANN output through a fixed partial analytical model (*Anal. model*) until it comes to a measured process variable (see Fig. 5). The proper choice of this *Anal. model* and the formulation of the error signal for network updating are discussed below. The procedure is based on the following assumptions:

(A1) Not all process states of model (1) are measured.

(A2) All kinetics coefficients are known, that is b and all entries of matrix K are available.

For more convenience, the model (1) is reformulated based on the following augmented vectors

$$X_{\text{aug}} = \begin{bmatrix} X \\ T \end{bmatrix}, \quad X_{\text{aug}} \in R^{n+1}, \quad K_{\text{aug}} = \begin{bmatrix} K \\ b \end{bmatrix}, \quad K_{\text{aug}} \in R^{(n+1)\times m}. \tag{7}$$

Then (1.1) is rewritten as

$$\frac{\mathrm{dX}_{\text{aug}}}{\mathrm{dt}} = K_{\text{aug}}\varphi\left(X_{\text{aug}}\right) - \bar{D}X_{\text{aug}} + U \text{ with } \bar{D} = \begin{bmatrix} D & 0 \\ 0 & d_0 \end{bmatrix}, \quad U = \begin{bmatrix} U_x \\ U_T \end{bmatrix} \tag{8}$$

Step 1: State vector partition A

The general dynamical model (8) represents a particular class of nonlinear state-space models. The nonlinearity lies in the kinetics rates $\varphi\left(X_{\text{aug}}\right)$ that are nonlinear functions of the state variables X_{aug}. These functions enter the model in the form $K_{\text{aug}}\varphi\left(X_{\text{aug}}\right)$, where K_{aug} is a constant matrix, which is a set of linear combinations of the same nonlinear functions $\varphi_1(X_{\text{aug}}), \ldots, \varphi_m(X_{\text{aug}})$. This particular structure can be exploited to separate the nonlinear part from the linear part of the model by a suitable linear state transformation. More precisely, the following nonsingular partition is chosen, Chen and Bastin [4].

$$LK_{\text{aug}} = \begin{bmatrix} K_a \\ K_b \end{bmatrix}, \ \text{rank}(K_{\text{aug}}) = l, \tag{9.1}$$

where $L \in R^{n \times n}$ is a quadratic permutation matrix, K_a is a lxm full row rank submatrix of K_{aug} and $K_b \in R^{(n-l) \times m}$. The induced partitions of vectors X_{aug} and U are

$$LX_{\text{aug}} = \begin{bmatrix} X_a \\ X_b \end{bmatrix}, \ LU = \begin{bmatrix} U_a \\ U_b \end{bmatrix}, \tag{9.2}$$

with $X_a \in R^l$, $U_a \in R^l$, $X_b \in R^{n-l}$, $U_b \in R^{n-l}$.

According to (9.1, 9.2), model (8) is also partitioned into two submodels

$$\frac{\mathrm{d}X_a}{\mathrm{d}t} = K_a \varphi(X_a, X_b) - \bar{D}X_a + U_a \tag{10}$$

$$\frac{\mathrm{d}X_b}{\mathrm{d}t} = K_b \varphi(X_a, X_b) - \bar{D}X_b + U_b \tag{11}$$

Based on (9.1, 9.2), a new vector $Z \in R^{n+1-l}$ is defined as a linear combination of the state variables

$$Z = A_0 X_a + X_b, \tag{12}$$

where matrix $A_0 \in R^{(n+1-l) \times l}$ is the unique solution of

$$A_0 K_a + K_b = 0, \tag{13}$$

that is

$$A_0 = -K_b K_a^{-1}, \tag{14}$$

Note that, a solution for A_0 exist if and only if K_a is not singular. Hence, a necessary and sufficient condition for the existence of a desired partition (9.1, 9.2), is that K_a is a pxm full rank matrix, which was the initial assumption. Then, the first derivative of vector Z is

$$\begin{aligned} \frac{\mathrm{d}Z}{\mathrm{d}t} &= A_0 \frac{\mathrm{d}X_a}{\mathrm{d}t} + \frac{\mathrm{d}X_b}{\mathrm{d}t} \\ &= A_0 [K_a \varphi(X_a, X_b) - \bar{D}X_a + U_a] + K_b \varphi(X_a, X_b) - \bar{D}X_b + U_b \\ &= (A_0 K_a + K_b) \varphi(X_a, X_b) - \bar{D}(A_0 X_a + X_b) + A_0 U_a + U_b \end{aligned} \tag{15}$$

Since matrix A_0 is chosen such that Eq. (13) holds, the term in (15) related with φ is canceled and we get

$$\frac{dZ}{dt} = -\bar{D}Z + A_0U_a + U_b \tag{16}$$

The state partition A results in a vector Z whose dynamics, given by Eq. (15), is independent of the kinetic rate vector φ. In general, (9) is not a unique partition and for any particular case a number of choices are possible.

Step 2: State vector partition B (measured and unmeasured states)
Now a new state partition is defined as subvectors of measured and unmeasured states X_1, X_2, respectively. The model (8) is also partitioned into two submodels

$$\frac{dX_1}{dt} = K_1\varphi(X_1, X_2) - \bar{D}X_1 + U_1 \tag{17.1}$$

$$\frac{dX_2}{dt} = K_2\varphi(X_1, X_2) - \bar{D}X_2 + U_2 \tag{17.2}$$

From state partitions A and B, vector Z can be represented in the following way

$$Z = A_0X_a + X_b = A_1X_1 + A_2X_2. \tag{18}$$

The first representation is defined in (12), then applying linear algebra transformations A_1 and A_2 are computed to fit the equality (18). **The purpose of state partitions *A* and *B* is to estimate the unmeasured states (vector X_2) independently of the kinetics rates (vector φ).** The recovery of X_2 is defined as state observer.

Step 3: State observer
Based on (16) and starting with known initial conditions, Z can be estimated as follow (in this work the estimations are denoted by hat)

$$\frac{d\hat{Z}}{dt} = -D\hat{Z} + A_0(F_{\text{in\_}a} - F_{\text{out\_}a}) + (F_{\text{in\_}b} - F_{\text{out\_}b}) \tag{19}$$

Then according to (18) the unmeasured states X_2 are recovered as

$$\hat{X}_2 = A_2^{-1}(\hat{Z} - A_1X_1) \tag{20}$$

Note that, estimates $\hat{X}_2$ exist if and only if A_2 is not singular, Bastin and Dochain [2]. Hence, **a necessary and sufficient condition for observability of the unmeasured states is that A_2 is a full rank matrix.**

Step 4: Error signal for NN training
The hybrid structure for NN training is shown in Fig. 6, where the adaptive hybrid model (AHM) is formulated as

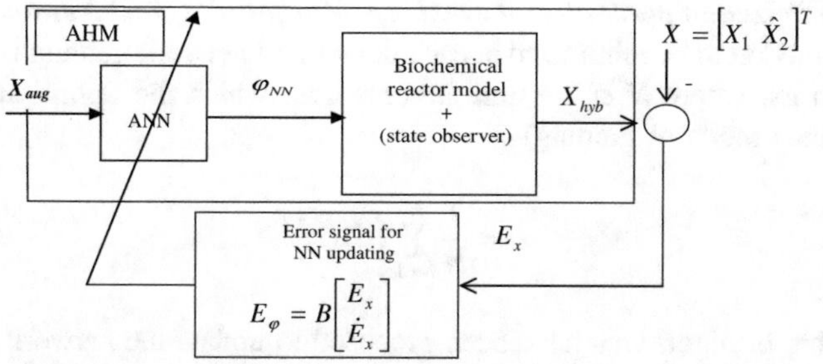

Fig. 6 Hybrid NN-based reaction rates identification structure

$$\frac{dX_{hyb}}{dt} = K_{aug}\varphi_{NN} - \bar{D}X_{hyb} + U + \Omega(X_{aug} - X_{hyb}) \tag{21}$$

The true (but unknown) process behavior is assumed to be represented by (8). Then the error dynamics is modeled as the difference between (8) and (21)

$$\frac{d(X_{aug} - X_{hyb})}{dt} = K_{aug}(\varphi - \varphi_{NN}) - \bar{D}(X_{aug} - X_{hyb}) + \Omega(X_{aug} - X_{hyb}) \tag{22}$$

The following definitions take place:
$E_x = (X_{aug} - X_{hyb})$ is termed as the observation error,
$E_\varphi = \varphi - \varphi_{NN}$ is the error signal for updating the ANN parameters.

X_{aug} consists of the measured (X_1) and the estimated ($\hat{X}_2$) states. Thus, (22) can be rearranged as follows

$$\frac{dE_x}{dt} = K_{aug}E_\varphi - (\bar{D} - \Omega)E_x \tag{23}$$

and from (23) the error signal for NN training is

$$E_\varphi = K_{aug}^{-1}\begin{bmatrix}\bar{D} - \Omega & 1\end{bmatrix}\begin{bmatrix} E_x \\ \dot{E}_x \end{bmatrix} = B\begin{bmatrix} E_x \\ \dot{E}_x \end{bmatrix}, \quad B = K_{aug}^{-1}\begin{bmatrix}\bar{D} - \Omega & 1\end{bmatrix} \tag{24}$$

Ω is a design parameter which defines the speed of the observation error convergence. The necessary identifiability condition for the kinetic rate vector is the nonsingularity of matrix K_{aug}. Note that, the error signal for updating the network parameters is a function of the observation error (E_x) and the speed of the observation error ($\dot{E}_x$). The intuition behind is that the network parameters are changed proportionally to their effect on the prediction of the process states and the prediction of their dynamics.

Step 5: Optimization porcesure—Levenberg–Marquardt Quasi-Newton algorithm The cost function to be minimized at each iteration of network training is the sum of squared errors, where N is the time instants over which the optimization is performed (batch mode of training)

$$J_k = \frac{1}{N} \sum_{i=1}^{N} \left[E_\varphi(i)\right]^2 \tag{25}$$

A number of algorithms have been proposed to update the network parameters (w). For this study the Levenberg–Marquardt (LM) Quasi-Newton method is the chosen algorithm due to its faster convergence than the steepest descent or conjugate gradient methods, Hagan et al. [6]. One (k) iteration of the classical Newton's method can be written as

$$w_{k+1} = w_k - H_k^{-1} g_k, \; g_k = \frac{\partial J_k}{\partial w_k}, \; H_k = \frac{\partial J_k^2}{\partial w_k \partial w_k} \tag{26}$$

where g_k is the current gradient of the performance index (25) H_k is the Hessian matrix (second derivatives) of the performance index at the current values (k) of the weights and biases. Unfortunately, it is complex and expensive to compute the Hessian matrix for a dynamical ANN. The LM method is a modification of the classical Newton method that does not require calculation of the second derivatives. It is designed to approach second-order training speed without having to compute directly the Hessian matrix. When the performance function has the form of a sum of error squares (25), at each iteration the Hessian matrix is approximated as

$$H_k = J_k^T J_k \tag{27}$$

where J_k is the Jacobian matrix that contains first derivatives of the network errors (e_k) with respect to the weights and biases

$$J_k = \frac{\partial E_{\varphi k}}{\partial w_k}, \tag{28}$$

The computation of the Jacobian matrix is less complex than computing the Hessian matrix. The gradient is then computed as

$$g_k = J_k E_{\varphi k} \tag{29}$$

The LM algorithm updates the network weights in the following way

$$w_{k+1} = w_k - \left[J_k^T J_k + \mu I\right]^{-1} J_k^T E_{\varphi k} \tag{30}$$

when the scalar μ □ is zero, this is just Newton's method, using the approximate Hessian matrix. When μ □ is large, this becomes gradient descent with a small step size. Newton's method is faster and more accurate near an error minimum, so the aim is to shift toward Newton's method as quickly as possible. Thus, μ □ is decreased after each successful step (reduction in performance function) and is increased only when a tentative step would increase the performance function. In this way, the performance function will always be reduced at each iteration of the algorithm.

5 Case Study—Estimation of Sugar Crystallization Growth Rate

Sugar crystallization occurs through mechanisms of nucleation, growth, and agglomeration that are known to be affected by several not well-understood operating conditions. The search for efficient methods for process description is linked both to the scientific interest of understanding fundamental mechanisms of the crystallization process and to the relevant practical interest of production requirements. The sugar production batch cycle is divided in several phases. During the first phase the pan is partially filled with a juice containing dissolved sucrose. The liquor is concentrated by evaporation, under vacuum, until the supersaturation reaches a predefined value. At this point seed crystals are introduced into the pan to induce the production of crystals (crystallization phase). As evaporation takes place further liquor or water is added to the pan. This maintains the level of supersaturation and increases the volume contents. The third phase consists of tightening which is controlled by the evaporation capacity, see Georgieva et al. [5] for more details. Since the objective of this paper is to illustrate the technique introduced in Sect. 4, the following assumptions are adopted:

i. Only the states that explicitly depend on the crystal growth rate are extracted from the comprehensive mass balance process model;
ii. The population balance is expressed only in terms of number of crystals;
iii. The agglomeration phenomenon is neglected.

The simplified process model is then

$$\frac{\mathrm{d}M_s}{\mathrm{d}t} = -k_1 G + F_f \rho_f B_f Pur_f \tag{31.1}$$

$$\frac{\mathrm{d}M_c}{\mathrm{d}t} = k_1 G \tag{31.2}$$

$$\frac{\mathrm{d}T_m}{\mathrm{d}t} = k_2 G + bF_f + cJ_{\mathrm{vap}} + d \tag{31.3}$$

$$\frac{\mathrm{d}m_0}{\mathrm{d}t} = k_3 G \tag{31.4}$$

where M_s is the mass of dissolved sucrose, M_c is the mass of crystals, T_m is the temperature of the massecuite, m_0 is the number of crystals. Pur_f and ρ_f are the purity (mass fraction of sucrose in the dissolved solids) and the density of the incoming feed. F_f is the feed flowrate, J_{vap} is the evaporation rate and *b, c, d* are parameters incorporating the enthalpy terms and specific heat capacities. They are derived as functions of physical and thermodynamic properties. The full state vector is

$X_{\mathrm{aug}} = [M_s \quad M_c \quad T_m \quad m_0]^T$, with $K_{\mathrm{aug}} = [-k_1 \quad k_1 \quad k_2 \quad k_3]^T$. Now we are in a position to apply the formalism developed in Sects. 2.2 and 2.3 for this particular reaction process.

We chose the following state partition A: $X_a = M_c$, $X_b = [M_s \quad T_m \quad m_0]^T$ and the solution of Eq. (13) is

$$A_0 = \begin{bmatrix} 1 & -\frac{k_2}{k_1} & -\frac{k_3}{k_1} \end{bmatrix}^T \tag{32}$$

M_c and T_m are the measured states, then the unique state partition B is

$$X_1 = [M_c \quad T_m]^T, X_2 = [M_s \quad m_0]^T,$$

Taking into account (32), the matrices of the second representation of vector Z in (18) are computed as

$$A_1 = \begin{bmatrix} 1 & -\frac{k_2}{k_1} & -\frac{k_3}{k_1} \\ 0 & 1 & 0 \end{bmatrix}^T, A_2 = \begin{bmatrix} 1 & 0 & 0 \\ 0 & 0 & 1 \end{bmatrix}^T$$

For this case $D = 0$, then the estimation of the individual elements of Z are

$$\hat{Z}_1 = M_c + \hat{M}_s, \quad \hat{Z}_2 = -\frac{k_2}{k_1} M_c + T_m, \quad \hat{Z}_3 = -\frac{k_3}{k_1} M_c + \hat{m}_0 \tag{33}$$

The analytical expression for the estimation of the unmeasured states is then

$$\begin{bmatrix} \hat{M}_s \\ \hat{m}_0 \end{bmatrix} = \begin{bmatrix} 1 & 0 & 0 \\ 0 & 0 & 1 \end{bmatrix} \left(\begin{bmatrix} \hat{Z}_1 \\ \hat{Z}_2 \\ \hat{Z}_3 \end{bmatrix} - \begin{bmatrix} 1 & 0 \\ -\frac{k_2}{k_1} & 1 \\ -\frac{k_3}{k_1} & 0 \end{bmatrix} \begin{bmatrix} M_c \\ T_m \end{bmatrix} \right) \tag{34}$$

The observation error is defined as

$$E_x = \begin{bmatrix} \hat{M}_s - M_{s\,\mathrm{hyb}} \\ M_c - M_{c\,\mathrm{hyb}} \\ \hat{m}_0 - m_{0\,\mathrm{hyb}} \\ T_m - T_{m\,\mathrm{hyb}} \end{bmatrix} \tag{35}$$

In the numerical implementation the first derivative of the observation error is computed as the difference between the current $E_x(k)$ and the previous value $E_x(k-1)$ of the observation error divided by the integration step (Δt)

$$\dot{E}_x = \frac{E_x(k) - E_x(k-1)}{\Delta t} \tag{36}$$

The three types of time accounting ANNs were trained with the same training data coming from six industrial batches (training batches). The physical inputs to all networks are (M_c, T_m,m_0,M_s), the network output is G_{NN}. Two of the inputs (M_c, T_m) are measurable, the others (m_0,M_s) are estimated. In order to improve the comparability between the different networks a linear activation function is located at the single output node (see Fig. 2, Layer 2—purelin) and hyperbolic tangent functions are chosen for the hidden nodes (Fig. 2, Layer 1—tansig). Though other S-shaped activation functions can be also considered for the hidden nodes, our choice was determined by the symmetry of the hyperbolic tangent function into the interval (−1, 1).

The hybrid models are compared with an analytical model of the sugar crystallization, reported in Oliveira et al. [12], where G is computed by the following empirical correlation

$$G = K_g \exp\left[-\frac{57000}{R(T_m + 273)}\right](S-1)\exp[-13.863(1-P_{\text{sol}})]\left(1+2\frac{V_c}{V_m}\right), \tag{37}$$

where S is the supersaturation, P_{sol} is the purity of the solution and V_c/V_m is the volume fraction of crystals. K_g is a constant, optimized following a nonlinear least-squares regression.

The performance of the different models is examined with respect to prediction quality of the crystal size distribution (CSD) at the end of the process which is quantified by two parameters—the final average (in mass) particle size (AM) and the final coefficient of particle variation (CV). The predictions given by the models are compared with the experimental data for the CSD (Table 1), coming from eight batches not used for network training (validation batches). The results with respect to different configurations of the networks are summarized in Tables 2, 3, and 4. All hybrid models (Eqs. 31 +RNN/TLNN/RCN) outperform the empirical model (37) particularly with respect to predictions of CV. The predictions based on TLFN and TCN are very close especially for higher reservoir dimension. Increasing the RCN hidden nodes (from 100 to 200) reduces the AM and CV prediction errors, however augmenting the reservoir dimension from 200 to 300 does not bring substantial improvements. The hybrid models with RNN exhibit the best performance though the successful results reported in Table 2 were preceded by a great number of unsuccessful (not converging) trainings. As with respect to learning efforts the RCN training takes in average few seconds on an Intel Core2 Duo processor based computer and by far is the easiest and fastest dynamical regressor.

Table 1 Final CSD—experimental data versus analytical model predictions

Batch no.	Experimental data		Analytical model (Eqs. 31 + 37)	
	AM (mm)	CV (%)	AM (mm)	CV (%)
1	0.479	32.6	0.583	21.26
2	0.559	33.7	0.542	18.43
3	0.680	43.6	0.547	18.69
4	0.494	33.7	0.481	14.16
5	0.537	32.5	0.623	24.36
6	0.556	35.5	0.471	13.642
7	0.560	31.6	0.755	34.9
8	0.530	31.2	0.681	27.39
Average error			13.7 %	36.1 %

Table 2 Final CSD—hybrid model predictions (Eqs. 31 + RNN)

RNN	Batch no.	AM (mm)	CV (%)
Exogenous input	1	0.51	29.6
Delay: 2	2	0.48	30.7
	3	0.58	33.6
Recurrent input	4	0.67	31.7
Delay: 2	5	0.55	29.5
	6	0.57	34.5
Total number of inputs: 14	7	0.59	29.6
Hidden neurons: 5	8	0.53	32.2
Average error (%)		*4.1*	*7.5*
exogenous input	1	0.59	30.7
Delay: 1	2	0.55	41.5
	3	0.59	39.3
Recurrent input	4	0.51	35.9
delay: 3	5	0.49	32.1
	6	0.58	31.7
Total number of inputs: 11	7	0.56	30.5
Hidden neurons: 5	8	0.53	36.8
Average error (%)		*5.2*	*9.2*
Exogenous input	1	0.51	30.9
Delay: 3	2	0.56	31.1
	3	0.59	37.2
Recurrent input	4	0.48	29.8
Delay: 1	5	0.52	34.8
	6	0.51	32.4
Total number of inputs: 17	7	0.59	30.6
Hidden neurons:5	8	0.50	33.5
Average error (%)		*3.6*	*6.9*

Table 3 Final CSD—hybrid model predictions (Eqs. 31 + TLNN)

TLNN	Batch no.	AM (mm)	CV (%)
Tap delay: 1	1	0.49	30.8
	2	0.51	37.1
	3	0.62	31.5
Total number of inputs: 8	4	0.60	35.5
Hidden neurons: 5	5	0.57	36.2
	6	0.52	28.7
	7	0.55	38.6
	8	0.54	32.4
Average error (%)		*6.02*	*11.0*
Tap delay: 2	1	0.51	37.5
	2	0.49	31.6
	3	0.59	34.6
	4	0.53	40.3
Total number of inputs: 12	5	0.60	35.2
Hidden neurons: 5	6	0.49	31.5
	7	0.51	29.6
	8	0.54	30.3
Average error (%)		*5.9*	*10.8*
Tap delay: 3	1	0.479	30.3
	2	0.559	41.2
	3	0.680	39.4
	4	0.494	35.7
Total number of inputs: 16	5	0.537	35.4
Hidden neurons: 5	6	0.556	30.3
	7	0.560	29.9
	8	0.530	28.3
Average error (%)		*5.8*	*10.3*

Table 4 Final CSD—hybrid model predictions (Eqs. 31 + RCN)

RCN	Batch no.	AM (mm)	CV (%)
Reservoir	1	0.53	31.2
Dimension: 100 nodes	2	0.49	28.1
	3	0.57	43.6
Total number of inputs: 4	4	0.61	41.7
	5	0.59	39.6
	6	0.60	36.1
	7	0.51	30.4
	8	0.54	40.2
Average error (%)		*6.8*	*12.0*
Reservoir	1	0.56	40.1
Dimension: 200 nodes	2	0.51	37.4
	3	0.61	36.2
	4	0.56	38.6
Total number of inputs: 4	5	0.49	28.9
	6	0.59	34.7
	7	0.61	30.4
	8	0.54	39.2
Average error (%)		*5.9*	*10.2*
Reservoir	1	0.59	33.9
Dimension: 300 nodes	2	0.48	28.8
	3	0.57	39.7
	4	0.51	29.6
Total number of inputs: 4	5	0.53	31.8
	6	0.51	33.9
	7	0.49	30.7
	8	0.57	36.9
Average error (%)		*5.9*	*9.8*

6 Conclusions

This work is focused on presenting a more efficient computational scheme for estimation of process reaction rates based on temporal ANN architectures. It is assumed that the kinetics coefficients are all known and do not change over the process run, while the process states are not all measured and therefore need to be estimated. It is a very common scenario in reaction systems with low or medium complexity.

The concepts developed here concern two aspects. On one side we formulate a hybrid (temporal ANN+ analytical) model that outperforms the traditional reaction rate estimation approaches. On the other side a procedure for ANN supervised training is introduced when target (reference) outputs are not available. The network is embedded in the framework of a first principle process model and the error signal for updating the network weights is determined analytically. According to the procedure, first the unmeasured states are estimated independently of the reaction rates and then the ANN is trained with the estimated and the measured data. Ongoing research is related with the integration of the hybrid models proposed in this work in the framework of a model-based predictive control.

Acknowledgments This work was financed by the Portuguese Foundation for Science and Technology within the activity of the Research Unit IEETA-Aveiro, which is gratefully acknowledged.

References

1. Antonelo, E.A., Schrauwen, B., Campenhout, J.V.: Generative modeling of autonomous robots and their environments using reservoir computing. Neural Process. Lett. **26**(3), 233–249 (2007)
2. Bastin, G., Dochain, D.: On-line Estimation and Adaptive Control of Bioreactors. Elsevier Science Publishers, Amsterdam (1990)
3. Bishop, C.M.: Pattern Recognition and Machine Learning. Springer (2006)
4. Chen, L., Bastin, G.: Structural identifiability of the yeals coefficients in bioprocess models when the reaction rates are unknown. Math. Biosci. **132**, 35–67 (1996)
5. Georgieva, P., Meireles, M.J., Feyo de Azevedo, S.: Knowledge based hybrid modeling of a batch crystallization when accounting for nucleation, growth and agglomeration phenomena. Chem. Eng. Sci. **58**, 3699–3707 (2003)
6. Hagan, M.T., Demuth, H.B., Beale, M.H.: Neural Network Design. PWS Publishing, Boston, MA (1996)
7. Haykin, S.: Neural Networks: A Comprehensive Foundation. Prentice Hall, NJ (1999)
8. Jaeger, H.: The "echo state" approach to analysing and training recurrent neural networks. Technical Report GMD Report 148, German National Research Center for Information Technology (2001)
9. Maass, W., Natschlager, T., Markram, H.: Real-time computing without stable states: a new framework for neural computation based on perturbations. Neural Comput. **14**(11), 2531–2560 (2002)

10. Mandic D.P., Chambers, J.A.: Recurrent Neural Networks for Prediction: Learning Algorithms, Architectures and Stability (Adaptive & Learning Systems for Signal Processing, Communications & Control). Wiley (2001)
11. Noykove, N., Muller, T.G., Gylenberg, M., Timmer J.: Quantitative analysis of anaerobic wastewater treatment processes: identifiably and parameter estimation. Biotechnol. Bioeng. **78** (1), 91–103 (2002)
12. Oliveira, C., Georgieva, P., Rocha, F., Feyo de Azevedo, S.: Artificial Neural Networks for Modeling in Reaction Process Systems, Neural Computing & Applications. Springer **18**, 15–24 (2009)
13. Principe, J.C., Euliano, N.R., Lefebvre, W.C.: Neural and Adaptive Systems: Fundamentals Through Simulations. Wiley, New York (2000)
14. Steil, J.J.:. Backpropagation-Decorrelation: Online recurrent learning with O(N) complexity. In: Proceedings of the International Joint Conference on Neural Networks (IJCNN), vol. 1, pp. 843–848
15. Walter, E., Pronzato, L.: Identification of Parametric Models from Experimental Data. Springer, London (1997)

Periodic Time-Varying Observer-Based Learning Control of A/F Ratio in Multi-cylinder IC Engines

Tomohiro Hara, Tielong Shen, Yasuhiko Mutoh and Yinhua Liu

Abstract This paper presents an air–fuel ratio control scheme via individual fuel injection for multi-cylinder internal combustion engines. The aim of presented control scheme is to improve air–fuel ratio precision by real-time compensation of the unknown offset in the fuel path of individual cylinder, which represents the effect of the cylinder-to-cylinder imbalance caused by the perturbations in each injector gain or disturbances in the dynamics of fuel injection path. First, the fueling-to-exhaust gas mixing system is treated as single-input single-output (SISO) periodic time-varying system where the input is fuel injection command for each cylinder and the output is the air–fuel ratio measured at each exhaust bottom dead center (BDC). Then, a periodic time-varying observer is presented that provides an estimation of the internal state of the system. Based on the presented observer, an iterative learning control strategy is proposed to compensate the unknown offset. The effectiveness of the learning control scheme will be demonstrated with the simulation and experiment results conducted on a commercial car used engine with six cylinders.

1 Introduction

For the internal combustion engines with multi-cylinders, individual cylinder actuation is an effective way to achieve high performance of the emission and the torque generation. Recently, due to the rapid progress in the technology of car electronics and electrical control unit (ECU), the individual cylinder control problem has begun to attract the attention of researchers in engine control community. For example, the literatures [1–3, 8, 9] addressed the air–fuel ratio control problem based on the estimation of the individual air–fuel ratio in each cylinder. The air–fuel ratio or torque balancing with individual fueling or variable valve timing is investigated by [4].

T. Hara · T. Shen (✉) · Y. Mutoh · Y. Liu
Department of Engineering and Applied Sciences, Sophia University,
Tokyo 102-8554, Japan
e-mail: tetu-sin@sophia.ac.jp

V. Sgurev et al. (eds.), *Recent Contributions in Intelligent Systems*,
Studies in Computational Intelligence 657,
DOI 10.1007/978-3-319-41438-6_5

Meanwhile, when an engine is operated on a static mode, the effect of cylinder-to-cylinder imbalance might be equivalently represented as offset in the level of actuation signal of each cylinder. If we focus on the air–fuel ratio control problem, the variation of the air–fuel ratio measured at the exhaust manifold, which caused by the perturbation of fuel mass injected into each cylinder, can be regarded as constant offset in each fuel injection path. As it is shown in [6], at a static mode, the dynamics of BDC-scaled air–fuel ratio under individually actuated fuel injection, which is denoted as single input delivered to each cylinder according the crank angle, can be represented as periodic time-varying linear system, where the periodic time-varying parameter is due to the difference between the characteristics of each cylinder. In general, the cylinder-to-cylinder imbalance represented by the offset is difficult to be calibrated exactly.

Motivated by the periodic time-varying characteristics of the fueling-to-exhaust gas mixing system, this paper proposes a periodic time-varying observer-based learning control scheme to real-time compensation of the unknown offset in the fueling path of each cylinder. At first, a periodic time-varying linear model is introduced from the physical observation to describe the dynamical behavior of the fueling-to-exhaust gas mixing system. Then, a state observer is designed based on the periodic time-varying model. By embedding the observer into the iterative learning control law, the air–fuel ratio controller with real-time unknown offset compensation is presented for a six cylinder gasoline engines. To demonstrate the effectiveness of the proposed control scheme, numerical simulation and experiments will be conducted on a full-scaled engine test bench.

This paper is organized as follows. Section 2 gives a detailed explanation of the targeted system and the control problem. Section 3 presents the periodic time-varying modeling for the engines, and in Sect. 4, the observer design will be given with the model. The learning control scheme embedded with the periodic time-varying observer is proposed in Sect. 5. Finally, Sect. 6 demonstrates the simulation and experiment results.

2 System and Problem

In multi-cylinder combustion engines, the combustion event in each cylinder occurs sequentially along the crankshaft angle, and the burnt gas is exhausted during the corresponding exhaust phase. Usually, in the multi-cylinder internal combustion engines, several cylinders share a common exhaust manifold where the burnt gases from different cylinders exhausted at different timing in crank angle are mixed and exhausted to the atmosphere passing through the tail pipe. To perform online air–fuel ratio control, air–fuel ratio sensor is usually equipped at the gas mixing point of the exhaust manifold, and the fuel mass injected to each cylinder is decided and delivered by ECU. As an ideal situation, if the air charge of each cylinder is same as the others, then the fuel injection command for all cylinders might take unified value to obtain the targeted air–fuel ratio. However, imbalance between the cylinders, for

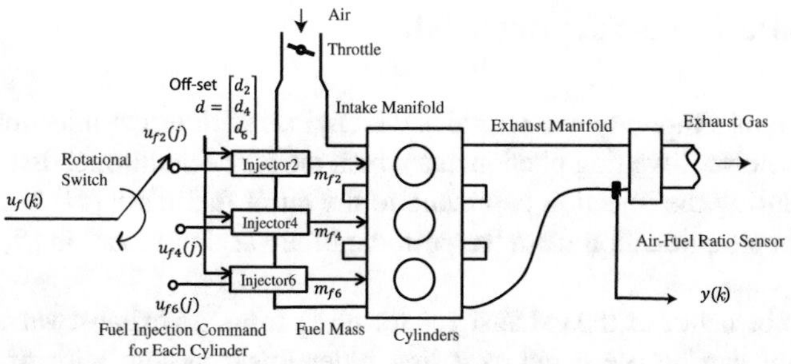

Fig. 1 Fueling-to-exhaust gas mixing system

instance, the injector gains and disturbances in the intake path of each cylinder, will unfortunately cause the errors of air–fuel ratio in each cylinder. As a result, it will damage the air–fuel ratio control performance.

For example, in the six-cylinder V-type engines, which is the main target of this paper, each of three cylinders shares a common exhaust manifold as shown in Fig. 1. We focus on one of the two exhaust manifolds, where the cylinders numbered as No.1, No.2, and No.3 share the same manifold, and for each cylinder the fuel injection command is decided individually, which is denoted with a unified signal $u_f(k)$ where k is the sampling index with BDC-scaled sampling period, i.e., $u_f(k)$ is assigned to the individual cylinder with the rotation $u_{f1}, u_{f2}, u_{f3}, u_{f1}, \ldots$, here u_{fi} $(i = 1, 2, 3)$ denotes the fuel injection command for the No. 1, No. 2, No. 3 cylinders. The sensor is mounted at the gas mixing point of exhaust manifold, where the burnt gas exhausted from each cylinder with BDC-scaled delay comes together. Hence, the measured air–fuel ratio, denoted as output $y(k)$, is of mixed gas from different cylinders. In order to represent the imbalance of air–fuel ratio between the cylinders, the unknown constant vector $d = [d_1, d_2, d_3]^T$ is introduced where $d_i (i = 1, 2, 3)$ denotes the unknown offset.

The control problem considered in this paper is to find a real-time control scheme that decides the fuel injection mass for each cylinder with the offset compensation without any previous information about the offset. As a result, the air–fuel ratio control precision is improved by the individual cylinder fuel injection. In the following, this problem is challenged with two steps: first, a state observer is designed for handling the internal state of the dynamical system from the unified fuel injection command $u(k)$ to the sensor output $y(k)$. Second, by embedding the state observer a learning control law is constructed to estimate the unknown offset and perform the compensation. We will start this challenging by introducing a periodic time-varying linear model for the targeted systems.

3 Periodic Time-Varying Model

For the sake of simplicity, we consider the engines with direct injection. It means that there is no wall-wetting phenomena which introduce additional first order delay dynamics during the injection command to the burnt fuel mass [6]. A brief review with a slight simplification from the periodic time-varying model in [5] is given as follows.

Suppose behavior of the exhaust gas traveling from the exhaust valve to the gas mixing point can be presented as a first order linear system with time constant $T_i (i = 1, 2, 3)$. Then, the rate of the air mass and the fuel mass flow measured at the sensor position, denoted as $\dot{m}_{sfi}$ and $\dot{m}_{sai}$ for No. i cylinder, are calculated as follows, respectively.

$$\dot{m}_{sai}(t') = \int_0^{t'} \frac{1}{T_i}\dot{m}_{ai}(\tau)e^{-\frac{1}{T_i}(t'-\tau)}\,d\tau \tag{1}$$

$$\dot{m}_{sfi}(t') = \int_0^{t'} \frac{1}{T_i}\dot{m}_{fi}(\tau)e^{-\frac{1}{T_i}(t'-\tau)}\,d\tau \tag{2}$$

where $t' = t - t_{BDC_i}$, t_{BDC_i} denotes the exhaust valve opening time of the No. i cylinder. Thus, the gas mixing including the air and the fuel is obtained as

$$\dot{m}_{sa}(t) = \sum_{i=1}^{N} \dot{m}_{sai}(t - t_{BDC_i}) \tag{3}$$

$$\dot{m}_{sf}(t) = \sum_{i=1}^{N} \dot{m}_{sfi}(t - t_{BDC_i}) \tag{4}$$

where $\dot{m}_{sa}(t)$ and $\dot{m}_{sf}(t)$ are the total mass of air and the fuel, respectively.

Therefore, the air–fuel ratio of the gas passing through the sensor head is given by

$$\eta(t) = \frac{\dot{m}_{sf}(t)}{\dot{m}_{sa}(t)} \tag{5}$$

Furthermore, taking the sensor's delay into account, which is usually represented as the first order system with time constant τ_s, the measured air–fuel ratio η_s is obtained as

$$\dot{\eta}_s(t) = \frac{1}{\tau_s}[-\eta_s(t) + \eta(t)] \tag{6}$$

Under the assumption that the exhaust gases exist in the manifold no longer that a combustion cycle, the system output $\eta_s(t)$ will be a periodic function with the period of combustion cycle time T when the engine is working at a static operation mode.

Note that the sampling period for individual cylinder injection is $2\pi/3$ in crank angle, i.e., the sampling and delivering the control signal are performed at each BTC. Denote the time interval between two adjacent BDCs as T_s, then $T = 3T_s$ for the exhaust system with three cylinders. Let j be the index of combustion cycle, we have the following relationship at each discrete sampling time kT_s

$$y(kT_s) = y(jT + (i-1)T_s) \tag{7}$$

Furthermore, for the direct injection engines, the fuel mass injected into the No. i cylinder is given by

$$\dot{m}_{fi}(t) = u_{fi}(t) + d_i, \quad (i = 1, 2, 3) \tag{8}$$

where d_i is introduced to represented the unknown offset in the fuel injection path. With the unified injection command $u_f(k) = u_{fi}(k)$, and $\mathrm{Mod}(k, 3) = i - 1$ in mind, we obtain the discrete gas mixing model as follows:

$$\eta(k) = \sum_{i=1}^{p} r_i(t) \;\; [u_{fi}(k - p + i) + d_i] \\ + \sum_{k+p+1}^{3} r_i(k)[u_{fi}(k - N - p + i) + d_i] \tag{9}$$

for $p = 1, 2, 3$ when $\mathrm{Mod}(k, 3) = p - 1$. For the sensor dynamics, we have

$$\eta_s(k+1) = g\eta_s(k) + (1-g)\eta(k) \tag{10}$$

with $g = 1 - T_s/\tau_s$.

In order to obtain a state-space model, define the state variables as

$$\begin{cases} x_1(k) = u_f(k-1) + m(k-1)d \\ x_2(k) = u_f(k-2) + m(k-2)d \\ x_3(k) = \eta_s(k) \end{cases} \tag{11}$$

with

$$d = \begin{bmatrix} d_1 \\ d_2 \\ d_3 \end{bmatrix}, \quad m(k) = \beta\,\Gamma^k$$

where β and Γ are defined by

$$\Gamma = \begin{bmatrix} 0\,1\,0 \\ 0\,0\,1 \\ 1\,0\,0 \end{bmatrix}, \quad \beta = \begin{bmatrix} 1\,0\,0 \end{bmatrix}$$

This leads to the following state equation:

$$\begin{bmatrix} x_1(k+1) \\ x_2(k+1) \\ x_3(k+1) \end{bmatrix} = A(k) \begin{bmatrix} x_1(k) \\ x_2(k) \\ x_3(k) \end{bmatrix} + B(k)[u(k) + m(k)d] \tag{12}$$

and the output equation

$$y(k) = C \begin{bmatrix} x_1(k) \\ x_2(k) \\ x_3(k) \end{bmatrix} \tag{13}$$

where

$$A(k) = \begin{bmatrix} 0 & 0 & 0 \\ 1 & 0 & 0 \\ (1-g)r_1(k) & (1-g)r_2(k) & g \end{bmatrix}$$

$$B(k) = \begin{bmatrix} 1 \\ 0 \\ (1-g)r_3(k) \end{bmatrix}, C = \begin{bmatrix} 0 & 0 & 1 \end{bmatrix}$$

and the parameter $r_i(k)$ $(i = 1, 2, 3)$ is a periodic time varying with the period 3, i.e.,

$$r_i(k+3) = r_i(k), \quad (i = 1, 2, 3) \tag{14}$$

holds for all $k \geq 0$.

4 Observer Design

In this section, a state observer will be designed based on the periodic time-varying model obtained in Sect. 3. For the sake of simplicity, the model (12) is rewritten as follows with a concise notation

$$\begin{cases} x(k+1) = A(k)x(k) + B(k)(u_f(k) + m(k)d) \\ y(k) = C(k)x(k) \end{cases} \tag{15}$$

where the periodic time-varying matrices are defined as

$$A(k) = \begin{bmatrix} 0 & 0 & 0 \\ 1 & 0 & 0 \\ a(k) & b(k) & g(k) \end{bmatrix}, B(k) = \begin{bmatrix} 1 \\ 0 \\ c(k) \end{bmatrix},$$
$$C(k) = \begin{bmatrix} 0 & 0 & 1 \end{bmatrix},$$

$$a(k) = (1 - g)r_1(k), \quad b(k) = (1 - g)r_s(k), \quad c(k) = (1 - g)r_3(k)$$

To construct an observer for the time-varying system with convergence of the state estimation error, the design process is focused on the case when $d = 0$. We start with some collection of fundamental concepts of time-varying system regarding the observer design.

Definition 1 System (15) is said to be completely reachable from the origin within n steps, if and only if for any $x_1 \in R^n$ there exists a bounded input $u(l)$ $(l = k, \ldots, k + n - 1)$ such that $x(0) = 0$ and $x(k + n) = x_1$ for all k.

Definition 2 System (15) is said to be completely observable within n steps, if and only if the state $x(k)$ can be uniquely determined from output $y(k), y(k + 1), \ldots, y(k + n - 1)$ for all k.

Under the definitions, the completely reachability can be concluded with the following conditions:

Lemma 1 *System (15) is completely reachable within n steps if and only if the rank of the reachability matrix defined below is n for all k.*

$$R(k) = \left[B_0(k)\, B_1(k) \;\cdots\; B_{n-1}(k) \right] \tag{16}$$

where,

$$\begin{aligned} B_0(k) &= B(k + n - 1) \\ B_1(k) &= A(k + n - 1)B(k + n - 2) \\ &= \Phi(k + n, k + n - 1)B(k + n - 2) \\ &\;\vdots \\ B_{n-1} &= \Phi(k + n, k + 1)B(k). \end{aligned}$$

and the state transition matrix of the system $\Phi(i, j)$ is defined as follows:

$$\Phi(i, j) = A(i - 1)A(i - 2) \cdots A(j) \quad i > j. \tag{17}$$

As shown in [7], the reachability matrix $R(k)$ can be rewritten as

$$R(k) = \left[b_1^0(k) \cdots b_m^0(k) | \cdots | b_1^{n-1}(k) \cdots b_m^{n-1}(k) \right] \tag{18}$$

with $b_i^r(k)$, the r-th column of $B_r(k)$, which is determined by

$$\begin{aligned} b_i^0(k) &= b_i(k+n-1) \\ b_i^1(k) &= A(k+n-1)b_i(k+n-2) \\ &= \Phi(k+n,k+n-1)b_i(k+n-2) \\ &\vdots \\ b_i^{n-1} &= \Phi(k+n,k+1)b_i(k) \end{aligned} \tag{19}$$

where $b_i(k)$ is the i-th column of $B(k)$.

Now, we construct the state observer for the system (15) as follows:

$$\hat{x}(k+1) = A(k)\hat{x}(k) + B(k)u(k) + H(k)(y(k) - C(k)\hat{x}(k)) \tag{20}$$

where $H(k)$ is the observer gain to be designed.

Then, the error of state observer is represented as follows:

$$e(k+1) = (A(k) - H(k)C(k))e(k). \tag{21}$$

where $e(k) = x(k) - \hat{x}(k)$. Therefore, our goal to finish the observer construction is to find a gain matrix $H(k)$ such that the error system is asymptotically stable. In the following, we will show a procedure to find a desired $H(k)$ under the condition of completely reachability, which can be done by routine work with the method presented in [7]. In principle, this is to perform the pole placement for an anti-causal system.

First, for the system (15), consider its anti-causal system

$$\xi(k-1) = A^T(k)\xi(k) + C^T(k)v(k). \tag{22}$$

where $\xi(k) \in R^n$, $v(k) \in R^p$. It has been shown in [7] that if the original system is observable, then there exists a time-varying vector function $\tilde{C}(k)$ such that the relative degree of the anti-causal system is n between the input v and the auxiliary output $\sigma(k)$ defined by

$$\sigma(k) = \tilde{C}(k+1)\xi(k). \tag{23}$$

By using the reachability matrix $\bar{R}(k-n)$ of the anti-causal system, the output matrix $\tilde{C}(k)$ can be constructed as follows:

Theorem 1 *If the anti-causal system is completely reachable in n steps, there exists a new output $\sigma(k)$ such that the relative degree from $v(k)$ to $\sigma(k)$ is n. And, such a $\tilde{C}(k)$ can be calculated by the following equation:*

$$\tilde{C}(k) = W\bar{R}^{-1}(k-n) \tag{24}$$

where $\bar{R}$ is the anti-causal reordered reachability matrix and W is constant matrix defined by

$$W = diag(w_1, w_2, \dots, w_m)$$
$$w_i = \begin{bmatrix} 0 \cdots 0 \; 1 \end{bmatrix} \in R^{1\times\mu_i} \tag{25}$$

Furthermore, for any given stable polynomials $q_i(z^{-1})$ $(i = 1, 2, \cdots, n)$, it is able to transform the system into the following:

$$\begin{bmatrix} q_1(z^{-1}) & & \\ & \ddots & \\ & & q_p(z^{-1}) \end{bmatrix} \sigma(k) = \Gamma(k)\xi(k) + \Lambda(k)v(k). \tag{26}$$

where $q^i(z^{-1})$ is the i-th ideal closed-loop characteristic polynomial from u_i to $\sigma_i(k)$

$$q^i(z^{-1}) = z^{-1\,\nu_i} + \overset{i}{\underset{\nu_i-1}{\alpha}} z^{-1\,\nu_i-1} + \cdots + \overset{i}{\underset{1}{\alpha}} z^{-1} + \overset{i}{\underset{0}{\alpha}} \tag{27}$$

where z^{-1} is the shift operator, ν_i is the reachability indice shown as [7].

It means that the anti-causal system can be stabilized by the state feedback

$$v(k) = \Lambda^{-1}(k)\Gamma(k)\xi(k). \tag{28}$$

Consequently, a desired observer gain matrix $H(k)$ is determined by

$$H(k) = \Gamma^T(k)\Lambda^{-T}(k). \tag{29}$$

For detailed algorithm to find $\Gamma(k)$ and $\Lambda(k)$ can be found in [7]. If we regard such an observer gain H(k) as a state feedback gain for the anti-causal system, then there exists a transformation matrix $Q(k)$ such that

$$Q^{-1}(k)(A^T(k) - C^T(k)H^T(k))Q(k+1) = A_a^{*T} \tag{30}$$

where A_a^{*T} is constant matrix which has the desired closed loop characteristic polynomial. Equivalently, we have

$$Q^T(k+1)(A(k) - H(k)C(k))Q^{-1^T}(k) = A_a{}^* \tag{31}$$

where $A(k) - H(k)C(k)$ has the desired characteristic polynomial to guarantee the asymptotical stability of the error system.

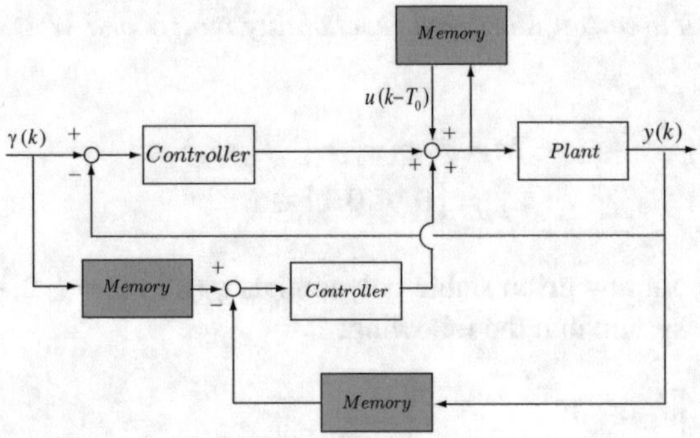

Fig. 2 PCCL block diagram

5 Learning Control Law

Figure 2 shows a typical structure of iterative learning control algorithm. $\gamma(k)$ is a given reference signal, $u(k)$ is the control signal and $u(k - T_0)$ is the input with T_0 step delay, which is to record the effective control signal used in the previous steps. The memory blocks play the role of learning.

For the periodic time-varying system discussed above, we introduce a learning mechanism to generate an estimation of the unknown offset $m(k)d$. Since $m(k)d$ takes the value d_1, d_2, d_3 periodically, which represent the offset of individual cylinder injector, for example,

$$m(k)d = m(k+3)d = \cdots = m(k+n) = d_1, \quad n = 1, 2, \ldots$$

we embed a memory with 3-steps delay in the estimation path of the offset as follows:

$$\hat{d}(k) = \hat{d}(k-3) + F(z^{-1})e(k) \tag{32}$$

with the learning gain function $F(z^{-1})$, which will be chosen later. Replacing the offset $m(k)d$ with this estimation in the observer, we have

$$\begin{aligned} \hat{x}(k+1) &= A(k)\hat{x}(k) + B(k)(u(k) + \hat{d}(k)) \\ &\quad + H(k)(y(k) - C(k)\hat{x}(k)), \\ \hat{d}(k) &= \hat{d}(k-3) + F(z^{-1})e(k) \end{aligned} \tag{33}$$

The block diagram of the observer-based learning algorithm is as shown in Fig. 3.

The learning gain function $F(k)$ can be simply chosen with current and previous learning error as

$$F(z) = K_c + K_p z^{-1} \tag{34}$$

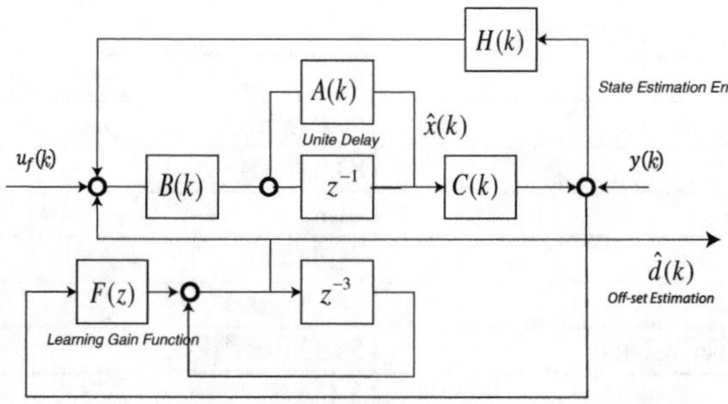

Fig. 3 Diagram of the observer-based learning algorithm

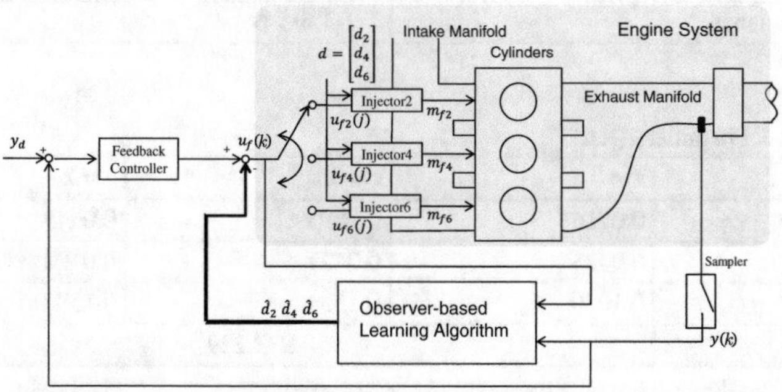

Fig. 4 Feedback control system with feedforward

Finally, by embedding the observer-based iterative learning algorithm to the fuel injection control loop, which provides a feedforward compensation with the estimation of the unknown offset, the whole control scheme is obtained as shown in Fig. 4. In the experiment shown in next section, the feedback control law of the fuel path is a conventional PI.

6 Numerical Simulation

To validate the learning algorithm in principle, numerical simulation is first conducted with a model identified from experiment data. The specification of the targeted engine is as shown in Table. 1.

Under the operation mode with the engine speed 1600 (rpm), the BDC-scaled sampling time is 25 (ms). The model parameter values obtained by the least square (LES) identification algorithm are given in Table 2. The observer-based iterative learning algorithm in open loop is tested with the model. The results are shown in Fig. 5. In the simulation, the offset is set to $d_1 = 1300, d_2 = 1000, d_3 = 1500$. It can

Table 1 Basic engine specifications

Number of cylinders	6
Bore	94.0 (mm)
Stroke	83.0 (mm)
Compression ratio	11.8
Crank radius	41.5 (mm)
Connection rod length	147.5 (mm)
Combustion chamber volume	53.33 (mm^3)
Displacement	3.456 (L)
Manifold volume	6.833 (L)
Fuel injection	D-4S
Injection sequence	1-2-3-4-5-6

Table 2 Model parameter($\times 10^{-3}$)

Mod(k,3) = r	r = 0	r = 1	r = 2
$a(k)$	0.0716	0.0695	0.0545
$b(k)$	0.0428	0.0427	0.0498
$c(k)$	0.1020	0.1534	0.1581
$g(k)$	932.229		

be observed that the estimation error of the state variables converge to zero, and the estimation of the offset convergence to the set value.

7 Experiment Results

The experiments are conducted on an engine test bench where a commercial production V6 gasoline engine with six cylinders is controlled by electrical control unit (ECU), which is bypassed via CAN bus so that the control law can be programmed on dSPACE prototype control environments. The system structure is shown in Fig. 6.

The engine in the test bench and the dynamometer are shown in lower image of Fig. 7. The upper image shows the control desk. The UEGO sensor placed at the gas mixing point of exhaust manifold provides the air–fuel ratio sampled with sampling period 240 (deg) in crank angle. The control input is the command delivered to each injector that inject the fuel directly into according cylinder.

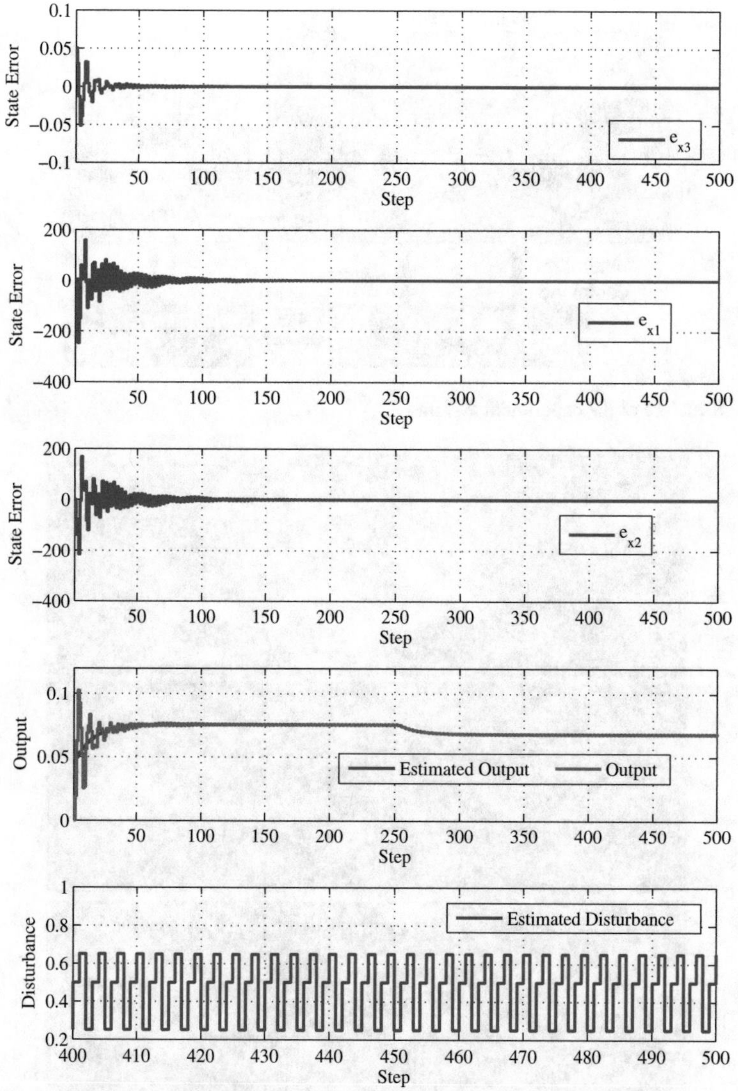

Fig. 5 Simulation results of iterative learning observer

7.1 Open-Loop Learning

To validate the effectiveness of the learning algorithm, we first conduct open-loop learning experiments, i.e., the observer-based learning algorithm is implemented without the air–fuel ratio feedback control loop.

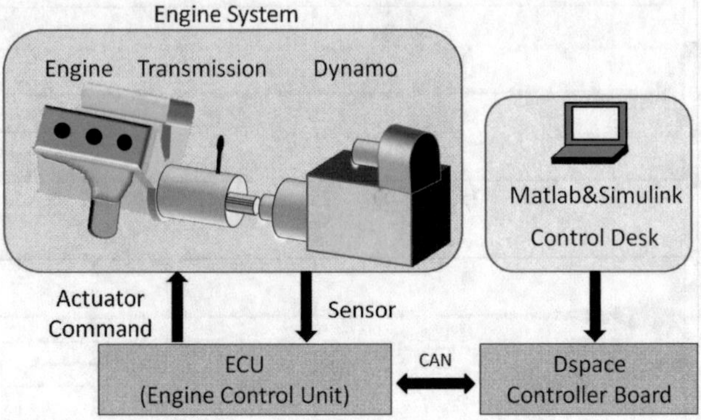

Fig. 6 Structure of the experiment system

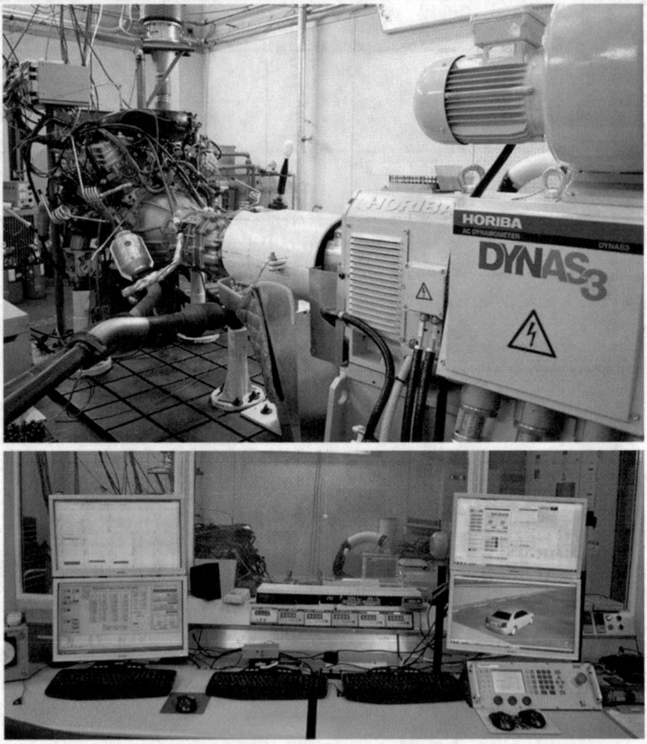

Fig. 7 Engine test bench and control desk

Case 1. The engine is under speed control set 1600 (rpm) and the load is changed from 60 to 70 (Nm). The response of the speed, state estimation, and the learning response of the offset are shown in Fig. 8.

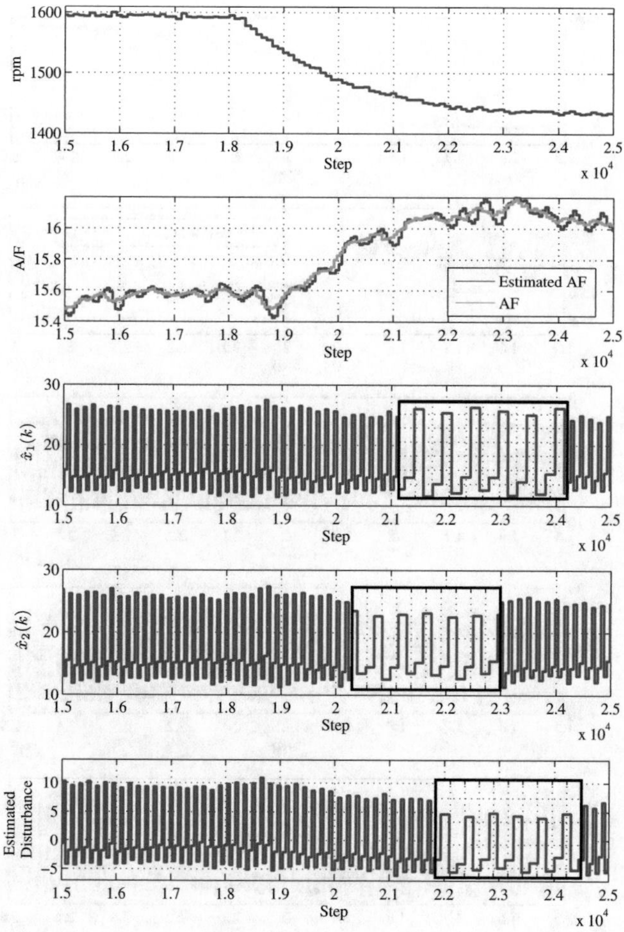

Fig. 8 Experiment results of iterative learning (acceleration)

Case 2. The same operation mode as the Case 1, but the load is changed from 70 to 60 (Nm). The response is shown in Fig. 9.

Case 3. The engine speed is operated at 1200 (rpm) and the load is changed from 80 to 70 (Nm).

Case 4. The engine speed is set to 1800 (rpm) with the load change from 70 to 80 (Nm).

It can be observed from the experiment results, zoomed window in the bottom of Fig. 8, the output of the learning algorithm converges to the individual offset value of each cylinder. The last two cases shows the learning algorithm has good robustness on the engine operating condition. Moreover, the experiment results operated at different speed ranges with the same model are shown in Figs. 10 and 11, which show that the model with fixed parameter value can work well at a rather wilder operation range.

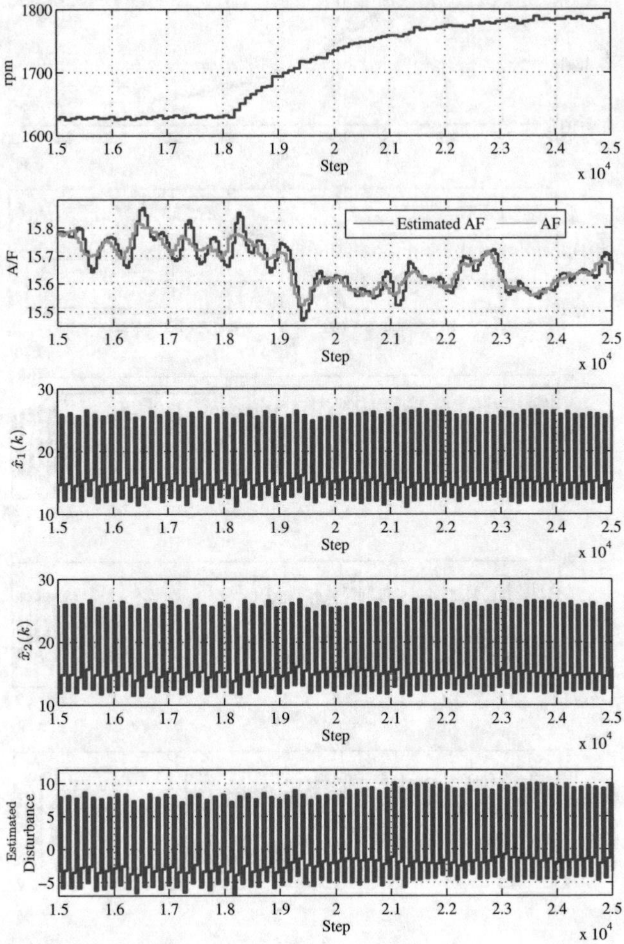

Fig. 9 Experiment results of iterative learning (deceleration)

7.2 *Learning-Based Control*

The feedback control with learning-based feedforward compensation presented in Sect. 5 is conducted. Figure 12 shows control responses data of the engine under the operating condition with speed 1600 (rpm), load 60 (Nm), and the nominal value of the injected fuel mass 17.26 (mm^3/str). In Fig. 12, the air–fuel ratio control error of the left bank is shown under the fuel injection commanded shown in the third figure, and the bottom shows the feedforward signal generated by the iterative learning algorithm. The learning signal converges to the three level which is equivalent to the offset value of the injector of individual cylinder.

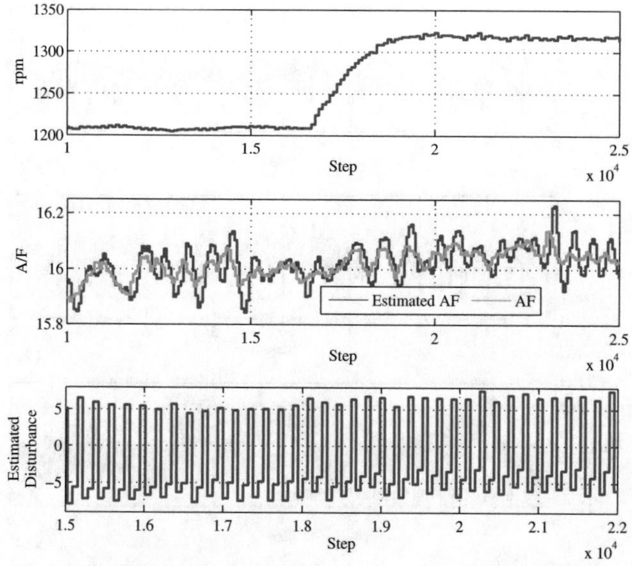

Fig. 10 Experiment results with speed 1200 (rpm)

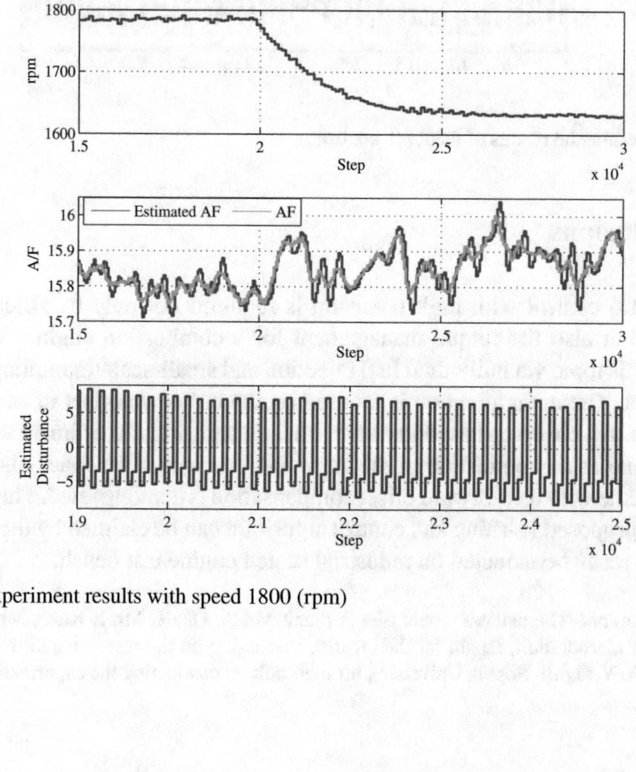

Fig. 11 Experiment results with speed 1800 (rpm)

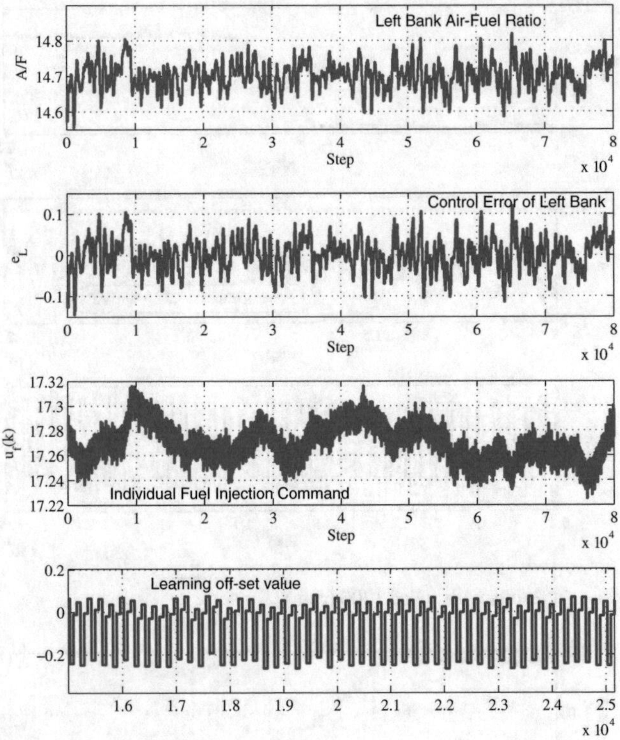

Fig. 12 Experimental results of feedback control

8 Conclusions

Air–fuel ratio control with high precision is required not only by strict emission constraint, but also the torque management for a combustion engine. This paper addressed this topic via individual fuel injection and small-scaled sampling and control decision. The main attention is focused on the unknown offset of each injector of cylinders. An observer-based iterative learning algorithm to estimate the offset is proposed, and as an application of the learning algorithm, a feedback air–fuel ratio control scheme with feedforward offset compensation is demonstrated. The effectiveness of the proposed learning and control algorithm can be claimed by the presented experiment results conducted on industrial scaled engine test bench.

Acknowledgments The authors would like to thank Mr. A. Ohata, Mr. J. Kako, Mr. K. Sata of Toyota Motor Corporation, Japan, for the creative discussion on the research and to thank Dr. J. Zhang and Mr. Y. Oguri, Sophia University, for their help in conducting the experiments.

References

1. Grizzle, J., Dobbins, K., Cook, J.: Individual cylinder air-fuel ratio control with a single EGO sensor. IEEE Trans. Veh. Technol. **40**(1) (1991)
2. Hasegawa, Y., Akazaki, S.: Individual Cylinder Air-Fuel Ratio Feedback Control Using an Observer. SAE Paper940376, SAE International, Warrendale, PA (1994)
3. He, B., Shen, T., Kako, J., Ouyang, M.: Input observer-based individual cylinder air-fuel ratio control, modelling, design and validation. IEEE Trans. Control Syst. Technol. **16**(5), 1057–1065 (2008)
4. Leroy, T., Chauvin, J., Petit, N.: Motion Planning for experimental air path control of variable-valve-timing spark ignition engine. Control Eng. Pract. **17**, 1432–1439 (2009)
5. Liu, Y.: Air-Fuel Ratio Control of SI Engine with Individual Cylinder Fuel Decision: Periodic Time-Vaeying Model-Based Approach. Ph.D. dissertation, Department of Mechanical Engineering, Sophia University, Tokyo, Japan (2013)
6. Liu, Y., Takayama, Y., Shen, T., Sata, K., Suzuki, K.: Exhaust gas flow mixing model and application to individual cylinder air-fuel ratio estimation and control. 6th IFAC Symposium on Advances in Automotive Control, Munich (2010)
7. Mutoh, Y., Hara, T.: A Design Method for Pole Placement and Observer of Linear Time-Varying Discrete MIMO Systems. IEEJ, vol. 132, No. 5, Sec. C (2012)
8. Smith, J.C., Schulte, C.W., Cabush, D.D.: Individual cylinder fuel control for imbalance diagnosis. SAE Int. J. Engines **3**(1), 28–34, 2010-01-0157 (2010)
9. Suzuki, K., Shen, T., Kako, J., Yoshida, S.: Individual A/F estimation and control with the fuel-gas ratio for multicylinder IC engine. IEEE Trans. Veh. Technol. **58**(9), 4757–4768 (2009)

Fuzzy T–S Model-Based Design of Min–Max Control for Uncertain Nonlinear Systems

Tatjana Kolemishevska-Gugulovska, Mile Stankovski, Imre J. Rudas, Nan Jiang and Juanwei Jing

Abstract The min–max robust control synthesis for uncertain nonlinear systems is solved using Takagi–Sugeno fuzzy model and fuzzy state observer. Existence conditions are derived for the output feedback min–max control in the sense of Lyapunov asymptotic stability and formulated in terms of linear matrix inequalities. The convex optimization algorithm is used to obtain the minimum upper bound on performance and the optimum parameters of min–max controller. The closed-loop system is asymptotically stable under the worst case disturbances and uncertainty. Benchmark of inverted pendulum plant is used to demonstrate the robust performance within a much larger equilibrium region of attraction achieved by the proposed design.

1 Introduction

The ever growing needs in control of nonlinear systems has also enhanced rapidly growing interest in applying fuzzy system-based control techniques [1]. Although in industrial practice many rather successful implementations exist heuristics-based fuzzy control designs employing both linguistic and fuzzy-set models still prevail along with empirical choice of membership functions [2, 3].

Initially, fuzzy control techniques employing semantic-driven fuzzy models [3] lacked a formal and systematic synthesis design methodology that would guarantee basic design requirements such as robust stability and acceptable system performance in closed loop. However, such a methodological approach has emerged through works based on Lyapunov stability theory via using specific fuzzy models, introduced in [4] by Takagi and Sugeno [4], by means of which any original nonlinear plant system can be emulated much more accurately. These models are known T–S fuzzy rule models on the grounds of which Tanaka and co-authors [5–8] have paved the way of Lyapunov based fuzzy control designs beginning with

T. Kolemishevska-Gugulovska (✉) · M. Stankovski · I.J. Rudas · N. Jiang · J. Jing
Skopje, Macedonia
e-mail: tanjakg@etf.ukim.edu.mk

V. Sgurev et al. (eds.), *Recent Contributions in Intelligent Systems*,
Studies in Computational Intelligence 657,
DOI 10.1007/978-3-319-41438-6_6

a stability condition that involved finding a common positive definite matrix P for the T–S models. For fuzzy control systems, being a specific kind of nonlinear systems, celebrated Lyapunov stability approach has been well established (e.g., see [9–17] and references therein) too. Furthermore, it has been into a fully elaborated LMI methodological approach [18].

Nowadays, control approaches employing fuzzy T–S models and Lyapunov stability theory are commonly viewed as conceptually amenable to design synthesis tasks of effective control for complex nonlinear systems [1, 3], switched fuzzy systems [17] inclusive. Recently, a number of methods to design fuzzy controllers for nonlinear systems based on T–S fuzzy models with embedded uncertainties were proposed. For it is feasible to decompose a complex plant system into several subsystems (via fuzzy rules employing T–S model), and then fairly simple control law synthesized for each subsystem so as to emulate the overall control strategy [4].

Techniques for fuzzy observer-based designs based on fuzzy T–S modeled nonlinear plants that guarantee robust stability [11, 12] are of particular importance since very seldom all plant states are measurable. In practice, measuring the states is physically difficult and costly, and often even impossible. Besides complicated sensors are often subject to noise, faults, and even complete failures. Thus asymptotic stability and quality performance are the two most important problems in control analysis and synthesis of nonlinear systems hence in fuzzy control systems too, as demonstrated in works [14–16]. In particular, Lin et al. [15] made a considerable improvement to the observer-based H^{∞} control of T–S fuzzy systems following the crucial contribution by Liu and Zhang [14]. In [12], Tong and Li gave a solution to observer-based robust fuzzy control of nonlinear systems in the presence of uncertainties. Works [19–21] studied and gave solutions to the fuzzy guaranteed cost control for nonlinear systems with uncertain and time delay via the LMI. An implicit assumption in most of those contributions is the measurability of state variables. Then controller design is based on a memory-less linear state feedback law. Fairly recently, a robust and non-fragile min–max control for a trailer truck represented by a derived T–S fuzzy model has been contributed in [22].

In the guaranteed cost optimum control synthesis only the existence of the disturbance is considered; the influence that disturbance may have actually is not dealt with. In contrast, the min–max control is a special kind of optimum control reaching beyond that of the guaranteed cost synthesis. In 1998, Kogan [22–24] was the first to find a solution and derive the sufficient condition for min–max control design of linear continuous-time systems; initially he termed it "minimax." Yet, he has not made it into the optimal parameter settings of his min–max controller. Recently, Yoon and co-authors studied the optimal min–max control for linear stochastic systems with uncertainty and gave a solution in [25]. However, they did not consider nonlinearities, which are most often present in plants-to-be-controlled (objects or processes) [1].

The chapter is largely based on the work given in [30]. We develop a new design synthesis of efficient min–max controller for nonlinear systems by employing fuzzy T–S models, which is solvable by using LMI toolbox of MATLAB [26, 27]. Since all the states of nonlinear systems usually are not measurable it is rather important to find a solution via employing a fuzzy state observer. Furthermore, in here the problem is solved under assumption of the worst uncertainty or disturbance [25] also accounting for possible state bias and control energy consumption too. Next, Sect. 2 introduces fuzzy T–S model construction for uncertain nonlinear systems and the definition of the min–max robust control. Existence conditions and design of min–max robust controller guaranteeing asymptotic Lyapunov stability by using LMI are investigated in Sect. 3. Section 4 presents application results to the inverted pendulum benchmark plant. Conclusion and references follow thereafter.

2 Plant Representation and Problem Statement

2.1 *Fuzzy Takagi–Sugeno Model of the Plant*

It is well known that fuzzy dynamic models crated by Takagi and Sugeno (T–S) are described by means of a finite set of specific fuzzy If-Then rules. Fuzzy T–S models involve rules that have math-analytic consequent part such that they interpolate the originally nonlinear system by local linear input–output relations [4].

Thus, a class of uncertain nonlinear systems can be described by the following T–S fuzzy model with parametric uncertainties:

$$\begin{aligned} &R^i: \quad \textbf{IF } z_1(k) \textbf{ is } F_1^i,\ z_2(k) \textbf{ is } F_2^i,\ \ldots,\ z_n(k) \textbf{ is } F_n^i, \\ &\textbf{THEN} \\ &\dot{x}_i(t) = (A_i + \Delta A_i)x_i(t) + (B_i + \Delta B_i)u(t) + D_i\omega(t) \qquad , \\ &y(t) = C_i x(t), \quad i = 1, \ldots, q \end{aligned} \tag{1}$$

Quantities in (1) denote: $F_j^i (j = 1, \ldots, n)$ is a fuzzy subset; $z(k) = [z_1(k), \ldots, z_n(k)]^{\mathrm{T}}$ is a vector of measurable quantities representing certain premise variables; $x(k) = [x_1(k),\ x_2(k),\ \ldots;\ x_n(k)]^T \in R^n$ is the state, $u(t) \in R^m$ is the control, and $y(t) \in R^l$ is the output vectors while $\omega(t)$ is the disturbance vector; matrices $A_i \in R^{n \times n}$, $B_i \in R^{n \times m}$, $C_i \in R^{l \times n}$ and $D_i \in R^{n \times m}$ represent the state, the input, the output, and the disturbance system matrices, respectively; ΔA_i and ΔB_i are constant matrices of appropriate dimensions representing the parametric uncertainties; and q is the number of rules of plant's fuzzy T–S model.

Application of Zadeh's inference [28] to this fuzzy rule-based model yields

$$\begin{aligned} \dot{x}(t) = &\sum_{i=1}^{q} h_i(z(t))[A_i x(t) + B_i u(t) + D_i\omega(t)] + \\ &+ \sum_{i=1}^{q} h_i(z(t))[\Delta A_i x(t) + \Delta B_i u(t)], \end{aligned} \tag{2}$$

$$y(t) = \sum_{i=1}^{q} h_i(z(t)) C_i x(t),\ i = 1, 2, \ldots, q \tag{3}$$

where

$$w_i(z(t)) = \prod_{j=1}^{n} F_j^i(z_j(t)),\ h_i(z(t)) = \frac{w_i(z(t))}{\sum_{i=1}^{q} w_i(z(t))} \tag{4}$$

Here, F_j^i is a fuzzy subset, h_i is the membership grade of $z_j(t)$ in F_j^i. Furthermore, $h_i(t)$ and $w_i(z(t))$ possess properties so as to satisfy the following relationships, respectively:

$$w_i(t) \geq 0,\ \sum_{i=1}^{q} w_i(t) > 0,\ i = 1, 2, \ldots, q; \tag{5}$$

$$h_i(z(t)) \geq 0,\ \sum_{i=1}^{q} h_i(z(t)) = 1,\ i = 1, \ldots, q. \tag{6}$$

It is realistic to assume existing uncertainties in the plant are bounded hence Assumption 1 is adopted.

Assumption 1 The parameter uncertainties are norm bounded hence represented as:

$$[\Delta A_i, \Delta B_i] = D_i F_i(t) [E_{i1}, E_{i2}] \tag{7}$$

where D_i, E_{i1}, E_{i2} are known real-valued constant matrices of appropriate dimensions, $F_i(t)$ is unknown matrix function with Lebesgue-measurable elements satisfying $F_i^T(t) F_i(t) \leq I$ with I an appropriate identity matrix.

Assumption 1 and Zadeh's inference enable deriving an equivalent representation to system (2)–(4) as follows:

$$\dot{x}(t) = \sum_{i=1}^{q} h_i(z(t)) [A_i x(t) + B_i u(t) + \bar{D}_i W(t)], \tag{8}$$

$$y(t) = \sum_{i=1}^{q} h_i(z(t)) C_i x(t),\ i = 1, \ldots, q. \tag{9}$$

Notice the new quantities in (8): $\bar{D}_i = [D_i \quad D_i \quad D_i]$, $W(t)^T = (v(t)^T, \omega(t)^T)^T$, $v(t)^T = (v_{i1}(t)^T, v_{i2}(t)^T)$, $v_{i1}(t) = F_i E_{i1} x(t)$, $v_{i2}(t) = F_i E_{i2} u(t)$.

2.2 Fuzzy Observer via T–S Fuzzy Model

Similarly, the fuzzy state observer [7, 11–14] for the T–S fuzzy model with parametric uncertainties (2)–(6) can be formulated in the rule-based form as follows:

$$\begin{aligned} &R^i\text{: } \textbf{IF } z_1(k) \textbf{ is } F_1^i,\ z_2(k) \textbf{ is } F_2^i,\ \ldots,\ z_n(k) \textbf{ is } F_n^i,\\ &\textbf{THEN}\\ &\dot{\tilde{x}}(t) = A_i\tilde{x}(t) + B_i u(t) + G_i[y - \tilde{y}],\\ &\tilde{y}(t) = C_i\tilde{x}(t), \quad i = 1, \ldots, q \end{aligned} \tag{10}$$

In here, $G_i \in R^{n \times l}$ is a constant observer gain matrix that is to be determined in due course of this synthesis. The output of the above observer is found as

$$\dot{\tilde{x}}(t) = \sum_{i=1}^{q} h(z(t))[A_i\tilde{x} + B_i u(t)] + \sum_{i=1}^{q} h_i(z(t))G_i[y - \tilde{y}] \tag{11}$$

$$\tilde{y}(t) = \sum_{i=1}^{q} h_i(z(t))C_i\tilde{x}(t),\ i = 1, \ldots, q. \tag{12}$$

The observation error traditionally is defined as follows:

$$e(t) = x(t) - \tilde{x}(t). \tag{13}$$

2.3 Task Problem Formulation

The adopted performance index (or criterion or else cost function) is the augmented one

$$J(u, W) = \int_0^\infty (\xi^T Q \xi + u^T u - W^T W)dt. \tag{14}$$

In here, $\xi^T = [x^T(t),\ e^T(t)]^T, e(t)$ is the observation error, $Q = Q^T \geq 0$, and $W^T W$ accounts for the admissible disturbances and uncertainty. It should be noted $\xi_0 = \xi(0)$.

Now, the task problem and objective of this research can be stated as follows: Via employing the above-defined fuzzy observer (12)–(13), synthesize a fuzzy feedback controller

$$u(t) = -\sum_{i=1}^{q} h_i(z(t))K_i\tilde{x}(t) \tag{15}$$

that reinforces the system (2)–(6) to the asymptotic stability equilibrium with the upper bound of performance cost (14) reaching its minimum under the worst case disturbances and uncertainty.

Representation model of the overall fuzzy control system described by (2)–(6), (11)–(13) and (15) is inferred and reformulated as follows:

$$\dot{x}(t) = \sum_{i=1}^{q} \sum_{j=1}^{q} h_i(z(t))h_j(t)\left[(A_i - B_iK_j)x(t) + B_iK_je(t) + \bar{D}_iW\right] \tag{16}$$

$$\dot{e}(t) = \sum_{i=1}^{q} \sum_{j=1}^{q} h_i(z(t))h_j(z(t))((A_i - G_iC_j)e(t) + \bar{D}_iW) \tag{17}$$

Definition 1 Consider the uncertain nonlinear system (2)–(6) and the performance cost (15) along with a positive number J^* representing its upper bound. Control $u^*(t)$ is called the min–max control law, if there exist a control law $u = u^*(t)$ and a positive number $J^* = \text{const}$ for the worst case disturbance and uncertainty such that the closed-loop system is asymptotically stable and the upper bound of the performance cost tends to its minimum ultimately $J \leq J^*$ is satisfied.

It should be noted, the bounds on performance in the specified min–max control are related to those of H^∞ control in the sense that the former are generally lower. A precise quantitative relationship has not been proved as yet [22].

3 Main New Results

The existence conditions for a mini–max controller design are explored in this section. It appeared, these ought to be investigated in two separate cases of control design synthesis. These new results address the optimal min–max control problem, which can be solved by convex optimization algorithms and the LMI tool [26]. For this purpose also Assumption 2 [5, 8] has to be observed.

Assumption 2 The initial value ξ_0 of the nonlinear system (16)–(17) is the zero mean random variable, satisfying $E\{\xi_0, \xi_0^T\} = I$, where $E(*)$ denotes the expectation operator.

Case 1 When $\tilde{D}_i\tilde{D}_j^T - \tilde{B}_i\tilde{B}_j^T < 0$, the existence conditions for the min–max controller are drawn from Theorem 1.

Theorem 1 *Consider system (2)–(6) and performance cost (14). If there exist common symmetric positive definite matrices X, Y, and Z_{ji} such that*

$$XA_i^T + A_iX + \Gamma_{ij} < 0, \ \ i;j = 1, \ldots, q \tag{18}$$

$$XA_i^T + A_iY - Z_{ji} - Z_{ji}^T + D_iD_j < 0, \ \ i;j = 1, \ldots, q, \tag{19}$$

where

$$X = P^{-1},\ Y = P_2^{-1},\ Z_{ji} = P_2^{-1} C_j G_i^T,\ \Gamma_{ij} = D_i D_j - B_i B_j; \tag{20}$$

then

$$u = -\sum_{i=1}^{q} h_i (B_i^T X \tilde{x}(t) + B_i^T X e(t)) \tag{21}$$

is a min–max output feedback control law for the plant system where state is estimated and guarantees asymptotic stability in the closed loop while the corresponding upper-bounded performance cost is

$$\bar{J} \leq \mathrm{Trace}(P) = J^*,$$

the value of which depends on the found matrix P.

Proof See Appendix 1.

Case 2 When $\tilde{D}_i \tilde{D}_j^T - \tilde{B}_i \tilde{B}_j^T \geq 0$, the existence conditions of min–max controller design are given by Theorem 2.

Theorem 2 *Consider system (2)–(6) and performance cost (14). If there exist the common symmetric positive definite matrices X, Y, and* Z_{ji} *such that*

$$XA_i^T + A_i X + 2\Gamma_{ij} < 0,\ i; j = 1, \ldots, q \tag{22}$$

$$XA_i^T + A_i Y - Z_{ji} - Z_{ji}^T + 2D_i D_j < 0,\ i; j = 1, \ldots, q; \tag{23}$$

then

$$u = -\sum_{i=1}^{q} h_i (B_i^T X \tilde{x}(t) + B_i^T X e(t)) \tag{24}$$

is a min–max output feedback control law for the plant system where state is estimated and guarantees asymptotic stability in the closed loop while the corresponding upper-bounded performance cost is

$$\bar{J} \leq \mathrm{Trace}(P) = J^*,$$

the value of which depends on the found matrix P.

Proof See Appendix 2.

Remark 1 If the above-presented existence condition is satisfied, then there exists a min–max output feedback control law that employs the fuzzy state observer for system (2). Furthermore, together Theorem 1 and Theorem 2 parameterize the min–max control law similarly to the results in [29]. Notice the upper bound of performance in fact depends on the selection of the min–max control law thus the choice of an appropriate min–max control law is crucial.

It is for that matter, the optimum parameter settings of the feedback controller and the minimum of the upper bound of performance index can be obtained by constructing and solving a convex optimization problem.

Theorem 3 *Consider the system (2)–(4) and performance cost (15), and assume*

$$M = P^{-1},\ N = P^{-1}\tilde{G}_i^T,\ \tilde{\Gamma}_{ij} = D_i D_j - B_i B_j, \tag{25}$$

$$\tilde{A}_i = \begin{bmatrix} A_i & 0 \\ 0 & A_i \end{bmatrix}, \tilde{G}_i = \begin{bmatrix} 0 & 0 \\ 0 & G_i C_j \end{bmatrix}, \tag{26}$$

and

$$\tilde{K} = I_0^T M I_0,\ I_0 = \begin{bmatrix} I \\ 0 \end{bmatrix}, \tag{27}$$

where I is the identity matrix of appropriate dimension.

If solvability condition for the following optimization problems are fulfilled

(i) When $\tilde{\Gamma}_{ij} \geq 0$,

$$\begin{aligned} &\min_{M, \tilde{M}, N} \ \mathrm{Trace}(\tilde{M}) \\ &\text{s.t} \quad M > 0, \\ &\qquad M\tilde{A}_i^T + \tilde{A}_i M - N - N^T + 2\tilde{\Gamma}_{ij} < 0,\ i,j = 1, \ldots, q, \\ &\qquad \begin{bmatrix} -\tilde{M} & I \\ I & -M \end{bmatrix} < 0 \end{aligned} \tag{28}$$

there exist a solution $(M, \tilde{M}, N)$;

(ii) When $\tilde{\Gamma}_{ij} < 0$,

$$\begin{aligned} &\min_{M, \tilde{M}, N} \ \mathrm{Trace}(\tilde{M}) \\ &\text{s.t} \quad M > 0, \\ &\qquad M\tilde{A}_i^T + \tilde{A}_i M - N - N^T + \tilde{\Gamma}_{ij} < 0,\ i,j = 1, \ldots, q \\ &\qquad \begin{bmatrix} -\tilde{M} & I \\ I & -M \end{bmatrix} < 0 \end{aligned} \tag{29}$$

there exist a solution $(M, \tilde{M}, N)$;

then

$$u = -\sum_{i=1}^{q} h_i (B_i^T \bar{K} \tilde{x}(t) + B_i^T \bar{K} e(t)) \tag{30}$$

is the optimal min–max control law for system (2)–(6) where plant state is estimated and guarantees asymptotic stability in the closed loop while the corresponding performance cost has an upper bound of optimum value.

Proof Given the proofs of the Theorem 1 (Appendix 1) and of Theorem 2 (Appendix 2), the proof of Theorem 3 readily inferred. □

4 Application Results for Inverted Pendulum Benchmark Example

The motion equations of the well-known benchmark example inverted pendulum [5, 12] are as follows:

$$\dot{x}_1 = x_2,$$

$$\dot{x}_2 = \frac{g\sin(x_1) - amlx_2^2\sin(2x_2)/2 - a\cos(x_1)u}{4l/3 - aml\cos^2(x_1)} + \omega_2(t).$$

In there, variables and coefficients denote: x_1 the angle (Rad) of the pendulum relative to the vertical; x_2 is the angular velocity (Rad/s); $g = 9.8$ (m/s^2) is Earth's gravity constant; $m = 2$ (kg) is the mass of the pendulum; $M = 8$(kg) is the mass of the cart; $l = 0.5$ (m) is the length from pendulum's center of mass to the shaft axis; u (N) is the force applied to the cart; $\omega_2(t)$ represents the disturbances; and $a = 1/(m + M)$ is the obvious accessory coefficient for pragmatic purpose. In order to design the fuzzy controller and the fuzzy observer, a fuzzy model that represents the dynamics of the nonlinear plant is needed. Therefore, first the plant system is represented with a T–S fuzzy model. To minimize the design effort and complexity, as few rules as possible should be employed. Notice that the given pendulum plant system is uncontrollable when $x_1 = \pm\pi/2$ [5].

In here, the plant system is approximated with the following two-rule fuzzy model:

Plant rule 1:
If x_1 is about 0 **Then** $\dot{x} = A_1 x(t) + B_1 u(t) + \omega(t)$.
Plant rule 2:
If x_1 is about $\pm 22\pi/45$ **Then** $\dot{x} = A_2 x(t) + B_2 u(t) + \omega(t)$.
where

$$A_1 = \begin{bmatrix} 0 & 1 \\ \frac{g}{\frac{4l}{3} - aml} & 0 \end{bmatrix}, \; B_1 = \begin{bmatrix} 0 \\ -\frac{a}{\frac{4l}{3} - aml} \end{bmatrix},$$

$$A_2 = \begin{bmatrix} 0 & 1 \\ \frac{2g}{\pi(\frac{4l}{3} - aml\beta^2)} & 0 \end{bmatrix}, \; B_2 = \begin{bmatrix} 0 \\ -\frac{a\beta}{\frac{4l}{3} - aml\beta^2} \end{bmatrix},$$

$$\beta = \cos(22\pi/45).$$

The adopted membership functions for this plant's rules are given as

$$h_1 = 1 - \frac{2}{\pi}|x_1|, \; h_2 = \frac{2}{\pi}|x_1|.$$

The corresponding model for the above dynamic T–S fuzzy system has been built up, and the parameter system matrices are found [12]:

$$A_1 = \begin{bmatrix} 0 & 1 \\ 17.294 & 0 \end{bmatrix}, B_1 = \begin{bmatrix} 0 \\ -0.1765 \end{bmatrix},$$

$$A_2 = \begin{bmatrix} 0 & 1 \\ 9.365 & 0 \end{bmatrix}, B_2 = \begin{bmatrix} 0 \\ -0.0349 \end{bmatrix}.$$

Using the LMI convex optimization algorithm [26] to solve the inequality (10), the feedback gain and the observer gain matrices can be obtained as follows:

$$K_1 = [\,1.2100 \quad -2.2027\,], \; K_2 = [\,0.3125 \quad -2.1007\,],$$

$$G_1 = G_2 = \begin{bmatrix} 3.8460 & 2.8071 \\ 5.0946 & 2.7978 \end{bmatrix}.$$

The experimental simulations were carried out under the inflicted persisting sine disturbance into the system and the following initial values for the state vectors:
$x(0) = [\,0.002 \quad 0\,]$, $\tilde{x}(0) = [\,0 \quad 0.001\,]$ (Figs. 1 and 2);
$x(0) = [\,0.609 \quad 0\,]$, $\tilde{x}(0) = [\,0 \quad 0.001\,]$, (Figs. 3 and 4).

The amplitude of the disturbances was suddenly increased from 0.01 to 1.0 but kept bounded. When the disturbance strength increased from 0.01 to 1.0, the maximum overshoot increased from 70 to about 160 (Figs. 1 and 3); and the stable control operating point also changed from 25 to 80 (Figs. 2 and 4). Apparently the control responses tend to stable steady states.

The simulation experiments when the initial values of x_1 is in 0.002 and 0.609 rad, respectively, showed that the system response curves remain stable still.

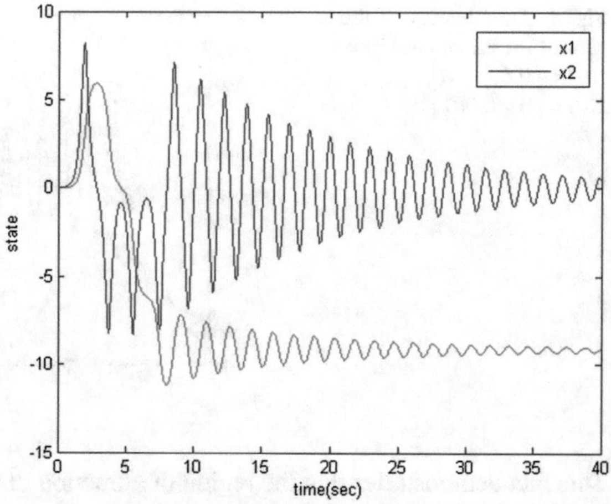

Fig. 1 Time history of the closed-loop state response for initial conditions $x(0) = [0.002 \quad 0]$, $\tilde{x}(0) = [0 \quad 0.001]$

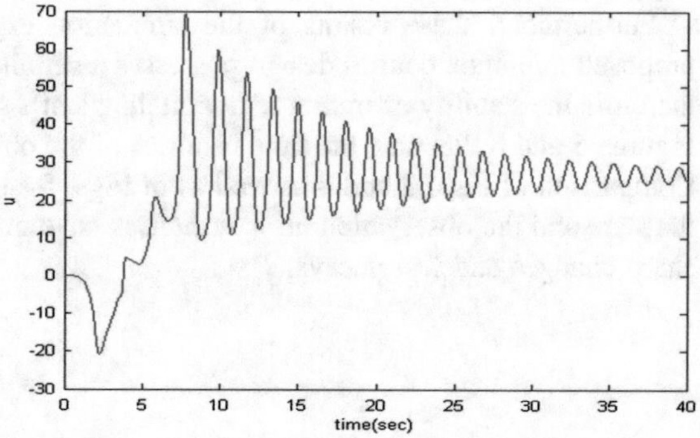

Fig. 2 Time history of the control signal for initial conditions $x(0) = [0.002 \quad 0]$, $\tilde{x}(0) = [0 \quad 0.001]$

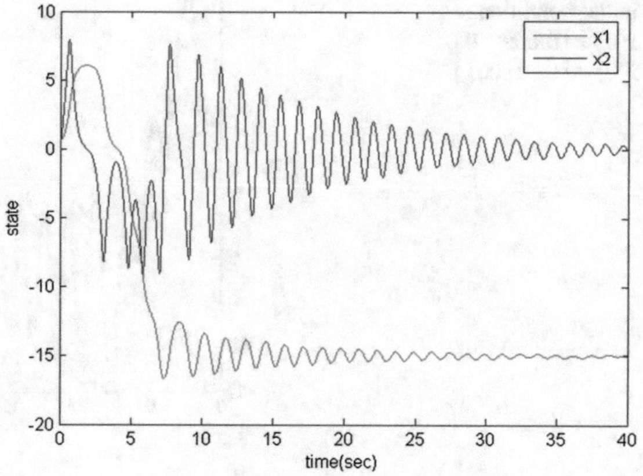

Fig. 3 Time history of the closed-loop state response for initial conditions $x(0) = [0.609 \quad 0]$, $\tilde{x}(0) = [0 \quad 0.001]$

Fig. 4 Time history of the control for initial conditions $x(0) = [0.609 \quad 0]$, $\tilde{x}(0) = [0 \quad 0.001]$

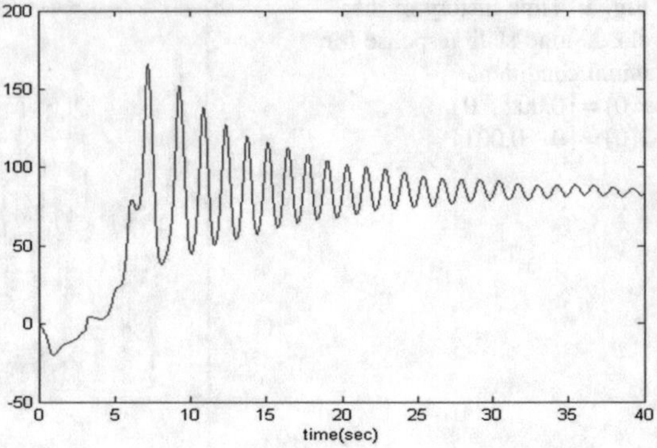

This fact demonstrates that the region of attraction of this balancing plant controlled by means of the proposed mini–max control design has been made considerably larger to reach (±34.89°).

Furthermore, these results of the simulation experiments illuminate that the proposed min–max control design possesses a certain robustness property [8]. And therefore the stability operating region in the plant's state space has been expanded. Figures 5 and 6 illustrate the time histories of the observation errors in the system. Comparison of Figs. 2 and 4 as well as of Figs. 5 and 6, respectively, demonstrate the state and the observation error responses have maintained largely their original sharp changes and fast decays.

Fig. 5 Time history of the observation error response for initial conditions $x(0) = [0.609 \quad 0]$, $\tilde{x}(0) = [0 \quad 0.001]$

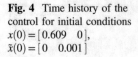

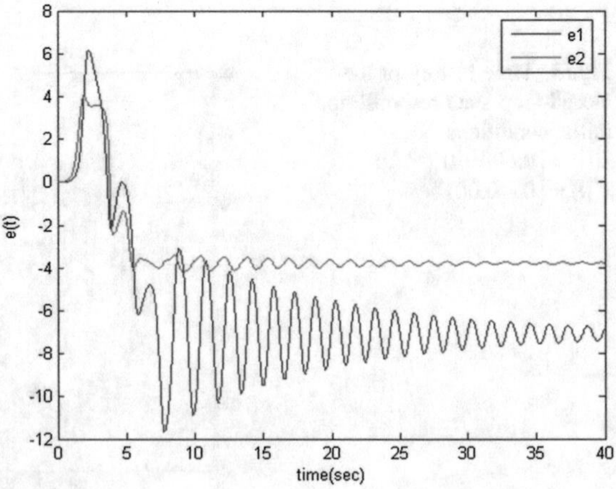

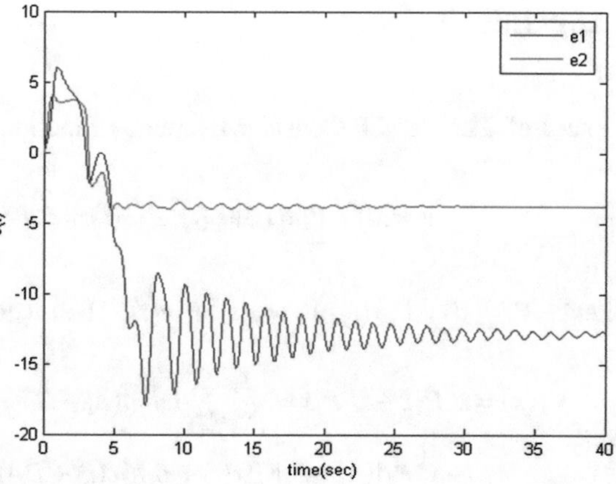

Fig. 6 Time history of the observation error response for initial conditions $x(0) = [0.609 \quad 0]$, $\tilde{x}(0) = [0 \quad 0.001]$

5 Concluding Remarks

A new design method for output feedback min–max control of nonlinear plants, following the original [23] and the works [22], based on a fuzzy observer [7, 11, 14], by using the linear matrix inequalities technique [18] was developed. The design procedure requires representing the original nonlinear plant system by means of a constructed fuzzy T–S model that can emulate it with considerable accuracy. Hence it has a potential for wide range of applications.

The proposed min–max control design minimizes the upper bound of performance index with low energy expenditure while guaranteeing asymptotic stability in the closed-loop system. Furthermore, as proved in [18, 22], the nonlinear closed-loop system possesses an operating equilibrium with quadratic stability property. This new control design is applied to motion control of an inverted pendulum to demonstrate the robust stability performance within a rather enlarged attraction domain of the equilibrium state than reported in so far despite the persisting disturbances. Refining improvements of transient control accuracy are the envisaged future research toward.

Acknowledgments The Authors gratefully acknowledge the crucial contribution by Professor Georgi M. Dimirovski in proving the theoretical results reported in this article.

Appendix 1

Proof of Theorem 1 Consider Lyapunov function candidate

$$V = x(t)^T P_1 x(t) + e(t)^T P_2 e(t) = \xi^T P \xi,\ P = \begin{bmatrix} P_1 & 0 \\ 0 & P_2 \end{bmatrix} \tag{31}$$

hence $V_1 = x(t)^T P_1 x(t)$, $V_2 = e(t)^T P_2 e(t)$. Then it follows:

$$\begin{aligned} \dot{V}_1(x) = \dot{x}^T P_1 x + x^T P_1 \dot{x} = & \sum_{i=1}^{q} \sum_{j=1}^{q} h_i h_j [[(A_i - B_i K_j)x(t) + B_i K_j e(t) + \bar{D}_i W]^T P_1 x \\ & + x^T P_1 [(A_i - B_i K_j)x(t) + B_i K_j e(t) + \bar{D}_i W]] = \\ = & \sum_{i=1}^{q} \sum_{j=1}^{q} h_i h_j [x^T (H_{ij}^T P_1 + P_1 H_{ij}) x + e^T(t)(B_i K_j)^T P_1 x(t) + \\ & + x^T(t) P_1 B_i K_j e(t) + 2x^T(t) P_1 \bar{D}_i W] \end{aligned} \tag{32}$$

and

$$\dot{V}_2(t) = \dot{e}^T P_2 e + e^T P_2 \dot{e} = \sum_{i=1}^{q} \sum_{j=1}^{q} h_i h_j [e^T (\Sigma_{ij}^T P_2 + P_2 \Sigma_{ij}) e + 2e^T(t) P_2 \bar{D}_i W], \tag{33}$$

where $H_{ij} = A_i - B_i K_j$. Hence

$$\begin{aligned} \dot{V} = \dot{V}_1 + \dot{V}_2 = & \sum_{i=1}^{q} \sum_{j=1}^{q} h_i h_j [\xi^T(t)(\Psi_{ij}^T P + P \Psi_{ij}) \xi(t) - \\ & - 2x^T(t) K_i^T B_j^T P_1 x(t) + 2e^T(t) K_i^T B_j^T P_1 x(t) + \\ & + 2x^T(t) P_1 \bar{D}_i W + 2e^T(t) P_2 \bar{D}_i W] = \\ = & \sum_{i=1}^{q} \sum_{j=1}^{q} h_i h_j [\xi^T(t)(\Psi_{ij}^T P + P \Psi_{ij}) \xi(t) + 2\xi^T(t) P \tilde{D}_i W - \\ & - 2x^T(t) K_i^T B_j^T P_1 x(t) + 2e^T(t) K_i^T B_j^T P_1 x(t)], \end{aligned} \tag{34}$$

where $\Sigma_{ij} = A_i - G_i C_j, \Psi_{ij} = \begin{bmatrix} A_i & 0 \\ 0 & \Sigma_{ij} \end{bmatrix}$, $\tilde{D}_i = \begin{bmatrix} \bar{D}_i \\ \bar{D}_i \end{bmatrix}$.

Next, the following local checking function:

$$\phi(t) = \dot{V} + u^T u - W^T W \tag{35}$$

is constructed. Substitution of (34) into the above expression yields

$$\begin{aligned}\varphi(t) = \sum_{i=1}^{q}\sum_{j=1}^{q} h_i h_j [\xi^T & \begin{bmatrix} A_i^T P_1 + P_1 A_i & 0 \\ 0 & \Sigma_{ij}^T P_2 + P_2 \Sigma_{ij} \end{bmatrix} \xi + \\ & + 2u^T(t)\, B_i P_1\, x(t) + 2x^T(t)\, P_1\, \bar{D}_i\, W + \\ & + 2e^T(t) P_2 \bar{D}_i W + u^T u - W^T W]\end{aligned} \tag{36}$$

Thus maximization of (36) about W yields

$$W^* = \sum_{i=1}^{q} h_i [\bar{D}_i^T P_1^T x(t) + \bar{D}_i^T P_2^T e(t)] = \sum_{i=1}^{q} h_i \tilde{D}_i^T P^T \xi(t) \tag{37}$$

Because of $\frac{\partial^2 \phi(t)}{\partial W^2} = -2 < 0$, (37) represents the parametric expression of the worst case disturbance [24–26] for system (16)–(17). Substitution of W^* into (36) gives:

$$\begin{aligned}\max_W \varphi(t) = \sum_{i=1}^{q}\sum_{j=1}^{q} h_i h_j [\xi^T & \begin{bmatrix} A_i^T P_1 + P_1 A_i & 0 \\ 0 & \Sigma_{ij}^T P_2 + P_2 \Sigma_{ij} \end{bmatrix} \xi + \\ & + 2x^T(t)\, P_1 B_i\, u(t) + 2x^T(t)\, P_1\, \bar{D}_i\, W + \\ & + \xi^T(t) P \tilde{D}_i \tilde{D}_j^T P \xi(t) + u^T(t) u(t) \\ & + 2e^T(t) P_2 \bar{D}_i W + u^T u - W^T W] \\ = \sum_{i=1}^{q}\sum_{j=1}^{q} & h_i h_j [\xi^T(t)(\Psi_{ij}^T P + P\Psi_{ij})\xi(t) + \\ & + 2x^T(t) P_1 B_i u(t) + u^T(t) u(t) + \xi^T(t) P \tilde{D}_i \tilde{D}_j^T P \xi(t)\end{aligned} \tag{38}$$

Minimization of the above expression about u yields

$$u^* = -\sum_{i=1}^{q} h_i B_i^T P_1 x(t) = -\sum_{i=1}^{q} h_i (B_i^T P_1 \tilde{x}(t) + B_i^T P_1 e(t)). \tag{39}$$

And for $\frac{\partial^2 (\max_W \phi(t))}{\partial u^2} = I > 0$, the (39) is the parametric expression of min–max controller for system (16)–(17) apparently. Next, the substitution of u^* into (38) gives

$$\begin{aligned}\min_u \max_W \varphi(t) = \sum_{i=1}^{q}\sum_{j=1}^{q} h_i h_j [\xi^T(t)(\Psi_{ij}^T P + P\Psi_{ij})\xi(t) - x^T(t)\, P_1^T B_i\, B_j^T P_1 x(t) + \\ + \xi^T(t) P \tilde{D}_i \tilde{D}_j^T P \xi(t)] \cdots = \sum_{i=1}^{q}\sum_{j=1}^{q} h_i h_j \xi^T(t)[\Psi_{ij}^T P + P\Psi_{ij} - P^T \tilde{B}_i \tilde{B}_j^T P + P \tilde{D}_i \tilde{D}_j^T P]\xi(t),\end{aligned}$$

where $\tilde{B}_i = \begin{bmatrix} B_i \\ 0 \end{bmatrix}$. Further, let it be denoted

$$\min_u \max_W \phi(t) = -\xi^T Q\xi(t) \tag{40}$$

Notice that if the following inequality is satisfied

$$\Psi_{ij}^T P + P\Psi_{ij} - P^T \tilde{B}_i \tilde{B}_j^T P + P\tilde{D}_i \tilde{D}_j^T P < 0, \tag{41}$$

then $Q > 0$ holds true. Substitution of (37) and (39) into (34) yields

$$\dot{V} = \sum_{i=1}^{q} \sum_{j=1}^{q} h_i h_j \xi^T(t) \left[\Psi_{ij}^T P + P\Psi_{ij} - 2P^T \tilde{B}_i \tilde{B}_j^T P + 2P\tilde{D}_i \tilde{D}_j^T P \right] \xi(t)$$

In turn, if $\tilde{D}_i \tilde{D}_j^T - \tilde{B}_i \tilde{B}_j^T < 0$ and (39) hold, then $\dot{V} < 0$ obviously. Thus the closed-loop system (16)–(17) is asymptotically stable and $\xi(\infty) = 0$. Pre- and post-multiplication of both sides of (41) by $diag(P_1^{-1}, P_2^{-1})$, and then application of Schur's complement yields

$$\begin{bmatrix} XA_i^T + A_i X + \Gamma & 0 \\ 0 & YA_i^T + A_i Y - Z - Z^T + D_i D_j \end{bmatrix} < 0 \tag{42}$$

Apparently, the above expression is equivalent to (18)–(19) in Theorem 1. Now the integral of (42) is calculated; after some appropriate transpose, to give:

$$\begin{aligned} \min_u \max_{v,\omega} J(u, W) &= \min_u \max_{v,\omega} \int_0^\infty (\xi^T Q\xi + u^T u - W^T W)dt \leq -\int_0^\infty \dot{V} dt \\ &= V(\xi(0)) - V(\xi(\infty)) = \xi(0)^T P\xi(0) \end{aligned} \tag{43}$$

Therefore, due to Assumption 2 and via considering the expected value of the performance cost, it follows

$$\bar{J} = E\{J\} \leq E\{\xi_0^T P\xi_0\} = Trace(P). \tag{44}$$

Nonetheless, notice the initial state of a plant system can hardly be accurately measured in real-world practice.

Appendix 2

Proof of Theorem 2 From the proof of Theorem 1 it is seen if the inequality

$$\Psi_{ij}^T P + P\Psi_{ij} - 2P^T \tilde{B}_i \tilde{B}_j^T P + 2P\tilde{D}_i \tilde{D}_j^T P < 0, \tag{45}$$

is satisfied then apparently $\dot{V} < 0$. That is, the closed-loop system (16)–(17) is asymptotically stable and $\xi(\infty) = 0$. Further, if $\tilde{D}_i \tilde{D}_j^T - \tilde{B}_i \tilde{B}_j^T \geq 0$, then

$$\Psi_{ij}^T P + P\Psi_{ij} - P^T \tilde{B}_i \tilde{B}_j^T P + P\tilde{D}_i \tilde{D}_j^T P < 0, \tag{46}$$

and thus $Q > 0$ is guaranteed. After pre- and post- multiplication of both sides of (23) by $diag(P_1^{-1}, P_2^{-1})$, the application of Schur's complement yields

$$\begin{bmatrix} XA_i^T + A_i X + 2\Gamma & 0 \\ 0 & YA_i^T + A_i Y - Z - Z^T + 2D_i D_j \end{bmatrix} < 0. \tag{47}$$

As seen, the above expression is equivalent to (21). Now recall the proof of Theorem 1. After the appropriate transposing and then calculating the integral, the investigation of expected value of performance index yields

$$\bar{J} = E\{J\} \leq E\{\xi_0^T P \xi_0\} = Trace(P). \tag{48}$$

References

1. Zak, S.H.: Systems and Control. Oxford University Press, New York (2003)
2. Driankov, D., Saffioti, A. (eds.): Fuzzy Logic Techniques for Autonomous Vehicle Navigataion. Physica-Verlag, Heidelberg, DE (2001)
3. Zadeh, L.A.: Is there a need for fuzzy logic? Inf. Sci. **178**, 2751–2779 (2008)
4. Takagi, T., Sugeno, M.: Stability analysis and design of fuzzy control systems. Fuzzy Sets Syst. **45**(2), 135–156 (1992)
5. Tanaka, K., Ikeda, T., Wang, H.O.: Robust stabilization of a class of uncertain nonlinear systems via fuzzy control: quadratic stabilizablilty, H^∞ control theory, and linear matrix inequalities. IEEE Trans. Fuzzy Syst. **4**(1), 1–13 (1996)
6. Wang, H.O., Tanaka, K., Griffin, M.F.: An approach to fuzzy control of nonlinear systems: Stability and design issues. IEEE Trans. Fuzzy Syst. **4**(1), 14–23 (1996)
7. Tanaka, K., Ikeda, T., Wang, H.O.: Fuzzy regulators and fuzzy observers: relaxed stability conditions and LMI-based designs. IEEE Trans. Fuzzy Syst. **6**(4), 250–265 (1998)
8. Leung, F.H.F., Lam, H.K., Tam, P.K.S.: Design of fuzzy controllers for uncertain nonlinear systems using stability and robustness analyses. Syst. Control Lett. **35**(3), 237–243 (1998)
9. Kim, E., Lee, T.H.: New approaches to relaxed quadratic stability condition on fuzzy control systems. IEEE Trans. Fuzzy Syst. **8**(6), 523–533 (2000)

10. Zheng, F., Wang, Q.-G., Lee, T.H.: Robust controller design for uncertain nonlinear systems via fuzzy modelling approach. IEEE Trans. Syst. Man Cybern. Part B Cybern. **34**(1), 166–178 (2002)
11. Tong, S.-C., Li, H.-H.: Observer-based robust fuzzy control of nonlinear systems with parametric uncertainties. Fuzzy Sets Syst. **131**(3), 165–184 (2002)
12. Kim, J., Park, D.: LMI-based design of stabilizing controllers for nonlinear systems described by Takagi-Sugeno model. Fuzzy Sets Syst. **122**(2), 73–82 (2003)
13. Liu, X., Zhang, Q.: New approaches to H^{∞} controller designs based on fuzzy observers for fuzzy system via LMI. Automatica **39**(12), 1571–1582 (2003)
14. Lin, C., Wang, Q.-G., Lee, T.H.: Improvement on observer-based H^{∞} control for T-S fuzzy systems. Automatica **41**(12), 1651–1656 (2005)
15. Guerra, T.M., Kruszewski, A., Vermeieren, L., Tirmant, H.: Conditions of output stabilization for nonlinear models in the Takagi-Sugeno's form. Fuzzy Sets Syst. **157**(9), 1248–1259 (2006)
16. Sala, A., Arino, C.: Asymptotically necessary and sufficient conditions for stability and performance in fuzzy control: applications of Polya's theorem. Fuzzy Sets Syst. **158**(24), 2671–2686 (2007)
17. Yang, H., Dimirovski, G.M., Zhao, J.: Switched fuzzy systems: modelling representation, stability, and control. Chap. 9. In: Kacprzyk, J. (ed.) Studies in Computational Intelligence, vol. 109, pp. 155–168. Springer, Heidelberg (2008)
18. Tanaka, K., Wang, H.O.: Fuzzy Control System Design and Analysis: A Linear Matrix Inequality Approach. Wiley (Canada) (2001)
19. Wu, H.-N., Cai, K.-Y.: H^2 guaranteed cost fuzzy control for uncertain nonlinear systems via linear matrix inequalities. Fuzzy Sets Syst. **148**(6), 411–429 (2004)
20. Wu, Z.-Q., Li, J., Gao, M.-J.: Guaranteed cost control for uncertain discrete fuzzy systems with delays. Fuzzy Syst. Math. **18**(3), 95–101 (2004)
21. Assawinchiachote, W., Nguang, S.K., Shi, P.: H^{∞} output feedback control design for uncertain fuzzy singularly disturbed systems: an LMI approach. Automatica **40**(12), 2147–2152 (2005)
22. Jing, Y., Jiang, N., Zheng, Y., Dimirovski, G.M.: Fuzzy robust and non-fragile minimax control of a trailer-truck model. In: Proceedings of the 46th IEEE Conference on Decision and Control, New Orleans, LA, USA, pp. 1221–1226 (2007)
23. Kogan, M.M.: Solution to the inverse problem of minimax control and minimax robust control. Avtomatika i Telemechanika **53**, 87–97 (1998). (in Russian)
24. Kogan, M.M.: Solution to the inverse problem of minimax control and worst case disturbance for linear continuous-time systems. IEEE Trans. Autom. Control **43**(5), 670–674 (1998)
25. Yoon, M., Utkin, V., Postletwhite, I.: On the worst-case disturbance of minimax optimal control. Automatica **41**(7), 847–855 (2005)
26. MathWorks.: Using Matlab—LMI Toolbox. The Mathworks Inc, Natick, NJ
27. MathWorks.: Using Simulink—Version 5. The Mathworks Inc, Natick, NJ
28. Zadeh, L.A.: Inference in fuzzy logic. IEEE Proc. **68**, 124–131 (1980)
29. Tuan, H.D., Apkarian, P., Narikiyo, T., Yamamoto, Y.: Parameterized linear matrix inequality techniques in fuzzy control system design. IEEE Trans. Fuzzy Syst. **9**(2), 324–332 (2001)
30. Kolemishevska-Gugulovska, T., Stankovski, M., Rudas, I.J., Jiang, N., Jing, J.: A min-max control synthesis for uncertain nonlinear systems based on fuzzy T-S model. In: Proceedings of the IEEE 6th International Conference on Intelligent Systems, IS 2012, Sofia, Bulgaria, 6–8 Sept 2012, pp. 303–310

Modeling Parallel Optimization of the Early Stopping Method of Multilayer Perceptron

Maciej Krawczak, Sotir Sotirov and Evdokia Sotirova

Abstract Very often, overfitting of the multilayer perceptron can vary significantly in different regions of the model. Excess capacity allows better fit to regions of high, nonlinearity; and backprop often avoids overfitting the regions of low nonlinearity. The used generalized net will give us a possibility for parallel optimization of MLP based on early stopping algorithm.

1 Introduction

In a series of papers, the process of functioning and the results of the work of different types of neural networks are described by Generalized Nets (GNs). Here, we shall discuss the possibility for training of feed-forward Neural Networks (NN) by backpropagation algorithm. The GN optimized the NN-structure on the basis of connections limit parameter.

The different types of neural networks [1] can be implemented in different ways [2–4] and can be learned by different algorithms [5–7].

M. Krawczak (✉)
Higher School of Applied Informatics and Management, Warsaw, Poland
e-mail: krawczak@ibs.pan.waw.pl

S. Sotirov · E. Sotirova
Asen Zlatarov University, Burgas 8000, Bulgaria
e-mail: ssotirov@btu.bg

E. Sotirova
e-mail: esotirova@btu.bg

V. Sgurev et al. (eds.), *Recent Contributions in Intelligent Systems*,
Studies in Computational Intelligence 657,
DOI 10.1007/978-3-319-41438-6_7

2 The Golden Sections Algorithm

Let the natural number N and the real number C be given. They correspond to the maximum number of the hidden neurons and the lower boundary of the desired minimal error.

Let real monotonous function f determine the error $f(k)$ of the NN with k hidden neurons.

Let function c: $R \times R \rightarrow R$ be defined for every $x, y \in R$ by

$$c(x,y) = \begin{cases} 0; \text{ if } \max(x;y) < C \\ \frac{1}{2}; \text{ if } x \leq C \leq y \\ 1; \text{ if } \min(x,y) > C \end{cases}$$

Let $\varphi = \frac{\sqrt{5}+1}{2} = 0.61$ be the Golden number.

Initially, let we put: $L = 1$; $M = [\varphi^2 : N] + 1$, where $[x]$ is the integer part of the real number $x \geq 0$.

The algorithm is the following:

1. If $L \geq M$ go to 5.
2. Calculate $c(f(L), f(M))$. If

$$c(x,y) = \begin{cases} 1 \text{ to go } 3 \\ \frac{1}{2} \text{ to go } 4 \\ 0 \text{ to go } 5 \end{cases}$$

3. $L = M + 1$; $M = M + [\varphi^2 \cdot (N-M)] + 1$ go to 1.
4. $M = L + [\varphi^2 \cdot (N-M)] + 1$; $L = L + 1$ go to 1.
5. End: final value of the algorithm is L.

3 Neural Network

The proposed generalized-net model introduces parallel work in learning of two neural networks with different structures. The difference between them is in neurons' number in the hidden layer, which directly reflects on the all network's properties. Through increasing their number, the network is learned with fewer number of epoches achieving its purpose. On the other hand, the great number of neurons complicates the implementation of the neural network and makes it unusable in structures with elements' limits [5].

Figure 1 shows abbreviated notation of a classic tree-layered neural network.

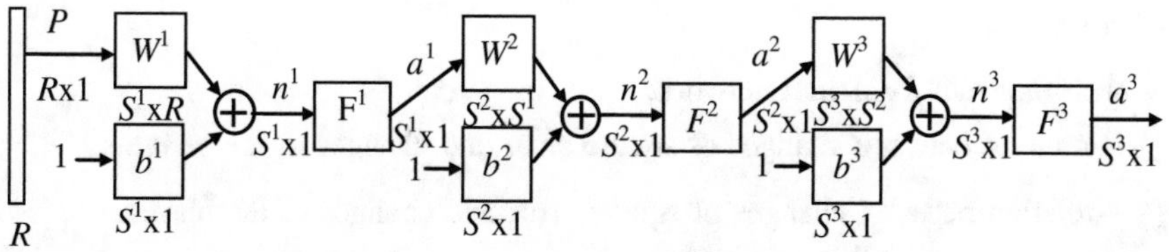

Fig. 1 xxx

In the many-layered networks, the one layer's exits become entries for the next one. The equations describing this operation are

$$a^3 = f^3\left(w^3 f^2\left(w^2 f^1\left(w^1 p + b^1\right) + b^2\right) + b^3\right), \tag{1}$$

where

- a^m is the exit of the *m*-layer of the neural network for $m = 1, 2, 3$;
- w is a matrix of the weight coefficients of the everyone of the entries;
- b is neuron's entry bias;
- f^m is the transfer function of the *m*-layer.

The neuron in the first layer receives outside the entries p. The neurons' exit from the last layer determine the neural network's exit a.

Because it belongs to the learning with teacher methods, the algorithm are submitted couple numbers (an entry value and an achieving aim—on the network's exit)

$$\{p_1, t_1\},\ \{p_2, t_2\},\ \ldots,\ \{p_Q, t_Q\}, \tag{2}$$

$Q \in (1 \ldots n)$, n—numbers of learning couple, where p_Q is the entry value (on the network entry), and t_Q is the exit's value replying to the aim. Every network's entry is preliminary established and constant, and the exit have to reply to the aim. The difference between the entry values and the aim is the error—$e = t - a$.

The "back propagation" algorithm [6] uses least-quarter error

$$\hat{F} = (t - a)^2 = e^2. \tag{3}$$

In learning the neural network, the algorithm recalculates network's parameters (W and b) to achieve least-square error.

The "back propagation" algorithm for *i*-neuron, for $k + 1$ iteration use equations

$$w_i^m(k+1) = w_i^m(k) - \alpha \frac{\partial \hat{F}}{\partial w_i^m}, \tag{4}$$

$$b_i^m(k+1) = b_i^m(k) - \alpha \frac{\partial \hat{F}}{\partial b_i^m}, \tag{5}$$

where

α—learning rate for neural network;

$\frac{\partial \hat{F}}{\partial w_i^m}$—relation between changes of square error and changes of the weights;

$\frac{\partial \hat{F}}{\partial b_i^m}$—relation between changes of square error and changes of the biases.

The overfitting [8] appears in different situations, which effect over trained parameters and make worse output results, as show in Fig. 2.

There are different methods that can reduce the overfitting—"Early Stopping" and "Regularization". Here we will use Early Stopping [9].

When multilayer neural network will be trained, usually the available data must be divided into three subsets. The first subset named "Training set" is used for computing the gradient and updating the network weighs and biases. The second subset is named "Validation set". The error on the validation set is monitored during the training process. The validation error normally decreases during the initial phase of training, as does the training set error. Sometimes, when the network begins to overfit the data, the error on the validation set typically begins to rise. When the validation error increases for a specified number of iterations, the training is stopped, and the weights and biases at the minimum of the validation error are returned [5]. The last subset is named "test set". The sum of these three sets has to be 100 % of the learning couples.

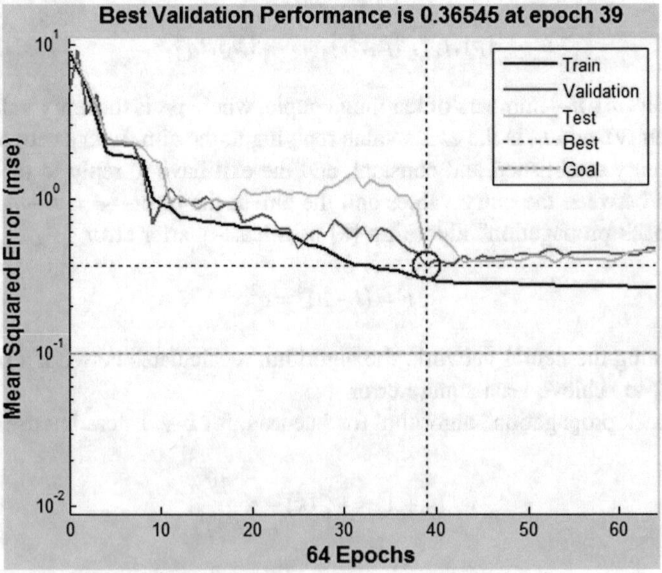

Fig. 2 xxx

When the validation error e_v increases (the changing de_v have positive value) the neural network learning stops when

$$de_v > 0 \tag{6}$$

The classic condition for the learned network is when

$$e^2 < E\max, \tag{7}$$

where Emax is maximum square error.

4 GN Model

All definitions related to the concept "GN" are taken from [10]. The network, describing the work of the neural network learned by "Backpropagation" algorithm [9], is shown in Fig. 3.

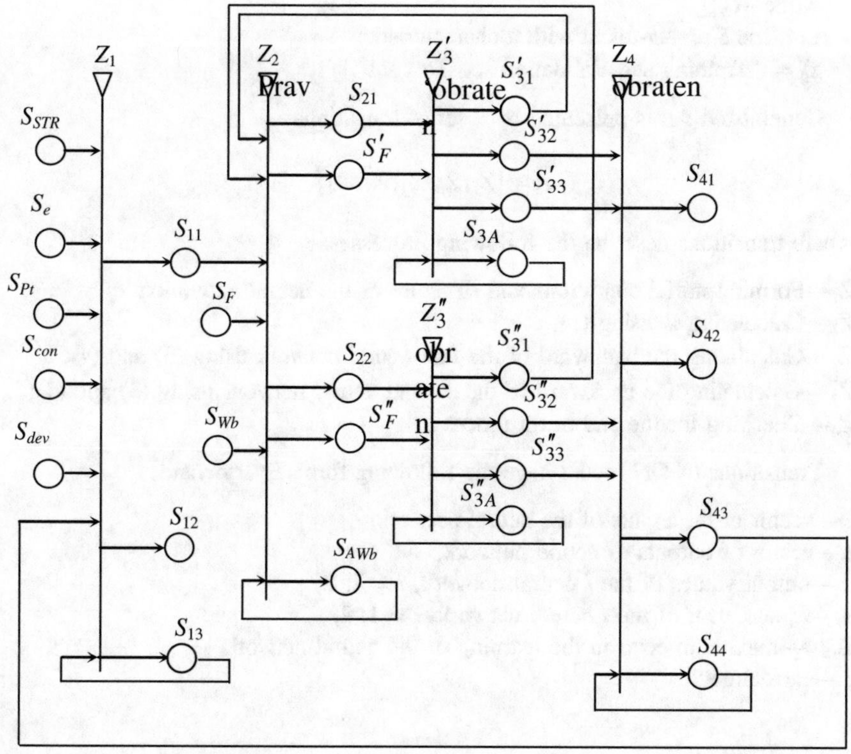

Fig. 3 xxx

The below constructed GN model is the reduced one. It does not have temporal components, the priorities of the transitions; places and tokens are equal, the place and arc capacities are equal to infinity.

Initially the following tokens enter in the generalized net:

- in place S_{STR}—α-token with characteristic
  x_0^{α} = "number of neurons in the first layer, number of neurons in the output layer";
- in place S_e—β-token with characteristic
  x_0^{β} = "maximum error in neural network learning Emax";
- in place S_{Pt}—γ-token with characteristic
  x_0^{γ} = "$\{p_1, t_1\}$, $\{p_2, t_2\}$, $\{p_3, t_3\}$";
- in place S_F—one δ-token with characteristic
  x_0^{δ} = "f^1, f^2, f^3".
  The token splits into two tokens that enters respectively in places S_F' and S_F'';
- in place S_{Wb}—ε-token having characteristics
  x_0^{ε} = "w, b";
- in place S_{con}—ξ-token with a characteristics
  x_0^{ξ} = "maximum number of the neurons in the hidden layer in the neural network—C_{max}".
- in place S_{dev}—ψ-token with a characteristics
  x_0^{ψ} = "Training set, Validation set, Test set".

Generalized net is presented by a set of transitions

$$A = \{Z_1, Z_2, Z_3', Z_3'', Z_4\},$$

where transitions describe the following processes:

Z_1—Forming initial conditions and structure of the neural networks;
Z_2—Calculating a_i using (1);
Z_3'—Calculating the backward of the first neural network using (3) and (4);
Z_3''—Calculating the backward of the second neural network using (3) and (4);
Z_4—Checking for the end of all process.

Transitions of GN model have the following form. Everywhere

p—vector of the inputs of the neural network,
a—vector of outputs of neural network,
a_i—output values of the i neural network, i = 1, 2,
e_i—square error of the i neural network, i = 1, 2,
$E_{\max}$—maximum error in the learning of the neural network,
t—learn target;

w_{ik}—weight coefficients of the i neural networks i = 1, 2 for the k iteration;
b_{ik}—bias coefficients of the i neural networks i = 1, 2 for the k iteration.

$$Z_1 = \langle\{S_{STR}, S_e, S_{Pt}, S_{\text{con}}, S_{\text{dev}}, S_{43}, S_{13}\}, \{S_{11}, S_{12}, S_{13}\}, R_1, \wedge(\vee(\wedge(S_e, S_{Pt}, S_{\text{con}}, S_{\text{dev}}), S_{13}), \vee(S_{STR}, S_{43}))\rangle,$$

where:

$R_1 =$

	S_{11}	S_{12}	S_{13}
S_{STR}	*False*	*False*	*True*
S_e	*False*	*False*	*True*
S_{Pt}	*False*	*False*	*True*
S_{con}	*False*	*False*	*True*
S_{dev}	*False*	*False*	*True*
S_{43}	*True*	*False*	*False*
S_{13}	$W_{13,11}$	$W_{13,12}$	*True*

and

$W_{13,11}$ = "the learning couples are divided into the three subsets";
$W_{13,12}$ = "is it not possible to divide the learning couples into the three subsets".

The token that enters in place S_{11} on the first activation of the transition Z_1 obtain characteristic

$$x_0^{\theta'} = "pr_1 x_0^{\alpha}, \left[1; x_0^{\xi}\right], pr_2 x_0^{\alpha}, x_0^{\gamma}, x_0^{\beta}".$$

Next it obtains the characteristic

$$x_{cu}^{\theta'} = "pr_1 x_0^{\alpha}, [l_{\min}; l_{\max}], pr_2 x_0^{\alpha}, x_0^{\gamma}, x_0^{\beta}",$$

where $[l_{min};l_{max}]$ is the current characteristics of the token that enters in place S_{13} from place S_{43}.

The token that enters place S_{12} obtains the characteristic $[l_{\min};l_{\max}]$.

$$Z_2 = \langle\{S_{31}', S_{31}'', S_{11}, S_F, S_{Wb}, S_{AWb}\}, \{S_{21}, S_F', S_{22}, S_F'', \} R_2 \vee(\wedge(S_F, S_{11}), \vee(S_{AWb}, S_{Wb}), (S_{31}', S_{31}''))\rangle,$$

where

$$R_2 = \begin{array}{c|ccccc} & S_{21} & S'_F & S_{22} & S''_F & S_{AWb} \\ \hline S'_{31} & True & False & False & False & True \\ S''_{31} & False & False & True & False & True \\ S_{11} & True & False & True & False & False \\ S_F & True & True & True & True & False \\ S_{Wb} & True & False & True & False & False \\ S_{12} & True & False & True & False & False \\ S_{AWb} & True & False & True & False & False \end{array}.$$

The tokens that enter places S_{21} and S_{22} obtain the characteristics respectively

$$x_{cu}^{\eta'} = "x_{cu}^{\varepsilon'}, x_0^{\gamma}, x_0^{\beta''}, a_1, pr_1 x_0^{\alpha}, [l_{\min}], pr_2 x_0^{\alpha}"$$

and

$$x_{cu}^{\eta''} = "x_{cu}^{\varepsilon'}, x_0^{\gamma}, x_0^{\beta''}, a_2, pr_1 x_0^{\alpha}, [l_{\max}], pr_2 x_0^{\alpha}".$$

$$Z_3' = \langle \{S_{21}, S_F', S_{3A}'\}, \{S_{31}', S_{32}', S_{3A}'\}, R_3', \wedge(S_{21}, S_F', S_{3A}')\rangle,$$

where

$$R_3' = \begin{array}{c|cccc} & S'_{31} & S'_{32} & S'_{33} & S'_{3A} \\ \hline S_{21} & False & False & False & True \\ S'_F & False & False & False & True \\ S'_{3A} & W'_{3A,31} & W'_{3A,32} & W'_{3A,33} & True \end{array},$$

and

$W'_{3A,31}$ = "$e_1 > E_{\max}$ or $de_{1v} < 0$";
$W'_{3A,32}$ = "$e_1 < E_{\max}$ or $de_{1v} < 0$";
$W'_{3A,33}$ = "($e_1 > E_{\max}$ and $n_1 > m$) or $de_{1v} > 0$";

where

n_1—current number of the first neural network learning iteration,
m—maximum number of the neural network learning iteration,
de_{1v}—validation error changing of the first neural network.

The token that enters place S'_{31} obtains the characteristic "first neural network: $w(k + 1)$, $b(k + 1)$", according (4) and (5). The λ'_1 and λ'_2 tokens that enter place S'_{32} and S'_{33} obtain the characteristic

$$x_0^{\lambda_1'} = x_0^{\lambda_2'} = ''l_{\min}''$$

$$Z_3'' = \langle \{S_{22}, S_F'', S_{A3}''\}, \{S_{31}'', S_{32}'', S_{33}'', S_{A3}''\}, R_3'' \wedge (S_{22}, S_F'', S_{A3}'')\rangle$$

where

$$R_3'' = \begin{array}{c|cccc} & S_{31}'' & S_{32}'' & S_{33}'' & S_{A3}'' \\ \hline S_{22} & False & False & False & True \\ S_{3F}'' & False & False & False & True \\ S_{A3}'' & W_{A3,31}'' & W_{A3,32}'' & W_{A3,33}'' & True \end{array},$$

and

$W_{3A,31}''$ = "$e_2 > E_{\max}$ or $de_{2v} < 0$",
$W_{3A,32}''$ = "$e_2 < E_{\max}$ or $de_{2v} < 0$",
$W_{3A,33}''$ = "($e_2 > E_{\max}$ and $n_2 > m$) or $de_{2v} > 0$",
where

n_2—current number of the second neural network learning iteration;
m—maximum number of the neural network learning iteration;
de_{2v}—validation error changing of the second neural network.

The token that enters place S_{31}'' obtains the characteristic "second neural network: $w(k + 1)$, $b(k + 1)$", according (4) and (5). The λ_1'' and λ_2'' tokens that enter place S_{32}'' and S_{33}'' obtain, respectively

$$x_0^{\lambda_1''} = x_0^{\lambda_2''} = ''l_{\max}''$$

$$Z_4 = \langle \{S_{32}', S_{33}', S_{32}'', S_{33}'', S_{44}\}, \{S_{41}, S_{42}, S_{43}, S_{44}\}, R_4, \wedge (S_{44} \vee (S_{32}', S_{33}', S_{32}'', S_{33}''))\rangle,$$

where

$$R_4 = \begin{array}{c|cccc} & S_{41} & S_{42} & S_{43} & S_{44} \\ \hline S_{32}' & False & False & False & True \\ S_{33}' & False & False & False & True \\ S_{32}'' & False & False & False & True \\ S_{33}'' & False & False & False & True \\ S_{44} & W_{44,41} & W_{44,42} & W_{44,43} & True \end{array},$$

and

$W_{44,41}$ = "$e_1 < E_{\max}$" and "$e_2 < E_{\max}$";
$W_{44,42}$ = "$e_1 > E_{\max}$ and $n_1 > m$" and "$e_2 > E_{\max}$ and $n_2 > m$";

$W_{44,43}$ = "($e_1 < E_{max}$ and ($e_2 > E_{max}$ and $n_2 > m$)) or ($e_2 < E_{max}$ and ($e_1 > E_{max}$ and $n_1 > m$))".

The token that enters place S_{41} obtains the characteristic

Both NN satisfied conditions—for the solution is used the network who wave smaller numbers of the neurons.

The token that enters place S_{42} obtain the characteristic

There is no solution (both NN not satisfied conditions).

The token that enters place S_{44} obtains the characteristic

the solution is in interval [l_{min}; l_{max}]—the interval is changed using the golden sections algorithm.

5 Conclusion

The proposed generalized-net model introduces the parallel work in the learning of the two neural networks with different structures. The difference between them is in the number of neurons in the hidden layer, which reflects directly over the properties of the whole network.

On the other hand, the great number of neurons complicates the implementation of the neural network.

The constructed GN model allows simulation and optimization of the architecture of the neural networks using golden section rule.

References

1. http://www.fi.uib.no/Fysisk/Teori/NEURO/neurons.html. Neural Network Frequently Asked Questions (FAQ), The information displayed here is part of the FAQ monthly posted to comp.ai.neural-nets (1994)
2. Krawczak, M.: Generalized net models of systems. Bull. Polish Acad. Sci. (2003)
3. Sotirov, S.: Modeling the algorithm Backpropagation for training of neural networks with generalized nets—part 1. In: Proceedings of the Fourth International Workshop on Generalized Nets, Sofia, 23 Sept, pp. 61–67 (2003)
4. Sotirov, S., Krawczak, M.: Modeling the algorithm Backpropagation for training of neural networks with generalized nets—part 2, Issue on Intuitionistic Fuzzy Sets and Generalized nets, Warsaw (2003)
5. Hagan, M., Demuth, H., Beale, M.: Neural Network Design. PWS Publishing, Boston, MA (1996)
6. Rumelhart, D., Hinton, G., Williams, R.: Training representation by back-propagation errors. Nature **323**, 533–536 (1986)
7. Sotirov, S.: A method of accelerating neural network training. Neural Process. Lett. Springer **22**(2), 163–169 (2005)
8. Bellis, S., Razeeb, K.M., Saha, C., Delaney, K., O'Mathuna, C., Pounds-Cornish, A., de Souza, G., Colley, M., Hagras, H., Clarke, G., Callaghan, V., Argyropoulos, C., Karistianos, C., Nikiforidis, G.: FPGA implementation of spiking neural networks—an initial step towards

building tangible collaborative autonomous agents, FPT'04. In: International Conference on Field-Programmable Technology, The University of Queensland, Brisbane, Australia, 6–8 Dec, pp. 449–452 (2004)

9. Haykin, S.: Neural Networks: A Comprehensive Foundation. Macmillan, NY (1994)
10. Atanassov, K.: Generalized Nets. World Scientific, Singapore (1991)
11. Gadea, R., Ballester, F., Mocholi, A., Cerda, J.: Artificial neural network implementation on a single FPGA of a pipelined on-line Backpropagation. In: 13th International Symposium on System Synthesis (ISSS'00), pp. 225–229 (2000)
12. Maeda, Y., Tada, T.: FPGA Implementation of a pulse density neural network with training ability using simultaneous perturbation. IEEE Trans. Neural Netw. **14**(3) (2003)
13. Geman, S., Bienenstock, E., Doursat, R.: Neural networks and the bias/variance dilemma. Neural Comput. **4**, 1–58 (1992)
14. Beale, M.H., Hagan, M.T., Demuth, H.B.: Neural Network Toolbox User's Guide R2012a (1992–2012)
15. Morgan, N.: H, pp. 630–637. Bourlard, Generalization and parameter estimation in feedforward nets (1990)

Intelligent Controls for Switched Fuzzy Systems: Synthesis via Nonstandard Lyapunov Functions

Jinming Luo and Georgi M. Dimirovski

Abstract This paper investigates the synthesis of intelligent control algorithms for switched fuzzy systems by employing nonstandard Lyapunov functions and combined techniques. Controlled plants are assumed nonlinear and to be represented by certain specific T–S fuzzy models. In one of case studies, a two-layer multiple Lyapunov functions approach is developed that yields a stability condition for uncontrolled switched fuzzy systems and a stabilization condition for closed-loop switched fuzzy systems under a switching law. State feedback controllers with the time derivative information of membership functions are simultaneously designed. In another case, Lyapunov functions approach is developed that yields a non-fragile guaranteed cost optimal stabilization in closed-loop for switched fuzzy systems provided a certain convex combination condition is fulfilled. Solutions in both cases are found in terms of derived linear matrix inequalities, which are solvable on MATLAB platform. In another case, a single Lyapunov function approach is developed to synthesize intelligent control in which a designed switching law handles stabilization of unstable subsystems while the accompanied non-fragile guaranteed cost control law ensures the optimality property. Also, an algorithm is proposed to carry out feasible convex combination search in conjunction with the optimality of intelligent control. It is shown that, when an optimality index is involved, the intelligent controls are capable of tolerating some uncertainty not only in the plant but also in the controller implementation. For both cases of intelligent

J. Luo (✉)
Signal State Key Laboratory of Synthetic Automation for Process Industries,
College of Information Science & Engineering, Northeastern University,
Shenyang 110819, Liaoning, People's Republic of China
e-mail: luojinming0830@163.com

G.M. Dimirovski
Departments of Computer and of Control & Automation Engineering,
School of Engineering, Dogus University, Acibadem,
Istanbul TR-34722, Turkey
e-mail: gdimirovski@dogus.edu.tr

G.M. Dimirovski
Institute of ASE, Faculty of Electrical Eng. & Information Technologies,
SS Cyril & Methodius University, Skopje MK-1000, Republic of Macedonia

V. Sgurev et al. (eds.), *Recent Contributions in Intelligent Systems*,
Studies in Computational Intelligence 657,
DOI 10.1007/978-3-319-41438-6_8

control synthesis solutions illustrative examples along with the respective simulation results are given to demonstrate the effectiveness and the achievable nonconservative performance of those intelligent controls in closed loop system architectures.

1 Introduction

Hybrid dynamic systems have been a scientific challenge for a fairly long time [1] in both communities of computer and of control sciences. A switched system is a special type of hybrid dynamical system that comprises several continuous-time or discrete-time subsystems (system models) and a rule governing the switching between them [13, 14, 20, 46]. The dynamics of switched systems is typically described by a set of differential equations, for continuous-time plants, or difference equations, for discrete-time plants, and a switching law that governs the switching among these system models. Such systems have drawn considerable attention from both academic and industrial-application-oriented researches during the past couple of decades [3, 9, 44, 46, 49]. In particular, in the study of switched systems, most works are focused on the stability and stabilization issues [14, 44, 45]. It has been shown in [9, 13, 49] that finding a common Lyapunov function guarantees stability in closed loop under arbitrary switching law. However, a common Lyapunov function may be too difficult to find and even may not exist, besides it may yield conservative results. The known effective design approaches for designing switching laws when no common Lyapunov function exists are the ones employing single Lyapunov function [13, 14, 47], multiple Lyapunov functions [4, 5, 15] and the average dwell-time [9, 15, 25].

On the other hand, it is well known that nonlinear real-world systems [44] can be rather well represented by means of the class of Takagi–Sugeno (T–S) fuzzy models [2, 26, 27], which interpolate a set of locally linear systems via fuzzy If-Then rules and inference in fuzzy-logic [41–43]. In turn, this fact has enabled many of the conventional linear system and control theories to be transformed in applications to a fairly general class of nonlinear systems, e.g., such as in [49]. Many synthesis design results by employing fuzzy T–S models have been reported, e.g., see [4, 5, 38, 47] for discrete-time systems and [29, 33, 36, 37, 39, 48] for various classes of plant systems via fuzzy models. Yet, the most widely used approach via fuzzy T–S models is the one seeking for common Lyapunov function that is applicable to each local linear model in order to obtain the overall fuzzy system stabilized with acceptable performance [31]; thus, the original nonlinear plant is stabilized as well. As the common Lyapunov function method is known to yield conservative results, successful attempts have made to use piecewise Lyapunov [11, 48] and switched Lyapunov function [6, 36, 48], fuzzy Lyapunov function [28] as well as multiple Lyapunov [3, 29] functions.

A switched system whose all subsystems are fuzzy systems is called a switched fuzzy system. Since this class of systems does exhibit features of both switched and fuzzy systems they possess useful properties. Namely, a switched fuzzy system may be viewed according to either the “hard switching” between its fuzzy subsystems or the “soft switching” among the linear models within a fuzzy T–S model. These two switching strategies and possibly their interaction may lead to very complex behaviors of switched fuzzy systems hence can represent the dynamics emanating from any nonlinear plant. Switched fuzzy systems too recently have attracted considerable research attention by both computer and control scientists, and by the community of intelligent control and decision, in particular [31, 36, 38].

Recent advances in switched fuzzy systems research contributed rather significant new results. Work [29] gave stability and smoothness conditions for the switched fuzzy system when a switching law is predetermined by a given partition of plant’s state space. Works [15, 33, 36, 37, 39, 48] proposed switched fuzzy system models by means of which the purposeful switching strategy between fuzzy subsystems can be designed. Most often the stabilization is achieved by designing a switching law via employing single Lyapunov function. In addition, work [21] investigated the impact of partitioning the state space fuzzy control designs while [22] explored the feasible performances of several typical controls for switched fuzzy systems.

The multiple Lyapunov functions method has also been established as an effective tool for designing switching laws to stabilize various systems. In particular, work [9] used multiple Lyapunov functions to solve the stabilization problem of networked continuous-time plants while [3, 36] of switched fuzzy systems. In there, each Lyapunov function of fuzzy subsystems is essentially a common Lyapunov function for every consequent linear model. Although the multiple Lyapunov functions method for switched systems relaxes the conservativeness, the fact of a common Lyapunov function for fuzzy subsystems still involves conservativeness. In this paper, the synthesis of intelligent controls of two types of plant stabilization problems for a class of nonlinear possibly uncertain plant systems based on switched fuzzy models have been solved.

In the first case study, synthesis of intelligent control the task is solved through exploring for a feasible solution via investigation employing Lyapunov functions which are multiple-ones for either the fuzzy subsystems or for the consequent local linear models. For this reason those Lyapunov functions referred to as two-layer multiple Lyapunov functions [17]. In addition, the state feedback controllers that have the first-time derivative information on the membership functions for the switched fuzzy system, which substantially increases the solvability of the investigated control problem, are also employed. The relevant literature does not seem to have exploited this idea hence it is believed that is, for the first time, used in this paper.

In the second case study synthesis of intelligent control, the task is solved through exploring for a feasible solution via investigation employing single

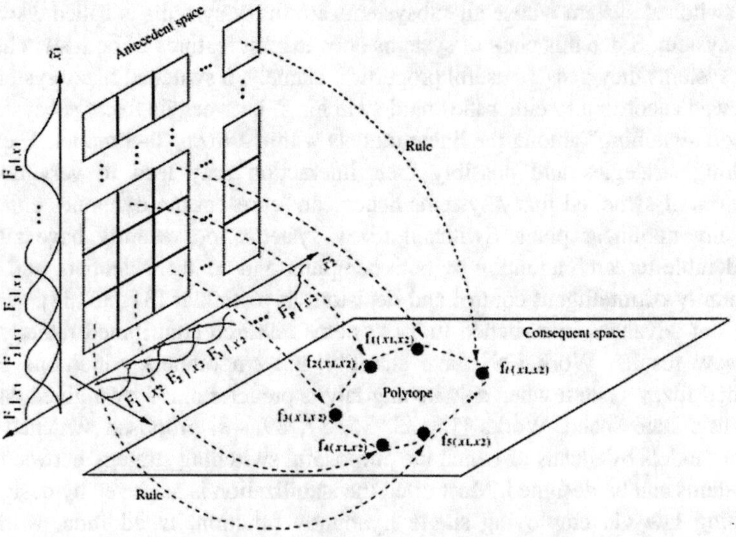

Fig. 1 The class of Takagi–Sugeno fuzzy systems models along with the involved rule-based polytopic mappings from the antecendent into the consequent space

Lyapunov function but in conjunction with certain optimality required in the sense of the non-fragile guaranteed cost control performance [10, 24]. It has been shown that such a solution for switched fuzzy systems exists provided a certain convex combination condition is also fulfilled [16]. This in turn imposes certain properties the rule based mappings, from the antecendent to the consequent spaces, have to constitute an "into" polytopic mapping operator (see Fig. 1).

Further this paper is organized as follows. In Sect. 2, the statement of the control synthesis tasks in both case studies and the needed preliminaries are given. In the sequel, the subsequent two sections present the first case of intelligent switched fuzzy control. Section 3 presents the solutions to stability analysis and controller design problems. In Sect. 4, a numerical example and its numerical and sample simulation results are given. Thereafter the next two sections present the second case of intelligent switched fuzzy control. In Sect. 5, the main synthesis solution results are derived. Then Sect. 6 presents the respective numerical example and its numerical and sample simulation results. Conclusion and references are given thereafter.

A note on the notation in this paper is given in the sequel. Namely, $X > O$ ($X \geq O$) means matrix X is positive definite (positive semi-definite) and $X < O$ ($X \leq O$) means that matrix X is negative definite (negative semi-definite). Capital letters I and O, respectively, represent the identity matrix and the zero matrix of appropriate dimensions as needed.

2 Mathematical Preliminaries and Statements of the Task Problems

Consider the specific Takagi–Sugeno (T–S) fuzzy model (Fig. 1) of switched fuzzy systems [38, 39], representing complex nonlinear and possibly uncertain plants. To the best of our knowledge, no results on using Lyapunov functions that are multiple either for each fuzzy subsystem or for every consequent local linear model have been found and were reported prior to our work [17]. This is the main motivation for the present investigation study.

In the case of the first task problem, one arbitrary inference rule [26, 42] is assumed to be represented as follows:

$$\begin{aligned} R_\sigma^i:\ &\text{if } x_1 \text{ is } \Omega^i_{\sigma 1},\ \text{is } \Omega^i_{\sigma 2},\ \ldots,\ x_n \text{ is } \Omega^i_{\sigma n}, \\ &\text{Then } \dot{x}(t) = A_{\sigma i}x(t) + B_{\sigma i}u_\sigma(t). \end{aligned} \tag{2.1}$$

where the constants, symbols, and variables denote:

R_σ^i	It denotes the *i*-th fuzzy inference rule
$x(t) = [x_1(t), x_2(t), \ldots, x_n(t)]^T \in R^n$	Vector of plant states variables, i.e., *n* component process variables
$\sigma: R_+ \to M = \{1, 2, \ldots, N\}$	A piecewise constant function that represents a switching signal
$\Omega^i_{\sigma n}$	It represents a fuzzy subset of the plant with a known fuzzy-set membership function
$u_\sigma(t)$	It represents the control input
$A_{\sigma i}$, $B_{\sigma i}$	These are known constant matrices of appropriate dimensions

Following Zadeh's fuzzy-logic inference [40, 41], by means of the product inference engine and center average defuzzification, the overall model of the system (2.1) is inferred as

$$\dot{x}(t) = \sum_{i=1}^{r_\sigma} h_{\sigma i}(x(t))[A_{\sigma i}x(t) + B_{\sigma i}u_\sigma(t)], \tag{2.2}$$

where r_σ is the number of inference rules. Thus the *l*-th fuzzy subsystem ($l \in M$) can be described by means of the model

$$\dot{x}_l(t) = \sum_{i=1}^{r_l} h_{li}(x(t))[(A_{li}x(t) + B_{li}u_l(t)], \tag{2.3}$$

where $h_{li}(x(t))$ is represented as

$$h_{li}(x(t)) = \frac{\Pi_{p=1}^{n} \mu_{lp}^{i}(x_p)}{\sum_{i=1}^{r_l} \Pi_{p=1}^{n} \mu_{lp}^{i}(x_p)}, \tag{2.4a}$$

satisfying

$$0 \le h_{li}(x(t)) \le 1, \ \sum_{i=1}^{r_l} h_{li}(x(t)) = 1. \tag{2.4b}$$

Here above $\mu_{lp}^{i}(x_p)$ denotes the membership function of x_p in the fuzzy subset Ω_{lp}^{i} for $p = 1, 2, \ldots, n$. For simplicity, the symbol $h_{li}(t)$ is used to denote $h_{li}(x(t))$ actually.

Assumption 2.1 The membership functions $h_{li}(t)$ are assumed to be C^1 functions.

This assumption implies

$$\sum_{i=1}^{r_l} \dot{h}_{li}(t) = 0, \tag{2.5}$$

which emanates directly from the membership functions. In turn, property (2.5) does imply that

$$\dot{h}_{lr_l}(t) = -\sum_{\rho=1}^{r_l - 1} \dot{h}_{l\rho}(t), \ \rho = 1, 2, \ldots, r_l - 1. \tag{2.6}$$

The state feedback controller $u_\sigma(t)$ and the considered system (2.1), respectively (2.2), employ the same premises as the original plant system. Thus, by means of the product inference engine along with center average defuzzification and via the parallel distributed compensation (PDC) scheme [29, 31], the global representation of the state feedback controller with the time derivative-information on fuzzy-set membership functions is inferred as follows:

$$u_\sigma(t) = \sum_{i=1}^{r_\sigma} h_{\sigma i}(t) K_{\sigma i} x(t) + \sum_{i=1}^{r_\sigma} \dot{h}_{\sigma i}(t) T_{\sigma i} x(t). \tag{2.7}$$

In here, quantities $K_{\sigma i}$ and $T_{\sigma i}$ are local feedback matrices to be designed. Naturally, the objectives are to solve the intelligent control synthesis task by solving both the analysis guaranteeing asymptotic stability and the control design, respectively.

It should be noted, in the case of switched fuzzy systems where both "hard switching" and "soft switching" are involved and coexist simultaneously, the investigation of any control synthesis becomes more complicated hence involved. Thus, the study of non-fragile guaranteed cost control and search for intelligent control synthesis that exploits its features becomes even more challenging. To the best of our knowledge, no results on the non-fragile guaranteed cost control for switched fuzzy systems have been reported prior to our work [16], which motivates the present investigation study. The non-fragile guaranteed cost control problem for uncertain switched fuzzy systems is studied via the single Lyapunov function method when a specific of convex combination exists. An algorithm to search for feasible solutions of the convex combination is also derived albeit this issue is not solved for the arbitrary case. In the present study, the presence of additive gain variation in the controller uncertainty is assumed and the existence of an upper bound on the guaranteed cost-function are assumed, which is quite a general case.

In this second task problem studied in here, a similar representation of one arbitrary inference rule is assumed:

$$R^i_\sigma\text{: if } x_1 \text{ is } \Omega^i_{\sigma 1}, \text{ is } \Omega^i_{\sigma 2}, \ldots, x_n \text{ is } \Omega^i_{\sigma n} \quad \text{Then } \dot{x}(t) = (A_{\sigma i} + \Delta A_{\sigma i}(t))x(t) + B_{\sigma i}u_\sigma(t). \tag{2.8}$$

where the constants, symbols, and variables denote:

R^i_σ	It denotes the i-th fuzzy inference rule
$x(t) = [x_1(t), x_2(t), \ldots, x_n(t)]^T \in R^n$	Vector of plant states variables, i.e., n component process variables
$\sigma: R_+ \to M = \{1, 2, \ldots, N\}$	A piecewise constant function representing a switching signal
$\Omega^i_{\sigma n}$	It represents a fuzzy subset of the plant with a known fuzzy-set membership function
$u_\sigma(t)$	It represents the control input
$A_{\sigma i}$, $B_{\sigma i}$	These are known constant matrices of appropriate dimensions
$\Delta A_{\sigma i}(t)$	Its is a time-varying matrix of appropriate dimensions representing uncertainty in the plant

Following Zadeh's fuzzy-logic inference [40, 41], by means of the product inference engine and center average defuzzification, the overall model of the system (2.8) is inferred as

$$\dot{x}(t) = \sum_{i=1}^{r_\sigma} h_{\sigma i}(x(t))[(A_{\sigma i} + \Delta A_{\sigma i}(t))x(t) + B_{\sigma i}u_\sigma(t)], \tag{2.9}$$

where r_σ is the number of inference rules. Thus the ℓ-th fuzzy subsystem ($l \in M$) is described by means of the model

$$\dot{x}_l(t) = \sum_{i=1}^{r_l} h_{li}(x(t))[(A_{\sigma i} + \Delta A_{\sigma i}(t))x(t) + B_{li}u_l(t)], \tag{2.10}$$

where $h_{li}(x(t))$ is represented as before by (2.4a) and satisfies (2.4b). Here above $\mu^i_{lp}(x_p)$ denotes the membership function of x_p in the fuzzy subset Ω^i_{lp} for $p=1,2,\ldots,n$. For simplicity, the symbol $h_{li}(t)$ is used to denote $h_{li}(x(t))$ actually.

The state feedback controller $u_\sigma(t)$ and the considered system (2.8), respectively (2.9), employ the same premises as the original plant system. Thus, by means of the product inference engine along with center average defuzzification and via the parallel distributed compensation (PDC) scheme [29, 31], the global representation of the state feedback control is inferred as follows:

$$u_\sigma(t) = \sum_{i=1}^{r_\sigma} h_{\sigma i}(t)(K_{li} + \Delta K_{li}(t))x(t). \tag{2.11}$$

In here, the quantity K_{li} is the controller gain for the ℓ-th fuzzy subsystem, which is to be designed, and ΔK_{li} represents the variation of this gain following gain drift due to the uncertainty. It should be noted

It should be noted, naturally, there may well appear in the plant both additive and multiplicative norm-bounded uncertainties. Hence also additive and multiplicative gain drifts may be considered, but this work is confined solely to the additive norm-bounded uncertainties.

Assumption 2.2 The norm-bounded uncertainties are $\Delta A_{li}(t) = D_{li}M_{li}(t)E_{li}$ and $\Delta K_{li}(t) = D_{\alpha li}M_{\alpha li}(t)E_{\alpha li}$, where D_{li}, E_{li} and $D_{\alpha li}$, $E_{\alpha li}$, respectively, are real-valued known constant while $M_{li}(t)$ and $M_{\alpha li}(t)$ are unknown time-varying matrices, all of appropriate dimensions, respectively satisfying $M_{li}(t)^T M_{li}(t) \le I$ and $M_{\alpha li}(t)^T M_{\alpha li}(t) \le I$.

Definition 2.1 The cost-function or the optimality index for the ℓ-th fuzzy control system (2.9), respectively (2.10), is defined as

$$J = \int_0^\infty [x(t)^T Q\, x(t) + u_\sigma(t)^T R\, u_\sigma(t)]dt, \tag{2.12}$$

where, following the optimal control theory, are positive definite weighting matrices.

Definition 2.2 For the fuzzy control system (2.9), respectively (2.10), and all admissible uncertainties, if there exist a state feedback control law $u_l = u_l(t)$ with $l \in M$ and $t \in [0, +\infty)$ for each subsystem, a switching law $\sigma = \sigma(t)$, and a positive real-valued scalar J^* such that the overall closed-loop system is asymptotically stable and the cost-function (2.12) satisfies $J \le J^*$, then the scalar index J^* is called a *non-fragile guaranteed cost* (NGC) and the state feedback control $u_l = u_l(t)$ is called a *non-fragile guaranteed cost* (NGC) *control*.

Remark 2.1 Notice that the above stated NGC control problem is different from the standard one [12, 34] because (a) the uncertainty time-varying gains in the control (2.11) affects the system matrices and (b) also enters the cost-function (2.12).

Solving this intelligent control synthesis task in the sense of the NGC control also belongs to the objectives of this paper.

3 Intelligent Control Synthesis Task One: Main New Results

In this section, first a stability condition for the uncontrolled switched fuzzy systems is derived that enables the existence of synthesis solution sought. Then, a synthesis design condition for a controller that renders the closed-loop switched fuzzy systems stable is presented.

Thus, the uncontrolled system (2.1) is investigated first and, since $u_\sigma(t) \equiv 0$, consequently the autonomous switched fuzzy system is derived:

$$\dot{x}_l(t) = \sum_{i=1}^{r_\sigma} h_{\sigma i}(t) A_{\sigma i} x(t). \tag{3.1}$$

This investigation yielded the next novel theoretical result for the investigated class of systems.

Theorem 3.1 Suppose that

$$\left|\dot{h}_{l\rho}(t)\right| \leq \Phi_{l\rho},\ l = 1,\ 2,\ \ldots, N,\quad \rho = 1,\ 2,\ \ldots,\ r_l - 1, \tag{3.2}$$

where $\Phi_{l\rho} \geq 0$. If there exist a set of symmetric positive definite matrices P_{li} $(i = 1, 2, \ldots, r_l)$ and P_{vq} $(v = 1, 2, \ldots, N,\ q = 1, 2, \ldots, r_v)$, and nonnegative constants β_{vl} such that

$$P_{l\rho} \geq P_{lr_l}, \tag{3.3}$$

and

$$\frac{1}{2}(A_{lj}^T P_{li} + P_{li} A_{lj} + A_{li}^T P_{lj} + P_{lj} A_{li}) + \sum_{\rho=1}^{r_l - 1} \Phi_{l\rho}(P_{l\rho} - P_{lr_l}) + \sum_{v=1, v\neq l}^{N} \beta_{vl}(P_{vq} - P_{li}) < 0,\ i \leq j, \tag{3.4}$$

for $l \in M$, then under the state dependent switching law

$$\sigma = \operatorname{argmin}\{x_l^T(t) P_{l} x_l(t)\}. \tag{3.5}$$

the autonomous system (3.1) is asymptotically stable when no operating sliding modes occur.

Proof Choose the following composite fuzzy function

$$V_l(x_l(t)) = x_l^T(t)P_l x_l(t),\ P_l = \sum_{i=1}^{r_l} h_{li}(t)P_{li},\ \ l=1,2,\ldots,N,\ \ i=1,2,\ \ldots,r_l. \ (13) \tag{3.6}$$

as the candidate Lyapunov function of the system (3.1). This kind of Lyapunov function candidate, defined indirectly via (1.4-a), is dependent on the plant's fuzzy sets' membership functions. Here it is called *two-layer multiple Lyapunov function* if the time derivative of $V_l(x_l(t))$ is negative at $x_l(t) \neq 0$ when all P_{li} are symmetric positive definite matrices. Notice that this is a novel kind of fuzzy-set dependent Lyapunov function.

Calculation of the time derivative along the trajectory of system state vector yields

$$\dot{V}_l(x(t)) = \sum_{i=1}^{r_l} h_{li}(t)[\dot{x}^T(t)P_{li}x(t) + x(t)P_{li}\dot{x}(t)] + \sum_{\rho=1}^{r_l} \dot{h}_{l\rho}(t)x^T(t)P_{l\rho}x(t).$$

Further, from the property (2.6), it follows

$$\dot{V}_l(x(t)) = \sum_{i,j=1}^{r_l} h_{li}(t)h_{lj}(t)x^T(t)(A_{lj}^T P_{li} + P_{li}A_{lj})x(t) + \sum_{\rho=1}^{r_l-1} \dot{h}_{l\rho}(t)x^T(t)P_{l\rho}x(t) + \dot{h}_{lr_l}(t)x^T(t)P_{lr_l}x(t)$$

$$= \sum_{i,j=1}^{r_l} h_{li}(t)h_{lj}(t)x^T(t)(A_{lj}^T P_{li} + P_{li}A_{lj})x(t) + \sum_{\rho=1}^{r_l-1} \dot{h}_{l\rho}(t)x^T(t)(P_{l\rho} - P_{lr_l})x(t)$$

$$= \sum_{i,j=1}^{r_l} h_{li}(t)h_{lj}(t)x^T(t)[\frac{1}{2}(A_{lj}^T P_{li} + P_{li}A_{lj} + A_{li}^T P_{lj} + P_{lj}A_{li})]x(t) + \sum_{\rho=1}^{r_l-1} \dot{h}_{l\rho}(t)x^T(t)(P_{l\rho} - P_{lr_l})x(t).$$

Taking into consideration the inequalities (2.2) (9) and (2.3) (10) into the last expression gives

$$\dot{V}_l(x(t)) \leq \sum_{i,j=1}^{r_l} h_{li}(t)h_{lj}(t)x^T(t)[\frac{1}{2}(A_{lj}^T P_{li} + P_{li}A_{lj} + A_{li}^T P_{lj} + P_{lj}A_{li})]x(t) + \\ + \sum_{\rho=1}^{r_l-1} \Phi_{l\rho}x^T(t)(P_{l\rho} - P_{lr_l})x(t).$$

Then the use of inequality (2.4a), (2.4b) on the right-hand side of this expression yields

$$\dot{V}_l(x(t)) \le \sum_{i,j=1}^{r_l}\sum_{q=1}^{r_v} h_{li}(t)h_{lj}(t)h_{vq}(t)x^T(t)[\frac{1}{2}({A_{lj}}^T P_{li} + P_{li}A_{lj} + {A_{li}}^T P_{lj} + P_{lj}A_{li}) +$$

$$+ \sum_{\rho=1}^{r_l-1} \Phi_{l\rho}(P_{l\rho} - P_{lr_l}) + \sum_{v=1, v\ne l}^{N} \beta_{vl}(P_{vq} - P_{li})]x(t) =$$

$$= \sum_{i,j=1}^{r_l} h_{li}(t)h_{lj}(t)x^T(t)[\frac{1}{2}({A_{lj}}^T P_{li} + P_{li}A_{lj} + {A_{li}}^T P_{lj} + P_{lj}A_{li}) +$$

$$+ \sum_{\rho=1}^{r_l-1} \Phi_{l\rho}(P_{l\rho} - P_{lr_l})]x(t) + \sum_{i=1}^{r_l}\sum_{q=1}^{r_v}\sum_{v=1, v\ne l}^{N} h_{li}(t)h_{vq}(t)\beta_{vl}x^T(t)(P_{vq} - P_{li})x(t) < 0.$$

Furthermore, by virtue of the switching law (2.5) (12), the following inequality

$$x^T(t)(P_v - P_l)x(t) = x^T(t)\sum_{q=1}^{r_v}\sum_{i=1}^{r_l} h_{vq}(t)h_{li}(t)(P_{vq} - P_{li})x(t) \ge 0$$

holds true. It is therefore that $\dot{V}_l < 0,\ \ \forall x \ne 0$, hence system (3.1) is asymptotically stable. Thus the proof is complete. □

Remark 3.1 Note that switching law (3.5) represents a set of LMI if $\Phi_{l\rho}$ is given beforehand. It should be noted, however, it is not an easy matter to select $\Phi_{l\rho}$.

In the sequel, the newly derived set of PDF controllers, which possess information on the time derivative of membership functions efficiently overcome this difficulty. This is presented in terms of the next theorem.

Theorem 3.2 If there exist a set of symmetric positive definite matrices $P_{li}\ (l = 1, 2, \ldots, N,\ i = 1, 2, \ldots, r_l)$ and $P_{vq}\ (v = 1, 2, \ldots, N, q = 1, 2, \ldots, r_v)$, matrices K_{li}, T_{li}, positive constants ε, γ, $s_{li}(s_{li} \ge 1)$, and nonnegative constants β_{vl} such that, for all $i \le j \le k,\ \rho = 1, 2, \ldots, r_l - 1,\ m = 1, 2,$ the inequalities

$$P_{li} \ge s_{li} I, \tag{3.7}$$

$$\mu_{l\rho m}(P_{l\rho} - P_{lr_l}) - [\frac{s_{li} + s_{lj} + s_{lk}}{3\varepsilon^2(r_l - 1)} + \frac{s_{li} + s_{lj}}{\gamma^2}]I + \frac{1}{6(r_l - 1)}\hat{U}_{lijk} + \frac{1}{2}\hat{M}_{lij\rho m} + \frac{1}{2}\hat{L}_{lijr_l m} + \sum_{v=1, v\ne l}^{N} \beta_{vl}(P_{vq} - P_{li}) < 0,$$

$$i < j < k,\ \rho = 1, 2,\ \ldots,\ r_l - 1,\ \ m = 1, 2 \tag{3.8}$$

where

$$\hat{U}_{lijk} = U_{lijk} + U_{likj} + U_{ljik} + U_{ljki} + U_{lkij} + U_{lkji}U_{lijk} = (\varepsilon {G_{ljk}}^T + \frac{1}{\varepsilon}P_{li})(\varepsilon {G_{ljk}}^T + \frac{1}{\varepsilon}P_{li})^T, \tag{3.9}$$

$$\hat{M}_{lij\rho m} = M_{lij\rho m} + M_{lji\rho m},\ M_{lij\rho m} = (\gamma\mu_{l\rho m}{H_{lj\rho}}^T + \frac{1}{\gamma}P_{li})(\gamma\mu_{l\rho m}{H_{lj\rho}}^T + \frac{1}{\gamma}P_{li})^T, \tag{3.10}$$

$$\hat{L}_{lijr_lm} = L_{lijr_lm} + L_{ljir_lm},\ L_{lijr_lm} = (\gamma\mu_{lr_lm}H_{ljr_l}{}^T - \frac{1}{\gamma}P_{li})(\gamma\mu_{lr_lm}H_{ljr_l}{}^T - \frac{1}{\gamma}P_{li})^T, \quad (3.11)$$

$$G_{ljk} = A_{lj} + B_{lj}K_{lk}\,,\ H_{lj\rho} = B_{lj}T_{l\rho}, \quad (3.12)$$

and where the set of $\mu_{l\rho m}$'s represent the maximum and the minimum of the $\dot{h}_{l\rho}(t)$, respectively, then the system (2.2) can be asymptotically stabilized by the synthesized control (2.7) when no operating sliding modes occur under the switching law (2.5).

Proof Choose (3.6) as the candidate Lyapunov functions for the system (2.2). Calculating the time derivative along the trajectory of the state vector yields:

$$\begin{aligned}
\dot{V}_l(x(t)) = {} & \sum_{\rho=1}^{r_l} \dot{h}_{l\rho}(t)x(t)^T P_{l\rho}x(t) + \sum_{i=1}^{r_l} h_{li}(t)[\dot{x}(t)^T P_{li}x(t) + x(t)^T P_{li}\dot{x}(t)] = \\
& = \sum_{\rho=1}^{r_l} \dot{h}_{l\rho}(t)x^T(t)P_{l\rho}x(t) + \\
& + \sum_{i=1}^{r_l} h_{li}(t)[x^T(t) \sum_{j,k=1}^{r_l} h_{lj}(t)h_{lk}(t)(A_{lj} + B_{lj}K_{lk} + \sum_{\rho=1}^{r_l} \dot{h}_{l\rho}(t)B_{lj}T_{l\rho})^T P_{li}x(t) + \\
& + x^T(t)P_{li} \sum_{j,k=1}^{r_l} h_{lj}(t)h_{lk}(t)(A_{lj} + B_{lj}K_{lk} + \sum_{\rho=1}^{r_l} \dot{h}_{l\rho}(t)B_{lj}T_{l\rho})x(t)] = \\
= {} & \sum_{\rho=1}^{r_l} \dot{h}_{l\rho}(t)x^T(t)P_{l\rho}x(t) + \sum_{i,j,k=1}^{r_l} h_{li}(t)h_{lj}(t)h_{lk}(t)x^T(t)[(A_{lj} + B_{lj}K_{lk})^T P_{li} + \\
& + P_{li}(A_{lj} + B_{lj}K_{lk}) + \sum_{\rho=1}^{r_l} \dot{h}_{l\rho}(t)((B_{lj}T_{l\rho})^T P_{li} + P_{li}B_{lj}T_{l\rho})]x(t) = \\
& = \sum_{i,j,k=1}^{r_l} h_{li}(t)h_{lj}(t)h_{lk}(t)x^T(t)[\sum_{\rho=1}^{r_l} \dot{h}_{l\rho}(t)(P_{l\rho} + (B_{lj}T_{l\rho})^T P_{li} + P_{li}B_{lj}T_{l\rho} + \\
& + (B_{li}T_{l\rho})^T P_{lj} + P_{lj}B_{li}T_{l\rho}) + (A_{lj} + B_{lj}K_{lk})^T P_{li} + P_{li}(A_{lj} + B_{lj}K_{lk})]x(t) = \\
= {} & \sum_{i,j,k=1}^{r_l} h_{li}(t)h_{lj}(t)h_{lk}(t)x^T(t)[\sum_{\rho=1}^{r_l} \dot{h}_{l\rho}(t)(P_{l\rho} + H_{lj\rho}^T P_{li} + P_{li}H_{lj\rho} + \\
& + H_{li\rho}^T P_{lj} + P_{lj}H_{li\rho}) + G_{ljk}^T P_{li} + P_{li}G_{ljk}]x(t) = \\
& = \sum_{i,j,k=1}^{r_l} h_{li}(t)h_{lj}(t)h_{lk}(t)x^T(t)[\sum_{\rho=1}^{r_l} \dot{h}_{l\rho}(t)(P_{l\rho} + J_{lij\rho}) + W_{lijk}]x(t) = \\
& = \sum_{i,j,k=1}^{r_l} h_{li}(t)h_{lj}(t)h_{lk}(t)x^T(t)[\sum_{\rho=1}^{r_l} \dot{h}_{l\rho}(t)(P_{l\rho} + \frac{1}{2}\hat{J}_{lij\rho}) + \frac{1}{6}\hat{W}_{lijk}]x(t),
\end{aligned}$$

where

$$W_{lijk}=G_{ljk}^{T}P_{li}+P_{li}G_{ljk},\quad \overset{\wedge}{W_{lijk}}=W_{lijk}+W_{likj}+W_{ljik}+W_{ljki}+W_{lkij}+W_{lkji},$$

$$J_{lij\rho}=H_{lj\rho}^{T}P_{li}+P_{li}H_{lj\rho},\quad \overset{\wedge}{J_{lij\rho}}=J_{lij\rho}+J_{lji\rho}:$$

On the grounds of the expression on property (2.6), it can be further derived:

$$\dot V_l(x(t))=$$

$$\sum_{i,j,k=1}^{r_l} h_{li}(t)h_{lj}(t)h_{lk}(t)x^T(t)[\sum_{\rho=1}^{r_l-1}\dot h_{l\rho}(t)(P_{l\rho}-P_{lr_l}+\frac{1}{2}\overset{\wedge}{J_{lij\rho}}-\frac{1}{2}\overset{\wedge}{J_{lijr_l}})+\frac{1}{6}\overset{\wedge}{W_{lijk}}]x(t)=$$

$$=\sum_{i,j,k=1}^{r_l} h_{li}(t)h_{lj}(t)h_{lk}(t)x^T(t)[\sum_{\rho=1}^{r_l-1}\sum_{m=1}^{2} a_{l\rho m}(t)\mu_{l\rho m}(P_{l\rho}-P_{lr_l})+\frac{1}{2}\overset{\wedge}{J_{lij\rho}}-\frac{1}{2}\overset{\wedge}{J_{lijr_l}})+\frac{1}{6}\overset{\wedge}{W_{lijk}}]x(t)=$$

$$=\sum_{i,j,k=1}^{r_l}\sum_{\rho=1}^{r_l-1}\sum_{m=1}^{2} h_{li}(t)h_{lj}(t)h_{lk}(t)a_{l\rho m}(t)x^T(t)[\mu_{l\rho m}(P_{l\rho}-P_{lr_l}+\frac{1}{2}\overset{\wedge}{J_{lij\rho}}-\frac{1}{2}\overset{\wedge}{J_{lijr_l}})+\frac{1}{6(r_l-1)}\overset{\wedge}{W_{lijk}}]x(t)=$$

$$\le\sum_{i,j,k=1}^{r_l}\sum_{\rho=1}^{r_l-1}\sum_{m=1}^{2} h_{li}(t)h_{lj}(t)h_{lk}(t)a_{l\rho m}(t)x^T(t)[\mu_{l\rho m}(P_{l\rho}-P_{lr_l})+\frac{1}{6(r_l-1)}(\overset{\wedge}{W_{lijk}}+$$

$$+\varepsilon^2G_{ljk}^TG_{ljk}+\varepsilon^2G_{lji}^TG_{lji}+\varepsilon^2G_{lik}^TG_{lik}+\varepsilon^2G_{lij}^TG_{lij}+\varepsilon^2G_{lki}^TG_{lki}+\varepsilon^2G_{lkj}^TG_{lkj})+$$

$$+\frac{1}{2}\mu_{l\rho m}(\overset{\wedge}{J_{lij\rho}}-\overset{\wedge}{J_{lijr_l}})+\frac{1}{2}(\gamma^2\mu_{l\rho m}^2H_{lj\rho}^TH_{lj\rho}+\gamma^2\mu_{l\rho m}^2H_{li\rho}^TH_{li\rho})+$$

$$+\frac{1}{2}(\gamma^2\mu_{lr_lm}^2H_{ljr_l}^TH_{ljr_l}+\gamma^2\mu_{lr_lm}^2H_{lir_l}^TH_{lir_l}]x(t)=$$

$$=\sum_{i,j,k=1}^{r_l}\sum_{\rho=1}^{r_l-1}\sum_{m=1}^{2} h_{li}(t)h_{lj}(t)h_{lk}(t)a_{l\rho m}(t)x^T(t)[\mu_{l\rho m}(P_{l\rho}-P_{lr_l})+\frac{1}{6(r_l-1)}\overset{\wedge}{S_{lijk}}+\frac{1}{2}\overset{\wedge}{N_{lij\rho m}}+\frac{1}{2}\overset{\wedge}{R_{lijr_lm}}]x(t)$$

where

$$S_{lijk}=U_{lijk}-\frac{1}{\varepsilon^2}P_{li}P_{li},\quad \overset{\wedge}{S_{lijk}}=S_{lijk}+S_{likj}+S_{ljik}+S_{ljki}+S_{lkij}+S_{lkji},$$

$$N_{lij\rho m}=M_{lij\rho m}-\frac{1}{\gamma^2}P_{li}P_{li},\quad \overset{\wedge}{N_{lij\rho m}}=N_{lij\rho m}+N_{lji\rho m},\quad R_{lijr_lm}=L_{lijr_lm}-\frac{1}{\gamma^2}P_{li}P_{li},$$

$$\overset{\wedge}{R_{lijr_lm}}=R_{lijr_lm}+R_{ljir_lm},\quad \dot h_{l\rho}(t)=\sum_{m=1}^{2}a_{l\rho m}(t)\mu_{l\rho m},\quad 0\le a_{l\rho m}(t)\le 1,\quad \sum_{m=1}^{2}a_{l\rho m}(t)=1.$$

Now taking into consideration the inequality (3.7), one obtains

$$\dot V_l(x(t))\le$$

$$\sum_{i,j,k=1}^{r_l}\sum_{\rho=1}^{r_l-1}\sum_{m=1}^{2} h_{li}(t)h_{lj}(t)h_{lk}(t)a_{l\rho m}(t)x^T(t)[\mu_{l\rho m}(P_{l\rho}-P_{lr_l})+\frac{1}{6(r_l-1)}\overset{\wedge}{S_{lijk}}+\frac{1}{2}\overset{\wedge}{N_{lij\rho m}}+\frac{1}{2}\overset{\wedge}{R_{lijr_lm}}]x(t)\le$$

$$\leq \sum_{i,j,k=1}^{r_l} \sum_{\rho=1}^{r_l-1} \sum_{m=1}^{2} h_{li}(t)h_{lj}(t)h_{lk}(t)a_{l\rho m}(t)x^T(t)[\mu_{l\rho m}(P_{l\rho} - P_{lr_l}) -$$

$$-(\frac{s_{li}+s_{lj}+s_{lk}}{3\varepsilon^2(r_l-1)} + \frac{s_{li}+s_{lj}}{\gamma^2})I + \frac{1}{6(r_l-1)}\hat{U}_{lijk} + \frac{1}{2}\hat{M}_{lij\rho m} + \frac{1}{2}\hat{L}_{lijr_l m}]x(t).$$

By making use of the inequality (3.8), it can be further found that the right-hand side of the above inequality yields

$$\sum_{i,j,k=1}^{r_l} \sum_{\rho=1}^{r_l-1} \sum_{m=1}^{2} \sum_{q=1}^{r_v} h_{li}(t)h_{lj}(t)h_{lk}(t)a_{l\rho m}(t)h_{vq}(t)x^T(t)\{\mu_{l\rho m}(P_{l\rho} - P_{lr_l}) -$$

$$-(\frac{s_{li}+s_{lj}+s_{lk}}{3\varepsilon^2(r_l-1)} + \frac{s_{li}+s_{lj}}{\gamma^2})I + \frac{1}{6(r_l-1)}\hat{U}_{lijk} + \frac{1}{2}\hat{M}_{lij\rho m} + \frac{1}{2}\hat{L}_{lijr_l m} + \sum_{v=1, v\neq l}^{N} \beta_{vl}(P_{vq} - P_{li})\}x(t)$$

$$= \sum_{i,j,k=1}^{r_l} \sum_{\rho=1}^{r_l-1} \sum_{m=1}^{2} h_{li}(t)h_{lj}(t)h_{lk}(t)a_{l\rho m}(t)x^T(t)[\mu_{l\rho m}(P_{l\rho} - P_{lr_l}) -$$

$$-(\frac{s_{li}+s_{lj}+s_{lk}}{3\varepsilon^2(r_l-1)} + \frac{s_{li}+s_{lj}}{\gamma^2})I + \frac{1}{6(r_l-1)}\hat{U}_{lijk} + \frac{1}{2}\hat{M}_{lij\rho m} + \frac{1}{2}\hat{L}_{lijr_l m}]x(t) +$$

$$+ \sum_{i=1}^{r_l} \sum_{q=1}^{r_v} \sum_{v=1, v\neq l}^{N} h_{li}(t)h_{vq}(t)\beta_{vl}x^T(t)(P_{vq} - P_{li})x(t) < 0.$$

Due to the switching law (3.5), the following inequality

$$x^T(t)(P_v - P_l)x(t) = x^T(t) \sum_{q=1}^{r_v} \sum_{i=1}^{r_l} h_{vq}(t)h_{li}(t)(P_{vq} - P_{li})x(t) \geq 0$$

is found to hold true. It is therefore that $\dot{V}_l < 0$, $\forall x \neq 0$, and the system (2.2), respectively (2.1), is asymptotically stable under the synthesized intelligent control, which combines the feedback and the switching control laws. Thus the proof is complete. □

Remark 3.2 In here a new kind of PDC controllers are proposed that employ the time derivative information on membership functions. These are valid in terms of the LMI for the case where the time derivative of $h_{l\rho}(t)$ can be calculated from the states and are not directly related to the control, which limits the applicability of these new synthesis designs.

Remark 3.3 The new PDC controllers have terms of $\dot{h}_{li}(t)$ and reduce to the original PDC controllers when $T_{li} = T$, i.e., when a single design matrix is used.

4 The First Intelligent Control Synthesis: Illustrative Example

As an illustrative example a switched fuzzy system composed of two subsystems is considered as in [17, 31]. These are given in terms of their system state

$$A_{11} = \begin{bmatrix} -5 & -4 \\ -1 & -2 \end{bmatrix}, A_{12} = \begin{bmatrix} -2 & -4 \\ 20 & -2 \end{bmatrix}, A_{21} = \begin{bmatrix} -5 & -4 \\ -1 & -2 \end{bmatrix}, A_{22} = \begin{bmatrix} -2 & -4 \\ 20 & -2 \end{bmatrix} \tag{4.1}$$

and input matrices:

$$B_{11} = \begin{bmatrix} 0 & 0 \\ 1 & 10 \end{bmatrix}, B_{12} = \begin{bmatrix} 0 & 0 \\ 1 & 3 \end{bmatrix}, B_{21} = \begin{bmatrix} 0 & 0 \\ 1 & 8 \end{bmatrix}, B_{22} = \begin{bmatrix} 0 & 0 \\ 1 & 5 \end{bmatrix}. \tag{4.2}$$

The actual nonlinear physical plant is assumed to be characterized by membership functions as follows:

$$h_{11}(x_1(t)) = h_{21}(x_1(t)) = \frac{1 + \sin x_1(t)}{2}, h_{12}(x_1(t)) = h_{22}(x_1(t)) = \frac{1 - \sin x_1(t)}{2}. \tag{4.3}$$

Following the technique developed in [29, 30] one may readily obtain:

$$\mu_{111} = \mu_{211} = 3.44,\ \mu_{121} = \mu_{221} = 3.68,\ \mu_{112} = \mu_{212} = -3.68,\ \mu_{122} = \mu_{222} = -3.44.$$

Solving the inequality (15) using the LMI toolbox of MATLAB [8] (also see [18, 19]) yields the following symmetric positive definite matrices

$$P_{11} = \begin{bmatrix} 4.0800 & 0.2222 \\ 0.2222 & 4.2513 \end{bmatrix}, P_{12} = \begin{bmatrix} 3.6174 & 0.2710 \\ 0.2710 & 4.0143 \end{bmatrix},$$

$$P_{21} = \begin{bmatrix} 3.9297 & 0.2613 \\ 0.2613 & 4.0545 \end{bmatrix}, P_{22} = \begin{bmatrix} 3.5241 & 0.2666 \\ 0.2666 & 4.0195 \end{bmatrix}$$

and feedback gain matrices

$$K_{11} = \begin{bmatrix} 0.0175 & 0.0760 \\ 0.0760 & -0.2327 \end{bmatrix}, K_{12} = \begin{bmatrix} -3.7911 & -5.4933 \\ -5.4933 & 1.1597 \end{bmatrix},$$

$$K_{21} = \begin{bmatrix} 0.0264 & 0.0890 \\ 0.0890 & -0.2679 \end{bmatrix}, K_{22} = \begin{bmatrix} -1.5140 & -3.7505 \\ -3.7505 & 0.3462 \end{bmatrix},$$

as well as the matrices

$$T_{11} = \begin{bmatrix} 0.0652 & 7.8165 \\ -7.8328 & 0.0406 \end{bmatrix},\ T_{12} = \begin{bmatrix} -0.0768 & 8.2736 \\ -8.2515 & -0.0300 \end{bmatrix},$$
$$T_{21} = \begin{bmatrix} -0.0106 & 8.2211 \\ -8.2168 & 0.0050 \end{bmatrix},\ T_{22} = \begin{bmatrix} 0.0042 & -5.2361 \\ 5.2353 & 0.0059 \end{bmatrix}.$$

Hence, the stabilizing control synthesis design according to Theorem 2, which renders system (2.1) asymptotically stable in closed loop, does exist and its implementation is feasible.

The resulting fuzzy control law synthesis is given as follows:

$$u_l(t) = \sum_{i=1}^{2} h_{li}(t)K_{li}x(t) + \sum_{i=1}^{2} \dot{h}_{li}(t)T_{li}x(t) \tag{4.4}$$

and it does complement operationally the switching law (2.5) in the sense of fuzzy-logic-based computational intelligence.

Computer simulation experiments were carried out by applying the control law (4.4) to the system representation (2.3) and by assuming $x(0) = [1, -1]^T$. A typical sample of the obtained simulation results are shown in Fig. 2. Apparently, the closed-loop system is asymptotically stable via the switching law (2.5), respectively (4.4).

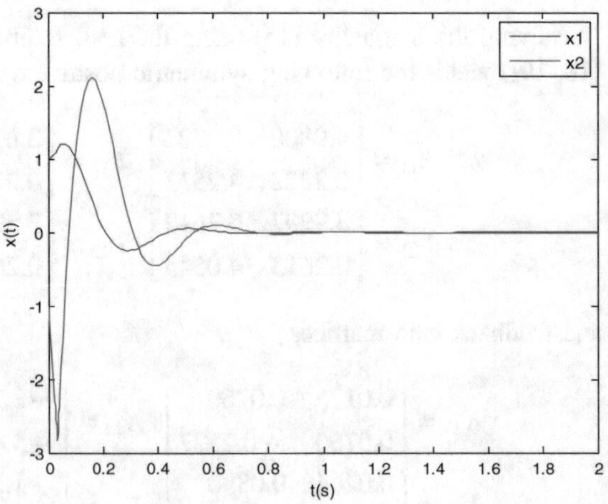

Fig. 2 The plant state responses in the designed closed-loop control system using the switched fuzzy system representation

5 Intelligent Control Synthesis Task Two: Main New Results

The second intelligent control synthesis task problem, due to its statement also requiring NGC control constraints, appeared to be subject to a certain solvability condition. It is this solvability condition that has been found using the single Lyapunov function approach. Yet, the control synthesis theorem appeared to imply computationally a hard NP (nondeterministic polynomial) problem hence some algorithm ought to be derived, that yielded a solvable case involving convexity condition. In what follows, the next two lemmas shall be needed.

Lemma 5.1 ([24]) *Let D, E, K, and $M(t)$ denote real-valued matrices of appropriate dimensions, and also let $M(t)^T M(t) \leq I$. Then for any scalar $\mu > 0$ and any positive real-valued matrix $W = W^T$ and $W - \mu DD^T > 0$, the following inequalities hold true*:

$$DM(t)E + (DM(t)E)^T \leq \mu^{-1} DD^T + \mu E^T E; \tag{5.1}$$

$$\begin{aligned}(K + DM(t)E)^T W^{-1}(K + DM(t)E) \leq K^T (W - \mu DD^T)^{-1} K + \\ + \mu^{-1} E^T E.\end{aligned} \tag{5.2}$$

Lemma 5.2 ([35]) *Let $X \in R^{m \times n}$ and $Y \in R^{m \times n}$ denote arbitrary real-valued matrices and $\varepsilon > 0$ any constant real-valued positive scalar. Then the following inequality holds true*:

$$X^T Y + Y^T X \leq \varepsilon X^T X + \varepsilon^{-1} Y^T Y \tag{5.3}$$

Both Lemmas 5.1 *and* 5.2 *are used to investigate the inequalities involved in establishing conditions under which the time derivative of Lyapunov function is guaranteed negative definite.*

Theorem 5.1 *For the considered control system (2.9) and for a given convex combination $\alpha_1, \alpha_2, \ldots, \alpha_N$, $\sum_{l=1}^{l=N} \alpha_i = 1$, if there exist a positive definite matrix P and a gain matrix K_{li} satisfying the following matrix inequalities*

$$\sum_{l=1}^{l=N} \alpha_l [\Pi_{lii}^* + T_{lii} + PS_{lii}P + K_{li}^T (R^{-1} - D_{\alpha li} D_{\alpha li}^T)^{-1} K_{li}] < 0 \ (7) \tag{5.4}$$

$$\begin{aligned}\sum_{l=1}^{l=N} \alpha_l [\Pi_{lii}^* + T_{lii} + PS_{lii}P + (RK_{lj})^T (I - D_{\alpha lj} D_{\alpha lj}^T)^{-1} (RK_{lj}) + \\ + K_{li}^T (I - D_{\alpha li} D_{\alpha li}^T)^{-1} K_{li}] < 0, \quad i < j, \qquad (8)\end{aligned} \tag{5.5}$$

where

$$\Pi^*_{lii} = PA_{li} + A^T_{li}P + PB_{li}K_{li} + K^T_{li}B^T_{li}P, \tag{5.6}$$

$$S_{lii} = D_{li}D^T_{li} + B_{li}D_{\alpha li}D^T_{\alpha li}B^T_{li}, \tag{5.7}$$

$$T_{lii} = 2E_{\alpha li}E^T_{\alpha li} + E_{li}E^T_{li} + Q, \tag{5.8}$$

$$\Pi^*_{lij} = PA_{li} + A^T_{li}P + PB_{li}K_{lj} + K^T_{lj}B^T_{li}P + PA_{lj} + A^T_{lj}P + PB_{lj}K_{li} + K^T_{li}B^T_{lj}P, \tag{5.9}$$

$$S_{lij} = D_{li}D^T_{li} + B_{li}D_{\alpha lj}D^T_{\alpha lj}B^T_{li} + D_{lj}D^T_{lj} + B_{lj}D_{\alpha li}D^T_{\alpha li}B^T_{lj}, \tag{5.10}$$

$$T_{lij} = 2E_{\alpha li}E^T_{\alpha li} + E_{li}E^T_{li} + 2E_{\alpha lj}E^T_{\alpha lj} + E_{lj}E^T_{lj} + 2Q, \tag{5.11}$$

Then there exists the NGC feedback control (2.11) such that the control system (2.10), respectively (2.9), is asymptotically stable under the switching law

$$\begin{aligned}\sigma(t) = \arg\min\{x(t)^T \sum_{i=1}^{i=r_l}\sum_{j=i}^{j=r_l} h_{li}(t)h_{li}(t)[(A_{li} + B_{li}K_{lj})^T + P(A_{li} + B_{li}K_{lj}) + \\ + PD_{li}D^T_{li}P + PB_{li}D_{\alpha lj}D^T_{\alpha lj}B^T_{li}P + E_{li}E^T_{li} + E_{\alpha lj}E^T_{\alpha lj}]x(t)\end{aligned} \tag{5.12}$$

and the cost-function possesses a non-fragile guaranteed optimum $J^* = x_0^T P x_0$ for any nonzero initial state $x(0) = x_0$.

Proof Substitution of the control law (2.11) into the system (2.10) yields

$$\dot{x}_l(t) = \sum_{i=1}^{r_l}\sum_{j=1}^{r_l} h_{li}(t)h_{lj}(t)[(A_{li} + \Delta A_{li}(t))x(t) + B_{li}(K_{lj} + \Delta K_{lj}(t))x(t)], \tag{5.13}$$

In conjunction with this representation of the closed-loop system consider next the analysis of the following overall system model:

$$\dot{x}_l(t) = \sum_{l=1}^{l=N} \alpha_l \sum_{i=1}^{r_l}\sum_{j=1}^{r_l} h_{li}(t)h_{lj}(t)[(A_{li} + \Delta A_{li}(t))x(t) + B_{li}(K_{lj} + \Delta K_{lj}(t))x(t)].$$

Further, it can be shown

$$\dot{x}_l(t) = \sum_{l=1}^{l=N} \alpha_l \sum_{i=1}^{r_l}\sum_{j=1}^{r_l} h_{li}(t)h_{lj}(t)[(A_{li} + \Delta A_{li}(t)) + B_{li}(K_{lj} + \Delta K_{lj}(t))]x(t)$$

hence

$$\dot{x}(t) = \sum_{l=1}^{l=N} \alpha_l \sum_{i=1}^{r_l} \sum_{j=1}^{r_l} h_{li}^2(t)[(A_{li} + \Delta A_{li}(t)) + B_{li}(K_{lj} + \Delta K_{lj}(t))]x(t) + \\ + \sum_{l=1}^{l=N} \alpha_l \sum_{i<j}^{r_l} \sum_{j=1}^{r_l} h_{li}(t)h_{lj}(t)[(A_{li} + \Delta A_{li}(t)) + B_{li}(K_{lj} + \Delta K_{lj}(t)) + \\ + (A_{lj} + \Delta A_{lj}(t)) + B_{lj}(K_{li} + \Delta K_{li}(t))]x(t). \tag{5.14}$$

On these grounds, now choose $V(x(t)) = x(t)^T P x(t)$ as the candidate Lyapunov function and calculate its time derivative $dV(t)/dt$ along the state trajectory of the system in closed loop using this equivalent representation model. Thus, it follows:

$$\dot{V}_l(x(t)) = \dot{x}(t)^T P_{li} x(t) + x(t)^T P_{li} \dot{x}(t)] =$$

$$= \sum_{l=1}^{l=N} \alpha_l \sum_{i=1}^{r_l} h_{li}^2(t)\{[(A_{li} + \Delta A_{li}(t)) + B_{li}(K_{li} + \Delta K_{li}(t))]^T P + P[(A_{li} + \Delta A_{li}(t)) + \\ + B_{li}(K_{li} + \Delta K_{li}(t))]\}x(t) + \sum_{l=1}^{l=N} \alpha_l \sum_{i<j}^{r_l} \sum_{j=1}^{r_l} h_{li}(t)h_{lj}(t)x(t)^T\{[(A_{li} + \Delta A_{li}(t)) + \\ + (B_{li}(K_{lj} + \Delta K_{lj}(t)) + (A_{lj} + \Delta A_{lj}(t)) + B_{lj}(K_{li} + \Delta K_{li}(t))]^T P + P[(A_{li} + \Delta A_{li}(t)) + \\ + B_{li}(K_{lj} + \Delta K_{lj}(t)) + (A_{lj} + \Delta A_{lj}(t)) + B_{lj}(K_{li} + \Delta K_{li}(t))]\}x(t);$$

This expression can be further re-arranged so as to yield:

$$\dot{V}_l(x(t)) = \\ = \sum_{l=1}^{l=N} \alpha_l \sum_{i=1}^{r_l} h_{li}^2(t)\{x(t)^T[Q + (K_{li} + \Delta K_{li}(t))^T R(K_{li} + \Delta K_{li}(t)) + PA_{li} + A_{li}^T P + \\ + PB_{li}K_{li} + K_{li}^T B_{li}^T P + P\Delta A_{li} + \Delta A_{li}^T P + PB_{li}\Delta K_{li}(t) + \Delta K_{li}(t)^T B_{li}^T P]x(t) - \\ - x(t)^T Qx(t) - x(t)^T (K_{li} + \Delta K_{li}(t))^T R(K_{li} + \Delta K_{li}(t))x(t)\} + \\ + \sum_{l=1}^{l=N} \alpha_l \sum_{i<j}^{r_l} \sum_{j=1}^{r_l} h_{li}(t)h_{lj}(t)x(t)^T\{[2Q + (K_{lj} + \Delta K_{lj}(t))^T R(K_{lj} + \Delta K_{lj}(t)) + \\ + (K_{lj} + \Delta K_{lj}(t))^T R(K_{li} + \Delta K_{li}(t)) + PA_{li} + A_{li}^T P + PB_{li}K_{lj} + K_{lj}^T B_{li}^T P + \\ + P\Delta A_{li} + \Delta A_{li}^T P + PB_{li}\Delta K_{lj}(t) + \Delta K_{lj}(t)^T B_{li}^T P + PA_{lj} + A_{lj}^T P + PB_{ji}K_{li} + \\ + K_{li}^T B_{lj}^T P + P\Delta A_{lj} + \Delta A_{lj}^T P + PB_{lj}\Delta K_{li}(t) + \Delta K_{li}(t)^T B_{lj}^T P]x(t) - 2x(t)^T Q x(t) - \\ - x(t)^T (K_{li} + \Delta K_{li}(t))^T R(K_{lj} + \Delta K_{lj}(t))x(t) - \\ - x(t)^T (K_{lj} + \Delta K_{lj}(t))^T R(K_{li} + \Delta K_{li}(t))x(t)\} \leq$$

$$\leq \sum_{l=1}^{l=N} \alpha_l \sum_{i=1}^{r_l} h_{li}^2(t)\{x(t)^T[Q + K_{li}^T(R^{-1} - D_{\alpha li}D_{\alpha li}^T)^{-1}K_{li} + E_{\alpha li}E_{\alpha li}^T + PA_{li} + A_{li}^T P +$$
$$+ PB_{li}K_{li} + K_{li}^T B_{li}^T P + PD_{li}D_{li}^T P + PB_{li}D_{\alpha li}D_{\alpha li}^T B_{li}^T P + E_{\alpha li}E_{\alpha li}^T + E_{li}E_{li}^T]x(t) -$$
$$- x(t)^T Qx(t) - x(t)^T(K_{li} + \Delta K_{li}(t))^T R(K_{li} + \Delta K_{li}(t))x(t)\} +$$
$$+ \sum_{l=1}^{l=N} \alpha_l \sum_{i<j}^{r_l} \sum_{j=1}^{r_l} h_{li}(t)h_{lj}(t)x(t)^T\{[2Q + (RK_{lj})^T(I - D_{\alpha lj}D_{\alpha lj}^T)^{-1}(RK_{lj}) +$$
$$+ K_{li}^T((I - D_{\alpha lj}D_{\alpha lj}^T)^{-1}K_{li} + E_{\alpha li}E_{\alpha li}^T + E_{\alpha lj}E_{\alpha lj}^T + PA_{li} + A_{li}^T P + PB_{li}K_{lj} +$$
$$+ K_{lj}^T B_{li}^T P + PD_{li}D_{li}^T P + PB_{li}D_{\alpha lj}D_{\alpha lj}^T B_{li}^T P + E_{\alpha lj}E_{\alpha lj}^T + E_{li}E_{li}^T + PA_{lj} + A_{lj}^T P +$$
$$+ PB_{lj}K_{li} + K_{li}^T B_{lj}^T P + PD_{lj}D_{lj}^T P + PB_{lj}D_{\alpha li}D_{\alpha li}^T B_{lj}^T P + E_{\alpha li}E_{\alpha li}^T + E_{lj}E_{lj}^T]x(t) -$$
$$- 2x(t)^T Qx(t) - x(t)^T(K_{li} + \Delta K_{li}(t))^T R(K_{lj} + \Delta K_{lj}(t))x(t) -$$
$$- x(t)^T(K_{lj} + \Delta K_{lj}(t))^T R(K_{li} + \Delta K_{li}(t))x(t)\}.$$

By making use of the two inequalities (5.4) and (5.5) in the statement of Theorem 5.1, one can show

$$\dot{V}_l(x(t)) < \sum_{l=1}^{l=N} \alpha_l \sum_{i=1}^{r_l} h_{li}^2(t)[-x(t)^T Qx(t) - x(t)^T(K_{li} + \Delta K_{li}(t))^T R(K_{li} + \Delta K_{li}(t))x(t)] +$$
$$+ \sum_{l=1}^{l=N} \alpha_l \sum_{i<j}^{r_l} \sum_{j=1}^{r_l} h_{li}(t)h_{lj}(t)[-2x(t)^T Qx(t) - x(t)^T(K_{li} + \Delta K_{li}(t))^T R(K_{lj} + \Delta K_{lj}(t))x(t) -$$
$$- x(t)^T(K_{lj} + \Delta K_{lj}(t))^T R(K_{li} + \Delta K_{li}(t))x(t)] = -\sum_{l=1}^{l=N} \alpha_l[x(t)^T Qx(t) + u_l(t)^T Ru_l(t)],$$

that is

$$\dot{V}(x(t)) = -\sum_{l=1}^{l=N} \alpha_l[x(t)^T Qx(t) + u_l(t)^T Ru_l(t)] < 0. \tag{5.15}$$

Integration of the left- and right-hind sides of (5.15) yields

$$V(x(\infty)) - V(x(0)) = -x_0^T Px_0 \leq -\sum_{l=1}^{l=N} \alpha_l \int_0^\infty [x(t)^T Qx(t) + u_l(t)^T Ru_l(t)]dt = -J. \tag{5.16}$$

hence it must be $J < x_0^T Px_0 = J*$ for any nonzero initial state vector.

Furthermore, it should be pointed out for any nonzero initial state x_0 there is at least one fuzzy system that satisfies

$$x(t)^T \sum_{i=j}^{r_l} \sum_{j=1}^{r_l} h_{li}(t)h_{lj}(t)\{[(A_{li}+\Delta A_{li}(t))+B_{li}(K_{lj}+\Delta K_{lj}(t))^T P+ \\ +P[(A_{li}+\Delta A_{li}(t))+B_{li}(K_{lj}+\Delta K_{lj}(t))^T]\}x(t)<0,$$

and therefore the arg min{•} switching law enforces the single Lyapunov function to decrease in the whole of the systems state space $\aleph^n$ as the time elapses. Thus the proof is complete. □

It should be noted, nonetheless, the above intelligent control synthesis solution id feasible for a family of gain coefficients α_l according to the convexity requirement, which turned out to be a NP-hard problem. For the general case, this is a research task for the future. For the particular case of the considered class of switched fuzzy systems that have two subsystems there is constructed in here a search algorithm, $\Im_2$, for finding coefficients α_l according to a feasible convex combination.

Algorithm 5.1 Assume the switched fuzzy systems possess two subsystems only, and let n_0 and n_e denote two given nonnegative integers. The search for a Hurwitz convex combinations of systems yields the necessary coefficients via the following iteration process $\Im_2$

Process $\Im_2(1, n_0, n_e)$

For $i = n_0 ; n_e$

$h_2 = \frac{1}{2^i}; s_{21} = \frac{1}{2}h_2; s_{22} = 1-\frac{1}{2}h_2; m_2 = s_{22}/s_{21}$

For $j = 0 : m_2$

$\alpha_1 = s_{21} + j \times h_2;\ \alpha_2 = 1-\alpha_2;$

If inequalities (7) and (8) are solvable, then α_1, α_2 represent a feasible convex combination, else continue Process $\Im_2$

end If

end For

end For

end $\Im_2$

Remark 5.1 In an analogous way, this process $\Im_2$ can be extended to switched system with $N(N > 2)$ subsystems; however, the solvability of the involved inequalities is an open issue yet to be fully explored.

6 The Second Intelligent Control Synthesis: Illustrative Example

As an illustrative example a switched fuzzy system composed of two subsystems is considered as in [16]. These are given by means of the respective matrices as follows:

$$A_{11}=\begin{bmatrix}-10 & 0.01\\ -9.3 & -1.0493\end{bmatrix},\ A_{12}=\begin{bmatrix}0 & 0.1\\ -32 & -4.529\end{bmatrix},$$
$$A_{21}=\begin{bmatrix}-10 & 0.1\\ 10 & -0.1\end{bmatrix},A_{22}=\begin{bmatrix}0 & 0.8\\ -8 & -0.9\end{bmatrix};\tag{5.17}$$

$$B_{11}=B_{12}=B_{21}=B_{22}=\begin{bmatrix}0 & 1\\ 1 & 0\end{bmatrix};\tag{5.18}$$

$$D_{11}=D_{12}=\begin{bmatrix}-0.1125 & 1\\ 1 & 0\end{bmatrix},\ D_{21}=D_{22}=\begin{bmatrix}0.01 & 1\\ 1 & 0\end{bmatrix};\tag{5.19}$$

$$E_{11}=E_{12}=\begin{bmatrix}1 & 0.2\\ 0 & 0\end{bmatrix},E_{21}=E_{22}=\begin{bmatrix}0.5 & 1\\ 0 & 0\end{bmatrix};\tag{5.20}$$

$$D_{\alpha 11}=D_{\alpha 12}=D_{\alpha 21}=D_{\alpha 22}=\begin{bmatrix}0.5 & 0\\ 0 & 0.5\end{bmatrix};\tag{5.21}$$

$$E_{\alpha 11}=E_{\alpha 12}=E_{\alpha 21}=E_{\alpha 22}=\begin{bmatrix}0.5 & 0\\ 0 & 0.5\end{bmatrix};\tag{5.22}$$

$$M_{11}(t)=M_{12}(t)=M_{21}(t)=M_{22}(t)=\begin{bmatrix}\sin t & 0\\ 0 & \cos t\end{bmatrix};\tag{5.23}$$

$$M_{\alpha 11}(t)=M_{\alpha 12}(t)=M_{\alpha 21}(t)=M_{\alpha 22}(t)=\begin{bmatrix}\sin t & 0\\ 0 & \cos t\end{bmatrix};\tag{5.24}$$

Also notice, in optimal control theory and applications it is assumed $Q=R=I$.

The actual nonlinear physical plant is assumed to be characterized by membership functions as follows:

$$h_{11}(x_1(t))=1-h_{12}(x_1(t))=1-\frac{1}{1+e^{-2x_1(t)}},\ h_{12}(x_1(t))=\frac{1}{1+e^{-2x_1(t)}};\tag{5.25}$$

$$h_{21}(x_1(t))=1-h_{22}(x_1(t))=1-\frac{1}{1+e^{-2(x_1(t)-0.3)}},\ h_{22}(x_1(t))=\frac{1}{1+e^{-2(x_1(t)-0.3)}};\tag{5.26}$$

Application of the search algorithm has yielded $\alpha_1 = 0.6$ and $\alpha_2 = 0.4$. Therefore the LMI toolbox of the MATLAB [8] (also see [18, 19]) can be used to solve the inequalities (5.4) and (5.5) in order to compute the symmetric positive definite matrix P as well as the respective feedback gain matrices. In turn, on can obtain

$$P = \begin{bmatrix} 33.4256 & -0.0000 \\ -0.0000 & 33.4256 \end{bmatrix} \tag{5.27}$$

and feedback gain matrices

$$\begin{aligned} K_{11} &= \begin{bmatrix} 16.6000 & 931.5000 \\ -1366.600 & 16.6000 \end{bmatrix}, K_{12} = \begin{bmatrix} 109.6370 & -309.5325 \\ -194.5343 & 109.6370 \end{bmatrix}, \\ K_{21} &= \begin{bmatrix} -20.3000 & -1468.6000 \\ 2001.5000 & -20.3000 \end{bmatrix}, K_{22} = \begin{bmatrix} -134.2774 & 398.9315 \\ 218.7505 & -134.2774 \end{bmatrix}. \end{aligned} \tag{5.28}$$

Hence, the intelligent control synthesis design combining the NGC feedback control of Theorem 5.1 and the $\arg\min\{\bullet\}$-switching control, which render plant system asymptotically stable in closed loop, does exist and its implementation is feasible.

The resulting fuzzy control law synthesis is given as follows:

$$u_l(t) = \sum_{i=1}^{2} h_{li}(t)(K_{li} + \Delta K_{li}(t)]x(t), \; \Delta K_{li}(t) = D_{\alpha li} M_{\alpha li}(t) E_{\alpha li} \tag{5.29}$$

It does complement the switching law (5.12) thus adding a blend of fuzzy-logic-based computational intelligence.

Computer simulation experiments were carried out by applying the synthesized intelligent control comprising the NGC-state feedback (5.29) and switching laws

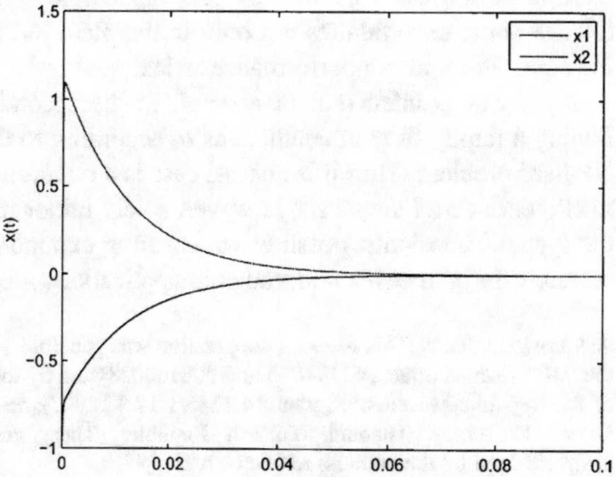

Fig. 3 The plant state responses in the designed closed-loop control system using the switched fuzzy system representation

(5.12) to the system representation (5.10) and by assuming $x(0) = [1, -1]^T$. A typical sample of the obtained simulation results are shown in Fig. 3. Apparently, in the closed loop, the plant system is enforced to operate with an asymptotically stable equilibrium state.

7 Conclusions

This work is focused on presenting novel synthesis design solutions for intelligent controls of complex plant processes, involving nonlinearities and uncertainties, using the synergy of fuzzy systems, switched systems and nontraditional Lyapunov functions approach. It is assumed that the physical parameters of the plant process to be controlled remain largely constant as time elapses. The minimum necessary performance requirement is the operating asymptotic stability in the closed loop.

A new intelligent control synthesis via two-layer multiple Lyapunov functions is proposed to solve the stabilization problem of arbitrary nonlinear continuous-time plants. It is derived on the grounds of switched fuzzy system models as developed via the theory of Takagi–Sugeno (T–S) fuzzy systems. A switching law and state feedback controller that use the time derivative information of membership functions are designed via the PDC scheme. The sufficient conditions for asymptotic stability are derived. The method is shown to be LMI solvable. Numerical and simulation results for an illustrative example are given to demonstrate the effectiveness and the control performance of the proposed synthesis design for nonlinear plant systems.

Also, for a class of uncertain switched fuzzy plant systems with unstable subsystems, the single Lyapunov function method is successfully applied to arrive at another new intelligent control synthesis, comprising a non-fragile guaranteed cost (NGC) feedback control law and a typical switching law, was derived. Furthermore, an algorithm is proposed so as to search a feasible convex combination. Numerical and simulation for another illustrative example are presented to demonstrate the feasible performance as well as that any employed controller should be able to tolerate some uncertainties not only in the plant but also in the implementation of the controller under a performance index.

As it was pointed out, in general, in the second intelligent control synthesis finding a family of gain coefficients α_l according to the convexity requirement is a NP-hard problem. Thus this general case is a research task for the future. In the first intelligent control synthesis, however, solely minor modifications so as to improve the systems transients, possibly via ensuring exponential stability, is left for future research. In both cases nonacademic applications are yet to be developed.

Acknowledgments This research cooperation was generously supported in part by the NSFC of the P.R. of China (grants 61174073 and 908160028) and by the Ministry of Education & Science of the Republic Macedonia (grant 14-3145/1-17.12.2007), and also by the Fund for Science of Dogus University, Istanbul, Turkish Republic. These respective supports are gratefully acknowledged by the authors, respectively.

References

1. Antsaklis, P (ed.): A brief introduction to the theory and applications of hybrid systems. In: Special Issue of the IEEE Proceedings—Hybrid Systems: Theory and Applications, vol. 8, no. 7, pp. 887–897 (2000)
2. Akar, M., Ozguner, U.: Decentralized techniques for the analysis and control of Takagi-Sugeno fuzzy systems. IEEE Trans. Fuzzy Syst. **8**(6), 691–704 (2000)
3. Branicky, M.S.: Multiple Lyapunov functions and other analysis tools for switched and hybrid systems. IEEE Trans. Autom. Control **43**(4), 475–482 (1998)
4. Branicky, M., Bokor, V., Mitter, S.: A unified framework for hybrid control: modal and optimal control theory. IEEE Trans. Autom. Control **43**(1), 31–45 (1998)
5. Choi, D.J., Park, P.G.: State-feedback controller design for discrete-time switching fuzzy systems. In: Proceedings of the 41st IEEE Conference on Decision and Control, NV, pp. 191–196 The IEEE, Piscataway, NJ, 10–13 Dec. (2002)
6. Daafouz, J., Riedinger, P., Iung, C.: Stability analysis and control synthesis for switched systems: a switched Lyapunov function approach. IEEE Trans. Autom. Control **47**(11), 1883–1887 (2002)
7. Fang, C.-H., Liu, Y.-S.: A new LMI-based approach to relaxed quadratic stabilization of T-S fuzzy control systems. IEEE Trans. Fuzzy Syst. **14**(3), 386–397 (2006)
8. Gahinet, P., Nemirovski, A., Laub, A.J., Chilali, M.: LMI Control Toolbox. The MathWorks, Natick, NJ (1995)
9. Hespanha, J.P., Morse, A.S.: Stability of switched systems with average dwell-time. In: Proceedings of the 38th IEEE Conference on Decision and Control, Phoenix, AZ, vol. 3, pp. 2655–2660. The IEEE, Piscataway, NJ, 7–10 Dec. (1999)
10. Jing, Y, Jiang, N., Zheng, Y. Dimirovski, G.M.: Fuzzy robust and non-fragile minimax control of a trailer-truck model. In: Proceedings of the 46th IEEE Conference on Decision and Control, New Orleans, LA, pp. 1221–1226. The IEEE, Piscataway, NJ, 12–14 Dec. (2007)
11. Johansson, M., Rantzer, A., Arzen, K.E.: Piecewise quadratic stability of fuzzy systems. IEEE Trans. Fuzzy Syst. **7**(6), 713–722 (1999)
12. Kogan, M.M.: Local-minimax and minimax control of linear neutral discrete system. *Avtomatika i Telemehainka*, vol. 11, pp. 33–44 (Automatic and Remote Control, English translations) (1997)
13. Liberzon, D., Morse, A.S.: Basic problems in stability and design of switched systems. IEEE Control Syst. Mag. **19**, 59–70 (1999)
14. Liberzon, D.: Switching in System and Control. Birkhauser, Boston, MA, USA (2003)
15. Liu, C.-H.: An LMI-based stable T-S fuzzy model with parametric uncertainties using multiple Lyapunov function approach. Proceedings of the 2004 IEEE Conference on Cybernetics and Intelligent Systems, pp. 514–519. The IEEE, Piscataway, NJ (2004)
16. Luo, J., Dimirovski, G.M., Zhao, J.: Non-fragile guaranteed cost control for a class of uncertain switched fuzzy systems. In: Proceedings of the 31st Chinese Control Conference, Hefei, Anhui, pp. 2112–2116. Chinese Association of Automation, Beijing, CN, 25–27 July (2012)
17. Luo, J., Dymirovsky, G.: A two-layer multiple Lyapunov functions stabilization control of switched fuzzy systems. In: Proceedings of the 6th IEEE Conference on Cybernetics Conference, Sofia, BG, vol. II, pp. 258–263. The IEEE and Bulgarian FNTS, Piscataway, NJ, and Sofia, BG, 6–8 Sept. (2012)
18. MathWorks, Using Matlab—Fuzzy Toolbox. The MathWorks, Natick, NJ, USA
19. MathWorks, Using Matlab—LMI Toolbox. The MathWorks, Natick, NJ, USA
20. Morse, S.: Supervisory control of families of linear set-point controllers. Part 1: exact matching. IEEE Trans. Autom. Control **41**(10), 1413–1431 (1996)
21. Ojleska, V.M., Kolemisevska-Gugulovska, T., Dimirovski, G.M.: Influence of the state space partitioning into regions when designing switched fuzzy controllers. Facta Univ. Series Autom. Control Robot. **9**(1), 103–112 (2010)

22. Ojleska, V.M., Kolemisevska-Gugulovska, T., Dimirovski, G.M.: Switched fuzzy control systems: Exploring the performance in applications. Int. J. Simul.—Syst. Sci. Technol. **12**(2), 19–29 (2012)
23. Ojleska, V., Kolemisevska-Gugulovska, T., Dymirovsky, G.: Recent advances in analysis and control design for switched fuzzy systems. In: Proceedings of the 6th IEEE Conference on Cybernetics Conference, Sofia, BG, vol. II, pp. 248–257. The IEEE and Bulgarian FNTS, Sofia, BG, 6–8 Sept. (2012)
24. Ren, J.S.: Non-fragile LQ fuzzy control for a class of nonlinear desriptor systems with time delay. Proceedings of the 4th International Conference on Machine Learning and Cybernetics, Guangzhou, pp. 18–25. Institute of Intellgent Machines, Chinese Academy of Sciences, Beijng, CN, 18–25 Aug. (2005)
25. Sun, X.M., Liu, G.P., Wang, W., Rees, D.: Stability analysis for networked control systems based on average dwell time method. Int. J. Robust Nonlinear Control **20**(15), 1774–1784 (2010)
26. Takagi, T., Sugeno, M.: Fuzzy identification of systems and its applications to modeling and control. IEEE Trans. Syst. Man & Cybern. **15**(1), 116–132 (1985)
27. Tanaka, K., Sugeno, M.: Stability analysis and design of fuzzy control systems. Fuzzy Sets Syst. **45**(2), 135–156 (1992)
28. Tanaka, K., Ikeda, T., Wang, H.O.: Fuzzy regulators and fuzzy observers: relaxed stability conditions and LMI-based designs. IEEE Trans. Fuzzy Syst. **6**(3), 250–265 (1998)
29. Tanaka, K., Iwasaki, M., Wang, H.O.: Stability and smoothness conditions for switching fuzzy systems. In: Proceedings of the 2000 American Control Conference, Chicago, IL, pp. 2474–2478. The AACC and IEEE Press, Piscataway, NJ (2000)
30. Tanaka, K., Hori, T., Wang, H.O.: Fuzzy Lyapunov approach to fuzzy control systems design. In: Proceedings of the 20th American Control Conference, Arlington, VA, pp. 4790–4795. The AACC and IEEE Press, Piscataway, NJ, June (2001)
31. Tanaka, K., Wang, H.O.: Fuzzy Control Systems Design and Analysis: A Linear Matrix Inequality Approach. Wiley, New York, NY (2001)
32. Wang, J.L., Shu, Z.X., Chen, L., Wang, Z.X.: Non-fragile fuzzy guaranteed cost control of uncertin nonlinear discrete-time systems. Proceedings of the 26th Chinese Control Conference, Zhangjiajie, Hunan, pp. 26–31. Chinese Association of Automation, Beijing, CN, 26–31 July (2007)
33. Wang, R.J., Lin, W.W., Wang, W.J.: Stabilizability of linear quadratic state feedback for uncertain fuzzy time-delay systems. IEEE Trans. Syst. Man Cybern. Part B: Cybern. **34**, 1288–1292 (2004)
34. Wang, R., Zhao, J.: Non-fragile hybrid guaranteed cost control for a class of uncertain switched linear systems. J. Control Theor. Appl. **23**, 32–37 (2006)
35. Wang, W.J., Kao, C.C., Chen, C.S.: Stabilization, estimation and robustness for large-scale time-delay systems. Control Theor. Adv. Technol. **7**, 569–585 (1991)
36. Yang, H., Dimirovski, G.M., Zhao, J.: Stability of a class of uncertain fuzzy systems based on fuzzy switching controller. In: Proceedings of the 25th American Control Conference, Minneapolis, MN, pp. 4067-4071. The AACC and Omnipress Inc., New York, NY, 14–16 June (2006)
37. Yang, H., Liu, H., Dimirovski, G.M., Zhao, J.: Stabilization control of a class of switched fuzzy systems. In: Proceedings of the 2007 IEEE Conference on Fuzzy Systems IEEE-FUZZ07, London, UK, pp. 1345-1350. The IEEE, Piscataway, NJ, July 23–28 (2007)
38. Yang, H., Dimirovski, G.M., Zhao, J.: Switched Fuzzy Systems: Representation modeling, stability and control design. In: Kacprzyk, J. (ed.) Chapter 9 in Studies in Computational Intelligence 109—Intelligent Techniques and Tool for Novel System Architectures, pp. 169–184. Springer, Berlin Heidelberg, DE (2008)
39. Yang, H., Zhao, J., Dimirovski, G.M.: State feedback H^{∞} control design fro switched fuzzy systems. In: Proceedings of the 4th IEEE Conference on Intelligent Systems, Varna, BG, Paper 4-2/pp. 1–6. The IEEE and Bulgarian FNTS, Piscataway, NJ, and Sofia, BG, 6–8 Sept. (2008)
40. Zadeh, L.A.: Inference in fuzzy logic. IEEE Proc. **68**, 124–131 (1980)

41. Zadeh, L.A.: The calculus of If-Then rules. IEEE AI Expert **7**, 23–27 (1991)
42. Zadeh, L.A.: Is there a need for fuzzy logic? Inf. Sci. **178**, 2751–2779 (2008)
43. Zak, S.H.: Systems and Control. Oxford University Press, New York, NY (2003)
44. Zhai, G.: Quadratic stabilizability of discrete-time switched systems via state and output feedback. In: Proceedings of the 40th IEEE Conference on Decision and Control, Orlando, FL, vol. 3, pp. 2165-2166. The IEEE, Piscataway, NJ, 4–7 Dec. (2001)
45. Zhai, G., Lin, H., Antsaklis, P.: Quadratic stabilizability of switched linear systems with polytopic uncertainties. Int. J. Control **76**(7), 747–753 (2003)
46. Zhang, L., Jing, Y., Dimirovski, G.M.: Fault-tolerant control of uncertain time-delay discrete-time systems using T-S models. In: Proceedings of the 2007 IEEE Conference on Fuzzy Systems IEEE-FUZZ07, London, UK 23–28, pp. 1339–1344. The IEEE, Piscataway, NJ, July (2007)
47. Zhang, L., Andreeski, C., Dimirovski, G.M., Jing, Y.: Reliable adaptive control for switched fuzzy systems. In: Proceedings of the 17th IFAC World Congress, Seoul, KO, pp. 7636–7641. The ICROS and IFAC, Seoul, KO, July 6–11 (2008)
48. Zhao, J., Spong, M.W.: Hybrid control for global stabilization of the cart-pendulum system. Automatica **37**(12), 1941–1951 (2001)
49. Zhao, J., Dimirovski, G.M.: Quadratic stability for a class of switched nonlinear systems. IEEE Trans. Autom. Control **49**(4), 574–578 (2004)

A New Architecture for an Adaptive Switching Controller Based on Hybrid Multiple T-S Models

Nikolaos A. Sofianos and Yiannis S. Boutalis

Abstract The scope of this chapter is to provide the reader with the latest advances in the field of switching adaptive control based on hybrid multiple Takagi-Sugeno (T-S) models. The method presented here proposes a controller which is based on some semi-fixed and adaptive T-S identification models which are updating their parameters according to a specified updating rule. The main target of this enhanced scheme—compared with the fixed and adaptive multiple models case—is to control efficiently a class of unknown nonlinear dynamical fuzzy systems. The identification models define the control signal at every time instant with their own state feedback fuzzy controllers which are parameterized by using the certainty equivalence approach. A performance index and an appropriate switching rule are used to determine the T-S model that approximates the plant best and consequently to pick the best available controller at every time instant. Three types of identification models are contained in the models bank: some semi-fixed T-S models which are redistributed during the control procedure, a free adaptive T-S model which is randomly initialized, and finally a reinitialized adaptive model which uses the parameters of the best semi-fixed model at every time instant. The asymptotic stability of the system and the adaptive laws for the adaptive models are given by using Lyapunov stability theory. The combination of these different model categories, offers many advantages to the control scheme and as it is shown by computer simulations, the semi-fixed models method enhances the system's performance and makes the initialization problem less significant than it is in the fixed models case.

N.A. Sofianos (✉) · Y.S. Boutalis
Department of Electrical and Computer Engineering, Democritus University of Thrace, 67100 Xanthi, Kimmeria, Greece
e-mail: nsofian@ee.duth.gr

Y.S. Boutalis
e-mail: ybout@ee.duth.gr

V. Sgurev et al. (eds.), *Recent Contributions in Intelligent Systems*,
Studies in Computational Intelligence 657,
DOI 10.1007/978-3-319-41438-6_9

1 Introduction

Controlling nonlinear plants with unknown or uncertain and time-varying parameters is unquestionably a very challenging problem. Amongst the various methods that have been proposed for this kind of problems, the multiple models control methods seem to exhibit very promising results.

The main advantage of using more than one identification models is that the identification procedure of the unknown parameters is more effective both in speed and accuracy terms. Also, in cases where the controllers are realized, the scheme is more reliable and is capable of producing the appropriate control signals even if one or some of the controllers are damaged or get out of order. As expected, all these advantages are achieved by increasing the computational burden or the cost of the control system if it is to be implemented in hardware. In [1, 2] the authors introduced for the first time the multiple model control methods. In these works the authors provided stability analysis and some simulation results which show the superiority of these controllers. In a more analytic work [3], Narendra and Balakrishnan proposed a number of various methods and techniques which are all based on multiple identification models. A common and novel constituent in all these methods—except from the fixed models case—is the inclusion of adaptive identification models. The adaptation algorithms and techniques that are commonly used can be found in [4, 5]. The main techniques which are described in [3] include fixed identification models, adaptive identification models, a combination of fixed and adaptive models, and finally an improved method which uses a reinitialized adaptive model. These control schemes are designed for linear plants. During the last decade the interest for multiple models control has been returned and some new notable works are available. In [6], the authors provide a scheme for simple nonlinear systems, and in [7] the authors present some new ideas and new perspectives in this field. In [8], the author uses some adaptive identification models and a backstepping technique in order to control a nonlinear plant with an unknown parameter vector but a known input vector. Ioannou and Kuipers in [9], propose a multiple models adaptive control scheme with mixing, using at the same time the powerful tools from linear time invariant theory and providing a robustness analysis. Recently, in [10, 11], the authors presented some new control schemes and techniques for linear systems which are based on multiple models logic. These techniques include the redistribution of some fixed models and the second level adaptation of these models. Although, all of the aforementioned methods are very effective, most of them are designed for linear plants. This fact, along with the observation that the single model adaptive control methods are not adequate enough in difficult control problems lead the authors in [12–14] to design reliable methods for nonlinear unknown systems utilizing the advantage of fuzzy theory to handle effectively the nonlinearities [15–18]. More specifically, T-S [14] fuzzy multiple models were used in order to express the nonlinear plants to be controlled. Depending on the control problem, the following three types of identification models were used: fixed models, free adaptive models, and reinitialized adaptive models. The proposed architectures have many advantages and provide better results than other methods which are based on single adaptive models. Moreover, the idea of combining adaptive and fixed models that change regularly has shown an improved performance [19]. In this chapter, the improved architecture which is mostly based on some redistributed fixed models (semi-fixed models), a free adaptive model and a reinitialized adaptive model is described. The semi-fixed models (SFMs) are changing their parameters toward a direction which minimizes a performance index. Also, the SFMs help the reinitialized adaptive model to be initialized according to a better point inside the uncertainty region and finally to approximate the real plant within a small time period. This strategy provides better results than the previous ones and in some cases reduces the complexity.

2 Problem Statement

In this section the objective of the controller is given and the expressions of the plant and the identification models are formulated in a suitable form for the designing procedure of the controller that follows in the next sections. Utilizing the following T-S fuzzy rules, one can describe a continuous-time nth order nonlinear plant,

$$\begin{aligned} &\textit{Rule } i: \text{IF } x_1(t) \text{ is } M_1^i \text{ and } x_2(t) \text{ is } M_2^i \text{ and } \ldots \\ &\ldots \text{ and } x_n(t) \text{ is } M_n^i \text{ THEN } \dot{x}(t) = A_i x(t) + Bu(t) \end{aligned}$$

where $i = 1, \ldots, l$ is the number of fuzzy rules, $M_p^i, p = 1, \ldots, n$ are the fuzzy sets, $x(t) \in R^n$ is the state vector, $u(t) \in R$ is the input vector, $A_i^{n\times n}$ are the state matrices which are considered to be unknown and $B^{n\times 1}$ is the known input matrix,

$$A_i = \begin{bmatrix} 0 & & & \\ 0 & & \mathbf{I}_{(\mathbf{n-1})} & \\ \vdots & & & \\ \alpha_n^i & \alpha_{n-1}^i & \ldots & \alpha_1^i \end{bmatrix}, \quad B = \begin{bmatrix} 0 \\ 0 \\ \vdots \\ 1 \end{bmatrix}$$

where $\mathbf{I}_{(\mathbf{n-1})}$ is an $(n-1) \times (n-1)$ identity matrix.
The following plant is used as a reference model,

$$\dot{x}_m = A_d x_m \tag{1}$$

where A_d is a $n \times n$ stable state matrix. The aim of the proposed control scheme is to enforce the state x of the unknown fuzzy system to track the desired state trajectory x_m ensuring at the same time the stability of the closed-loop system and an improved performance compared to former similar control schemes.

The dynamical fuzzy system which results from a fuzzy blending of the linear models in the consequent parts of the rules can be described as follows:

$$\dot{x}(t) = \frac{\sum_{i=1}^{l} h_i(x(t))(A_i x(t) + Bu(t))}{\sum_{i=1}^{l} h_i(x(t))} \tag{2}$$

where $h_i(x(t)) = \prod_{p=1}^{n} M_p^i(x_p(t)) \geq 0$ and $M_p^i(x_p(t))$ is the grade of membership of $x_p(t)$ in M_p^i for all $i = 1, \dots, l$ and $p = 1, \dots, n$. The state-space parametric model (SSPM) expression of (2) is given in (3).

$$\dot{x}(t) = A_d x(t) + \frac{\sum_{i=1}^{l} h_i(x(t))((A_i - A_d)x(t) + Bu(t))}{\sum_{i=1}^{l} h_i(x(t))} \tag{3}$$

The series parallel model (SPM) [20] of (2), is given as follows:

$$\dot{\hat{x}}(t) = A_d \hat{x}(t) + \frac{\sum_{i=1}^{l} h_i(x(t))((\hat{A}_i - A_d)x(t) + Bu(t))}{\sum_{i=1}^{l} h_i(x(t))} \tag{4}$$

where $\hat{A}_i$, are the estimations of A_i matrices.

3 The SFMs Based Switching Fuzzy Control Scheme

The proposed switching fuzzy control scheme is described in detail in this section along with some useful tools that are used to ensure a satisfactory performance and the stability of the system.

3.1 Multiple T-S Identification Models Architecture

A hybrid bank of fuzzy identification models is the main part of the proposed control scheme. The role of these fuzzy models is to approximate the behavior of the real plant. The identification models consist of N-2 semi-fixed T-S identification mod-

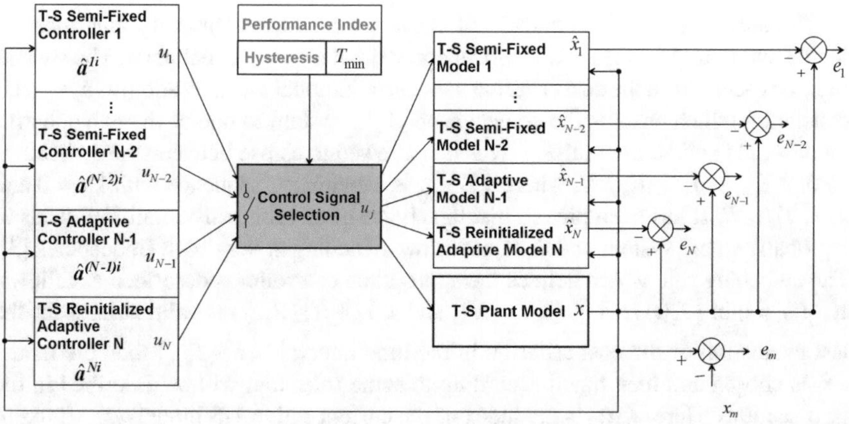

Fig. 1 The multiple T-S estimation models switching control scheme

els $\{\mathscr{M}_{sf}\}_{sf=1}^{N-2}$, one free running adaptive model $\mathscr{M}_{ad}$ and one reinitialized adaptive model $\mathscr{M}_{adr}$. Every identification model is connected with its own adaptive controller by using the certainty equivalence approach for the adaptation of the controllers' parameters. The main parts of the control scheme are given in Fig. 1. The objective of the control scheme is to drive the reference model tracking error, $e_m = x - x_m$, to zero. The SPM formulation (4) is used to describe all the identification models $\{\mathscr{M}_k\}_{k=1}^{N}$ whose initial parameter values are different. The estimations for A_i are denoted as $\hat{A}_{ki}$ and the uncertain parameters of A_i are denoted as $\mathscr{E}_A \in \Xi \subset R^n$, where Ξ is a compact space indicating the region defined by all possible parameter values combinations and n is equal to the number of the unknown parameters. The critical point here is that the initial estimations are not picked randomly but they are distributed uniformly over a lattice in Ξ. Although N controllers are used in the proposed scheme, only one of them defines the control signal u, which is finally applied to all the T-S models and the real fuzzy model of the plant and an output $\hat{x}_k$ is produced for every model. The identification error which defines how "close" the models are to the real plant is given as $e_k = x - \hat{x}_k$. A feedback linearization controller $\mathscr{C}_k$ with an output u_k corresponds to identification model $\mathscr{M}_k$. The semi-fixed controllers $\mathscr{C}_{sf}$ and the adaptive controllers $\mathscr{C}_{ad}$ and $\mathscr{C}_{adr}$ are updated in an indirect way using the certainty equivalence approach. The controller's $\mathscr{C}_k$ signal is designed so that when applied to the corresponding T-S plant $\mathscr{M}_k$, the output is given by a state equation identical to that of the reference model (1). At every time instant, the appropriate controller is chosen according to a switching rule, which is based on a cost criterion J_k, where

$$J_k(t) = \int_0^t e_k^T(r)e_k(r)dr \tag{5}$$

The cost criterion (5) is calculated at every time instant for every identification model and finally defines which model approximates best the real plant. The switching rule except from the cost criterion uses two additional tools; an additive hysteresis constant h which ensures the convergence of the system to one of the two adaptive models [21] and an interval T_{min} which is allowed to elapse between the switchings so that $T_{i+1} - T_i \geq T_{\min}, \forall i$ where $\left\{T_i\right\}_{i=0}^{\infty}$ is a switching sequence with $T_0 = 0$ and $T_i < T_{i+1}, \forall i$. It has been proved that the choice of an arbitrarily small T_{min} leads in a globally stable system and does not allow switching at very high frequencies [3]. The switching rule which defines the appropriate controller is described as follows: If $J_j(t) = \min_{k \in \Lambda} \left\{J_k(t)\right\}$, $\Lambda = \{1, \ldots, N\}$, and $J_j(t) + h \leq J_{cr}(t)$ is valid at least for the last evaluation of the cost criterion in the time interval $[t, t + T_{min}]$ then the model $\mathscr{M}_j$ is chosen and then tuned according to some rules that will be described in the next sections. Here, $J_{cr}(t)$ is the index of the current active T-S model $\mathscr{M}_{cr}$. It has to be noted that the algorithm step is smaller than T_{min} and thus there will be more than one evaluations in the time interval $[t, t + T_{min}]$. For example, if T_{min} is equal to three algorithm's steps then the inequality $J_j(t) + h \leq J_{cr}(t)$ should be valid at least during the third step in order to change the controller. If the aforementioned inequality is not valid, the controller $\mathscr{C}_{cr}$ remains active, meaning that it is the ideal controller for the time instant $t + T_{min}$. Note that $\mathscr{M}_j$, i.e., the model with the minimum cost criterion may change during the evaluations in the time interval $[t, t + T_{min}]$. The above procedure is repeated at every step. In case $\mathscr{M}_j \in \left\{\mathscr{M}_{sf}\right\}_{f=1}^{N-2}$, i.e., $\mathscr{M}_j$ is a semi-fixed model, the model $\mathscr{M}_{adr}$ reinitializes its parameter vector value, its cost criterion value, and its state vector value according to the corresponding values of the dominant semi-fixed model $\mathscr{M}_j$ and the entire adaptation process is continued. It is clear that the better the location of the semi-fixed models inside the uncertainty region Ξ is, the better will be the initialization of the adaptive model $\mathscr{M}_{adr}$. Every T-S identification model $\mathscr{M}_k$ is described by the following fuzzy rules:

T-S Identification Model $\mathscr{M}_k$
Rule i :
IF $x_1(t)$ is M_1^{ki} *and* $x_2(t)$ is M_2^{ki} and … and $x_n(t)$ is M_n^{ki}
THEN $\dot{\hat{x}}_k(t) = A_d \hat{x}_k(t) + (\hat{A}_{ki} - A_d)x(t) + Bu(t)$

where $k \in \Lambda = \{1, \ldots, N\}$ and $i = 1, \ldots, l$. The final form of every T-S model is given by the following equation:

$$\dot{\hat{x}}_k(t) = A_d \hat{x}_k(t) + \frac{\sum_{i=1}^{l} h_{ki}(x)((\hat{A}_{ki} - A_d)x(t) + Bu(t))}{\sum_{i=1}^{l} h_{ki}(x)} \tag{6}$$

where $h_{ki}(x) = \prod_{p=1}^{n} M_p^{ki}(x_p(t)) \geq 0$ and $M_p^{ki}(x_p(t))$ is the grade of membership of $x_p(t)$ in M_p^{ki}, $k \in \Lambda = \{1, \ldots, N\}$, $i = 1, \ldots, l$ and $p = 1, \ldots, n$. Also, $M_p^{ki}(x_p(t)) = M_p^{i}(x_p(t))$ and $h_{ki}(x) = h_i(x)$ for all k, i, p. The matrices of all the T-S models are of the following form:

$$\hat{A}_{ki} = \begin{bmatrix} 0 & & & \\ 0 & & \mathbf{I}_{(\mathbf{n-1})} & \\ \vdots & & & \\ \hat{a}_n^{ki} & \hat{a}_{n-1}^{ki} & \cdots & \hat{a}_1^{ki} \end{bmatrix}$$

3.2 Controller Design

Using a feedback linearization technique and supposing that $\mathcal{M}_j$ is the superior T-S model, the control signal for the plant is identical to the control signal of the controller $\mathcal{C}_j$ and is given by the following equation:

$$u(t) = u_j(t) = \frac{\sum_{i=1}^{l} h_{ji}(x)(\mathbf{a}^d - \hat{\mathbf{a}}^{ji})^T x(t)}{\sum_{i=1}^{l} h_{ji}(x)} \tag{7}$$

where $(\hat{\mathbf{a}}^{ji})^T = \begin{bmatrix} \hat{a}_n^{ji} \; \hat{a}_{n-1}^{ji} \; \cdots \; \hat{a}_2^{ji} \; \hat{a}_1^{ji} \end{bmatrix}$ and $(\mathbf{a}^d)^T = \begin{bmatrix} a_n^d \; a_{n-1}^d \; \cdots \; a_2^d \; a_1^d \end{bmatrix}$ are the nth rows of the identification and reference model state matrices respectively. Applying the control input $u(t)$ to $\mathcal{M}_j$ and taking into account that $h_{ji}(x) = h_{ki}(x) = h_i$ we obtain:

$$\dot{\hat{x}}_j(t) = A_d\hat{x}_j(t) + \frac{1}{\sum_{i=1}^{l} h_i}\left(\sum_{i=1}^{l} h_i\left((\hat{A}_{ji} - A_d)x(t) + B\frac{\sum_{i=1}^{l} h_i(\mathbf{a}^d - \hat{\mathbf{a}}^{ji})^T x(t)}{\sum_{i=1}^{l} h_i}\right)\right)$$

$$= A_d\hat{x}_j(t) + \frac{1}{\sum_{i=1}^{l} h_i}\left\{\left[\begin{bmatrix} 0 & & & \\ 0 & & \mathbf{I}_{(\mathbf{n-1})} & \\ \vdots & & & \\ \sum_{i=1}^{l} h_i\hat{a}_n^{ji} & \sum_{i=1}^{l} h_i\hat{a}_{n-1}^{ji} & \cdots & \sum_{i=1}^{l} h_i\hat{a}_1^{ji} \end{bmatrix}\right.\right.$$

$$
-\begin{bmatrix} 0 & & \mathbf{I}_{(\mathbf{n-1})} & \\ 0 & & & \\ \vdots & & & \\ \sum_{i=1}^{l} h_i a_n^d & \sum_{i=1}^{l} h_i a_{n-1}^d & \cdots & \sum_{i=1}^{l} h_i a_1^d \end{bmatrix}
$$

$$
\left. + \begin{bmatrix} 0 \\ 0 \\ \vdots \\ \sum_{i=1}^{l} h_i \end{bmatrix} \frac{1}{\sum_{i=1}^{l} h_i} \left[\sum_{i=1}^{l} h_i (a_n^d - \hat{a}_n^{ji}) \; \sum_{i=1}^{l} h_i (a_{n-1}^d - \hat{a}_{n-1}^{ji}) \cdots \sum_{i=1}^{l} h_i (a_1^d - \hat{a}_1^{ji}) \right] \right\} x(t)
$$

$$
= A_d \hat{x}_j(t) + \frac{1}{\sum_{i=1}^{l} h_i} \left\{ \begin{bmatrix} 0 & & \mathbf{0}_{(\mathbf{n-1})} & \\ 0 & & & \\ \vdots & & & \\ \sum_{i=1}^{l} h_i (\hat{a}_n^{ji} - a_n^d) & \sum_{i=1}^{l} h_i (\hat{a}_{n-1}^{ji} - a_{n-1}^d) & \cdots & \sum_{i=1}^{l} h_i (\hat{a}_1^{ji} - a_1^d) \end{bmatrix} \right.
$$

$$
\left. + \begin{bmatrix} 0 \\ 0 \\ \vdots \\ 1 \end{bmatrix} \left[\sum_{i=1}^{l} h_i (a_n^d - \hat{a}_n^{ji}) \; \sum_{i=1}^{l} h_i (a_{n-1}^d - \hat{a}_{n-1}^{ji}) \cdots \sum_{i=1}^{l} h_i (a_1^d - \hat{a}_1^{ji}) \right] \right\} x(t).
$$

where $\mathbf{I}_{(\mathbf{n-1})}$ is a $(n-1) \times (n-1)$ zero matrix.

From the above equations it follows that:

$$
\dot{\hat{x}}_j(t) = A_d \hat{x}_j(t) \tag{8}
$$

From (8) it is obvious that when $u_j(t)$ is applied to $\mathscr{M}_j$, this model is linearized and has an identical behavior to that of the desired reference model (1). When the numerator of the control signal (7) equals to zero, then the control signal will not be able to control the system. In this case, the switching rule is enriched with the following rule:

$$
\begin{gathered}
\text{If } \; \mathrm{C}_j \text{ is the ideal controller} \\
\text{and } \sum_{i=1}^{l} h_{ji}(x)(\mathbf{a}^d - \hat{\mathbf{a}}^{ji})^T x(t) = 0 \text{ and } x(t) \neq 0 \\
\text{Then } \; u(t) = u_{j(nbc)}(t), \; \text{where } j(nbc) = \arg \min_{k \in \Lambda, k \neq j} \left\{ J_k(t) \right\}.
\end{gathered}
$$

This rule ensures that the best alternative controller will undertake the control procedure for this time instant and the control signal will not be zero.

4 Learning Rules, Redistribution of Semi-Fixed T-S Models, and Stability Analysis

In this section a learning rule for the adaptive models, that leads the system to the desired behavior ensuring at the same time its stability, is given, and in addition a rule for the updating of the semi-fixed models is formulated. According to the previous sections, the time derivative of the identification error e_k of every T-S model is given by the following equation:

$$\dot{e}_k = \dot{x} - \dot{\hat{x}}_k = A_d e_k - \frac{\sum_{i=1}^{l} h_i \tilde{A}_{ki}}{\sum_{i=1}^{l} h_i} x \tag{9}$$

$$= A_d e_k - \frac{\sum_{i=1}^{l} h_i \left[\mathbf{0_{n\times(n-1)}} \; \tilde{\mathbf{a}}^{ki} \right]^T}{\sum_{i=1}^{l} h_i} x \tag{10}$$

where $\widetilde{A}_{ki} = \hat{A}_{ki} - A_i$, $\tilde{\mathbf{a}}^{ki} = \left[\tilde{a}^{ki}_n \; \tilde{a}^{ki}_{n-1} \; \cdots \; \tilde{a}^{ki}_1 \right]^T$ is a $n \times 1$ vector, $\widetilde{a}^{ki}_p = \hat{a}^{ki}_p - a^i_p$, $\mathbf{0_{n\times(n-1)}}$ is a $n \times (n-1)$ zero matrix and $p = 1, \ldots, n$. Consider the following Lyapunov function candidates:

$$V_k(e_k, \tilde{\mathbf{a}}^{ki}, \tilde{b}^{ki}) = e_k^T P_k e_k + \sum_{i=1}^{l} \frac{(\tilde{\mathbf{a}}^{ki})^T \tilde{\mathbf{a}}^{ki}}{r^{ki}} \tag{11}$$

where $r^{ki} > 0$ is the learning rate constant, $V_k \geq 0$, and $P_k = P_k^T$ is the solution of the Lyapunov equation:

$$A_d^T P_k + P_k A_d = -Q_k \tag{12}$$

where Q_k is an $n \times n$ positive definite matrix. The time derivative of V_k is given as follows:

$$\dot{V}_k = \dot{e}_k^T P_k e_k + e_k^T P_k \dot{e}_k + \sum_{i=1}^{l} 2 \frac{(\dot{\tilde{\mathbf{a}}}^{ki})^T \tilde{\mathbf{a}}^{ki}}{r^{ki}}$$

$$= e_k^T (A_d^T P_k + P_k A_d) e_k + \sum_{i=1}^{l} 2 \frac{(\dot{\hat{\mathbf{a}}}^{ki})^T \tilde{\mathbf{a}}^{ki}}{r^{ki}}$$

$$-\frac{\sum_{i=1}^{l} h_i x^T \tilde{\mathbf{a}}^{ki}}{\sum_{i=1}^{l} h_i} P_{ks}^T e_k - e_k^T P_{ks} \frac{\sum_{i=1}^{l} h_i (\tilde{\mathbf{a}}^{ki})^T x}{\sum_{i=1}^{l} h_i}$$

$$= -e_k^T Q_k e_k + \sum_{i=1}^{l} 2 \frac{(\dot{\tilde{\mathbf{a}}}^{ki})^T \tilde{\mathbf{a}}^{ki}}{r^{ki}} - 2 \frac{\sum_{i=1}^{l} h_i P_{ks}^T e_k x^T \tilde{\mathbf{a}}^{ki}}{\sum_{i=1}^{l} h_i}$$

where $P_{ks} \in R^{n \times 1}$ is the nth column of P_k. The semi-fixed models may not change their parameters at every step—this depends on the algorithm—so whenever $\mathscr{M}_k \in \left\{\mathscr{M}_{sf}\right\}_{f=1}^{N-2}$, i.e., $\mathscr{M}_k$ is a semi-fixed model, $(\dot{\tilde{\mathbf{a}}}^{ki})^T = 0$. In that case,

$$\dot{V}_k = -e_k^T Q_k e_k - 2 \frac{\sum_{i=1}^{l} h_i P_{ks}^T e_k x^T \tilde{\mathbf{a}}^{ki}}{\sum_{i=1}^{l} h_i} \tag{13}$$

Since the semi-fixed models are not used for stability purposes and, as it can be seen from (13), when only semi-fixed models are used in the control scheme, there is no guarantee for the boundedness of e_k, the duty of stabilizing the system is assigned to the adaptive models. In case $\mathscr{M}_k \in \left\{\mathscr{M}_{ad}, \mathscr{M}_{adr}\right\}$, i.e., $\mathscr{M}_k$ is an adaptive model, the updating law for $(\tilde{\mathbf{a}}^{ki})^T$ is given as follows:

$$(\dot{\tilde{\mathbf{a}}}^{ki})^T = r^{ki} \frac{h_i}{\sum_{i=1}^{l} h_i} P_{ks}^T e_k x^T \tag{14}$$

On the other hand, the semi-fixed models will update their parameters too, toward the region of the uncertainty space that is closer to the real plant. From (10) one has:

$$\dot{e}_k = A_d e_k - B \frac{\sum_{i=1}^{l} h_i (\tilde{\mathbf{a}}^{ki})^T}{\sum_{i=1}^{l} h_i} x \tag{15}$$

By solving (15) and assuming $x(0) = \hat{x}_k(0)$, it follows that

$$e_k = -\int_0^t e^{A_d(t-r)} B x^T(r) dr \frac{\sum_{i=1}^{l} h_i \tilde{\mathbf{a}}^{ki}}{\sum_{i=1}^{l} h_i} \quad (16)$$

The identification errors can be expressed as follows:

$$e_k(t) = -K(t) \frac{\sum_{i=1}^{l} h_i \tilde{\mathbf{a}}^{ki}}{\sum_{i=1}^{l} h_i} \quad (17)$$

where $K(t) = \int_0^t e^{A_d(t-r)} B x^T(r) dr$ can be computed at every time instant since its parameters are known and accesible. It is obvious that $e_k(t)$ and $\tilde{\mathbf{a}}^{ki}$ are linearly related. The performance index can be expressed as follows:

$$J_k(t) = \int_0^t \left(\frac{\sum_{i=1}^{l} h_i (\tilde{\mathbf{a}}^{ki})^T}{\sum_{i=1}^{l} h_i} K^T(r) K(r) \frac{\sum_{i=1}^{l} h_i \tilde{\mathbf{a}}^{ki}}{\sum_{i=1}^{l} h_i} \right) dr$$

$$= \frac{\sum_{i=1}^{l} h_i (\tilde{\mathbf{a}}^{ki})^T}{\sum_{i=1}^{l} h_i} \left(\int_0^t K^T(r) K(r) dr \right) \frac{\sum_{i=1}^{l} h_i \tilde{\mathbf{a}}^{ki}}{\sum_{i=1}^{l} h_i} = \frac{\sum_{i=1}^{l} h_i (\tilde{\mathbf{a}}^{ki})^T}{\sum_{i=1}^{l} h_i} C(t) \frac{\sum_{i=1}^{l} h_i \tilde{\mathbf{a}}^{ki}}{\sum_{i=1}^{l} h_i}$$

The above expression for $J_k(t)$ shows that the performance indices are quadratic functions of the unknown vectors $\tilde{\mathbf{a}}^{ki}$. If matrix $C(t)$ is positive definite for $t \geq T$, then the performance indices are parts of a quadratic surface in which the minimum corresponds to the real plant, where the performance index is equal to zero. Based on this fact, an algorithm for the redistribution of the semi-fixed models is used in this paper [10]. The main objective of this algorithm is to change the parameter values of the semi-fixed models toward a direction which is close to the real plant. The algorithm can be expressed as follows: If $J_j(t) = \min_{k \in \Lambda} \{J_k(t)\}$ where $\Lambda = \{1, \ldots, N\}$ and $j \in [1, \ldots, N-2]$, then all the semi-fixed models—except $\mathcal{M}_j$—redistribute their parameters according to the following equation:

$$\hat{\mathbf{a}}^{ki}_{new} = \frac{\sqrt{J_k}}{\sqrt{J_j}+\sqrt{J_k}}\hat{\mathbf{a}}^{ji} + \frac{\sqrt{J_j}}{\sqrt{J_j}+\sqrt{J_k}}\hat{\mathbf{a}}^{ki} \tag{18}$$

where $k \neq j$ and $k \in [1, \ldots, N-2]$. The redistribution does not take place at every algorithm step and is defined according to the problem specifications. It is obvious that when the dominant model is an adaptive model the redistribution does not take place.

The following Theorem ensures that the dynamical fuzzy system (2) is asymptotically stable and its state follows the state of the reference model (1).

Theorem 1 *The proposed indirect switching control architecture for the dynamical fuzzy system (2), along with the control law (7), the adaptive law (14) and the reference model (1), guarantees that:* $\hat{\mathbf{a}}^{ki}$, $e_k(t)$ *are bounded,* $[e_k(t), \dot{\hat{\mathbf{a}}}^{ji}(t), e_m(t)] \to 0$ *as* $t \to \infty$ *for all* $i, j = 1, \ldots, l$, $k \in \{ad, adr\}$ *and finally the controller converges to one of the two available adaptive controllers (except from the case where one of the semi-fixed controllers matches the plant exactly). The stability results of Theorem 1 are valid only when adaptive models are used in the control architecture.*

Proof Considering the adaptive law (14), the time derivative of V_k is given in as follows:

$$\dot{V}_k = -e_k^T e_k \tag{19}$$

Therefore, $\dot{V}_k = -e_k^T e_k \leq 0$, for $k \in \{ad, adr\}$ and $t \geq 0$. Consequently, the function V_k is a Lyapunov function for the systems (10), (14) when $k \in \{ad, adr\}$. This implies that e_k, $\tilde{\mathbf{a}}^{ki}$, $\hat{\mathbf{a}}^{ki} \in L_\infty$. Due to the fact that V_k is bounded from below ($V_k \geq 0$) and nonincreasing with time ($\dot{V}_k \leq 0$), the following equation holds,

$$\lim_{t\to\infty} V_k(e_k(t), \tilde{\mathbf{a}}^{ki}(t)) = V(\infty) < \infty \tag{20}$$

and

$$\int_0^\infty e_k^T e_k dr \leq -\int_0^\infty \dot{V}_k dr = (V_k(0) - V_k(\infty)) \tag{21}$$

where $V_k(0) = V_k(e_k(0), \tilde{\mathbf{a}}^{ki}(0))$. Consequently, $e_k \in L_2 \cap L_\infty$. At any time instant, only one controller $\mathscr{C}_j$ is chosen in order to control the plant. Applying $u_j(t)$ from (7) in $\mathscr{M}_j$, and taking into account the expression (3), the time derivative of the identification error for $\mathscr{M}_j$ is given as follows:

$$\dot{e}_j(t) = \dot{x}(t) - \dot{\hat{x}}_j(t) = A_d e_j(t) + \frac{\sum_{i=1}^{l} h_i(x(t))((A_i - A_d)x(t) + Bu(t))}{\sum_{i=1}^{l} h_i(x(t))} \tag{22}$$

The time derivative of the reference model tracking error $e_m = x - x_m$, is given in (23):

$$\dot{e}_m(t) = \dot{x}(t) - \dot{x}_m(t) = A_d e_m(t) + \frac{\sum_{i=1}^{l} h_i(x(t))((A_i - A_d)x(t) + Bu(t))}{\sum_{i=1}^{l} h_i(x(t))} \tag{23}$$

From (22), (23) one has:

$$\dot{e}_j(t) - \dot{e}_m(t) = A_d(e_j(t) - e_m(t)) \tag{24}$$

Taking into account that $e_j \in L_2 \cap L_\infty$ for all $j \in \{ad, adr\}$, and that equation (24) is satisfied at any instant, it follows that $e_m(t)$ is bounded and consequently $x,\ u \in L_\infty$ when one of the adaptive models ends up to be the dominant model. This is the case in the proposed control architecture with the particular performance index (5). The performance indices J_{ad} or J_{adr} of the adaptive models are bounded while the performance indices of the semi-fixed models grow in an unbounded way due to the fact that they are formulated using the integral terms of the squared errors. A semi-fixed model may never approximate the parameters or the behavior of the controlled plant. Therefore, there exists a finite time t_{ad} such that the system switches to one of the available adaptive controllers and stays there for all $t \geq t_{ad}$ [6]. The role of the semi-fixed models is very significant because they provide to the reinitialized adaptive model the best possible starting point. Moreover, they revise this starting point many times through the control procedure by adapting their own parameters. On the other hand, the reinitialized adaptive model takes the control role—if necessary—from a semi-fixed model very quickly and leads the system in stability, in case the free adaptive model does not have a satisfactory parameters initialization. Consequently, the main role of the semi-fixed models is to adapt their parameters towards a direction which minimizes their performance indices and provide the best possible approximation for the plant's parameters in the reinitialized adaptive model during the first steps of the control procedure. Considering these facts, equation (10) implies that $\dot{e}_k \in L_\infty$ and if combined with $e_k \in L_2 \cap L_\infty$, we obtain that $e_k \to 0$ asymptotically. Using (14) we conclude that $\dot{\hat{\mathbf{a}}}^{ji} \to 0$. Finally from (24) it follows that $\lim_{t\to\infty} e_m(t) \to 0$ which is the objective of the proposed controller architecture.

Remark 1 The computational cost of the proposed control scheme mainly depends on the number of the T-S identification models. The number of T-S models is specified by the controller designer and is associated with the size of the uncertainty region. Compared to the case where only adaptive T-S models are used, we conclude that this method is less complicated and requires a reduced computational effort. On the other hand, compared with approaches employing only fixed models that do not change their parameters, the method appears more complicated and is computationally more demanding. Nevertheless, the redistribution of the fixed models speeds up the control procedure and leads the system to the desired point faster than all the previous methods. In order to obtain all the advantages of the proposed method, the designer should compromise between the performance and the computational requirements of the controller.

5 Simulation Results

The effectiveness and the advantages of the proposed controller are demonstrated by utilizing a known mechanical system as the plant to be controlled. It should be noted that the theoretical analysis of the proposed method supports its use for the control of the T-S fuzzy system model. However, in the simulation carried out, the derived control signal is applied on the real system demonstrating with simulations the robustness of the method. Three methods are tested and compared here: (i) the control scheme that uses only adaptive multiple models (AMs) [14], (ii) the hybrid scheme where fixed and adaptive multiple models are used together (FMs) [13], and (iii) the proposed controller scheme (SFMs). These three controllers are applied to the mass–spring system [22], which is depicted in Fig. 2 and is described by the following equations in state-space form:

$$\begin{bmatrix} \dot{x}_1 \\ \dot{x}_2 \end{bmatrix} = \begin{bmatrix} 0 & 1 \\ -k - ka^2 x_1{}^2 & -c \end{bmatrix} \begin{bmatrix} x_1 \\ x_2 \end{bmatrix} + \begin{bmatrix} 0 \\ 1 \end{bmatrix} u(t) \tag{25}$$

where $x_1 \in [-d, d]$ is the displacement of the mass from the stability point, u is the control signal, $a^2 = 0.90$, $m = 1Kg$, and k, c are the uncertain constants. More specifically $k \in [0.4, 1.4]$ and $c \in [0.1, 1.3]$. The nonlinear term $x_1{}^2$ can be approximated by using two fuzzy rules with membership functions, $M_1 = (d^2 - x_1{}^2)/d^2$ and $M_2 = x_1{}^2/d^2$.

The fuzzy rules are given as follows:

Rule 1 : IF $x_1(t)$ is about 0 THEN $\dot{x}(t) = A_1 x(t) + B_1 u(t)$

Rule 2 : IF $x_1(t)$ is about $\pm d$ THEN $\dot{x}(t) = A_2 x(t) + B_2 u(t)$,

where

$$A_1 = \begin{bmatrix} 0 & 1 \\ -k & -c \end{bmatrix}, B_1 = \begin{bmatrix} 0 \\ 1 \end{bmatrix}$$

Fig. 2 The mass–spring mechanical system

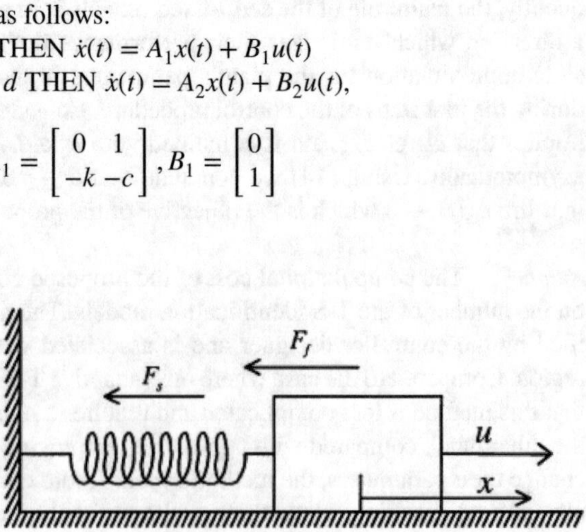

$$A_2 = \begin{bmatrix} 0 & 1 \\ -k - ka^2d^2 & -c \end{bmatrix}, B_2 = \begin{bmatrix} 0 \\ 1 \end{bmatrix}$$

and $d = 1$. The control objective is to force the system (25) to follow the reference model (1) where:

$$A_d = \begin{bmatrix} 0 & 1 \\ -5 & -5 \end{bmatrix}$$

Using $Q_k = \mathbf{I}_{2\times 2}$ in Lyapunov equation (12) one obtains:

$$P_k = \begin{bmatrix} 0.1 & 0.1 \\ 0.1 & 0.12 \end{bmatrix}, P_{ks} = \begin{bmatrix} 0.1 & 0.12 \end{bmatrix}^T.$$

The first control scheme consists of ten adaptive models $\{\mathscr{M}_k\}_{k=1}^{10}$ along with their corresponding fuzzy controllers. The second control scheme consists of nine fixed models $\{\mathscr{M}_k\}_{k=1}^{9}$, one free adaptive model $\mathscr{M}_{10}$ and one reinitialized adaptive model $\mathscr{M}_{11}$, and the third control scheme consists of nine semi-fixed fuzzy models $\{\mathscr{M}_k\}_{k=1}^{9}$, one free adaptive fuzzy model $\mathscr{M}_{10}$ and one reinitialized adaptive fuzzy model $\mathscr{M}_{11}$ along with their corresponding controllers. The initial estimates $\hat{\mathbf{a}}^{ki}$ in the fixed models, the semi-fixed models, and the free adaptive models are the same

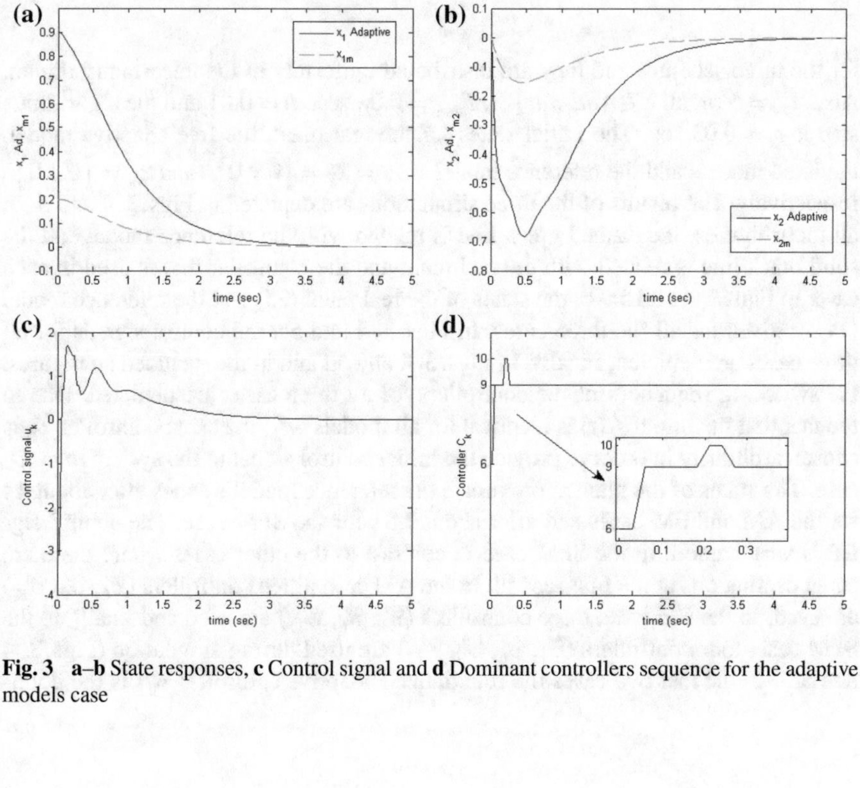

Fig. 3 **a–b** State responses, **c** Control signal and **d** Dominant controllers sequence for the adaptive models case

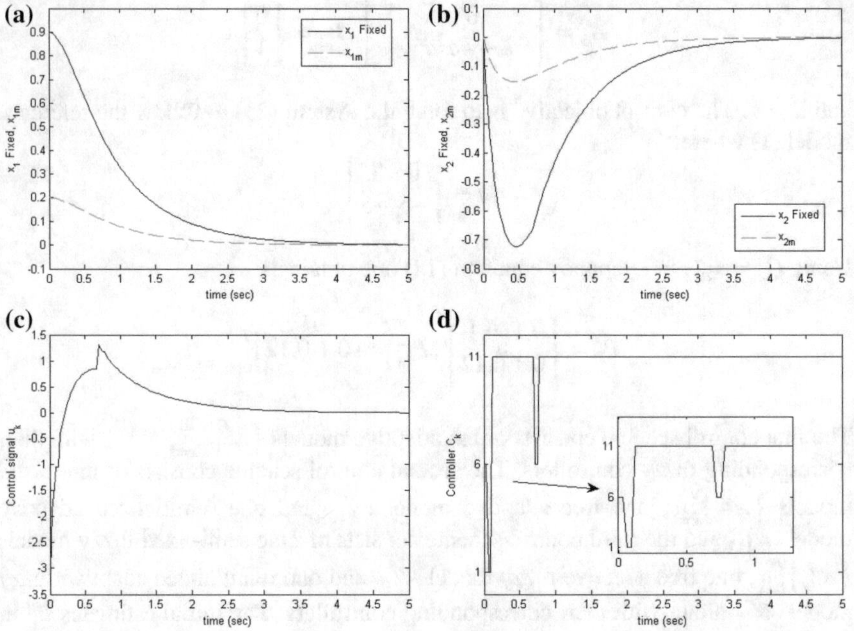

Fig. 4 **a–b** State responses, **c** Control signal and **d** Dominant controllers sequence for the fixed and adaptive models case

for the three schemes and they are distributed uniformly in the uncertainty region. Also, $r^{ki} = 5$ for all $k \in \{ad, adr\}$, i, $T_{\min} = 0.06$ sec., $h = 0.01$ and the algorithm's step is $t_s = 0.03$ sec. The initial states for the real plant, the free adaptive model, the fixed models and the reference model are $x = \hat{x}_k = \begin{bmatrix} 0.9 & 0 \end{bmatrix}^{\mathrm{T}}$ and $x_m = \begin{bmatrix} 0.2 & 0 \end{bmatrix}^{\mathrm{T}}$ respectively. The results of the three simulations are depicted in Figs. 3, 4 and 5. In all these figures, the dashed green line is related with the reference model and the solid black line is related with the real plant and the controller that is used in each case. In Figs. 3, 4 and 5a–b, the states of the real plant (25) and the reference model (1) are given for all the three cases. In Figs. 3, 4 and 5c, the control signals for all three cases are depicted. Finally, in Figs. 3, 4 and 5d and in the included subfigures, the switching sequences of the controllers for all three cases are depicted. Due to the fact that the initial $J_k(t)$ is identical for all models $\mathscr{M}_k$, the same controller $\mathscr{C}_6$ is chosen arbitrarily in order to provide the initial control signal to the system in every case. The states of the plant approximate the reference model's states after about 4 s for the AM and FM cases and after about 3.5 s for the SFM case. The control signal is very smooth in the SFM case in contrast to the other cases where there are some oscillations at the first second. In the AM case, three controllers ($\mathscr{C}_6$, $\mathscr{C}_9$, $\mathscr{C}_{10}$) are used, in the FM case, three controllers ($\mathscr{C}_1$, $\mathscr{C}_6$, $\mathscr{C}_{11}$) are used and finally in the SFM case, four controllers ($\mathscr{C}_1$, $\mathscr{C}_6$, $\mathscr{C}_9$, $\mathscr{C}_{11}$) are used during simulation (Figs. 3, 4 and 5d). In the last two cases the reinitialized adaptive controller $\mathscr{C}_{11}$ is the domi-

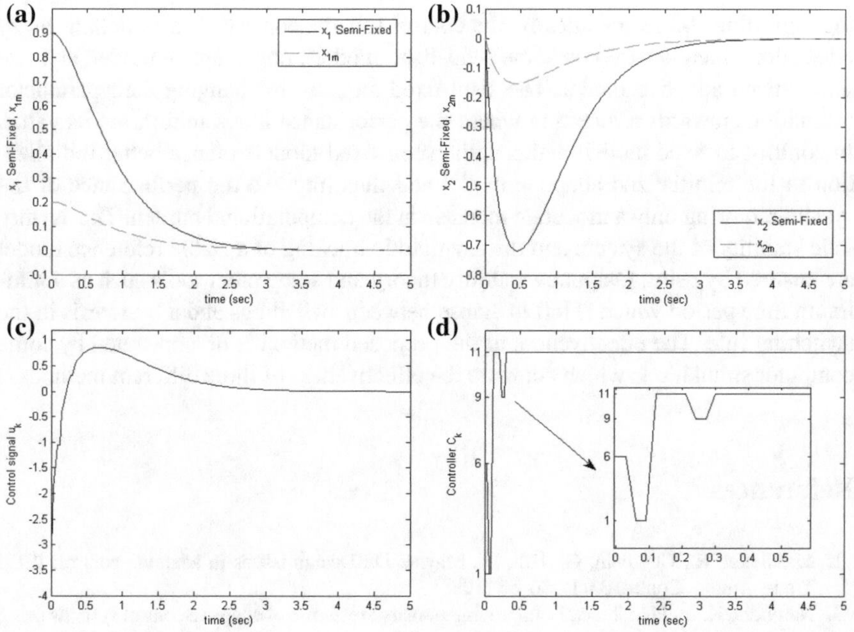

Fig. 5 **a–b** State responses, **c** Control signal and **d** Dominant controllers sequence for the semi-fixed and adaptive models case

nant controller. It is obvious that in SFM case, the semi-fixed model $\mathscr{M}_1$ provides a good reinitialization in the adaptive model $\mathscr{M}_{11}$ at about 0.1 s and after that period of time the controller $\mathscr{C}_{11}$ gains the domination. On the other hand in the FM case, the reinitialized controller is not ideally parameterized by the fixed models and this causes a small problem in the control signal at about 0.7 s. Consequently, although all three methods are very reliable, there is an improvement in the system's dynamics when the SFM method is used due to the fact that the reinitialized adaptive controller exploits a better initialization from the semi-fixed models and thus a faster convergence and better performance is achieved. Also the SFM control scheme reduces the computational burden compared with the AM case.

6 Conclusions

Multiple models control architectures which contain tunable fixed models are very effective and offer an enhanced performance compared with the architectures which do not use that kind of models. The usual methodologies locate the fixed models uniformly inside the uncertainty region. This chapter describes a technique where the fixed models are transformed to semi-fixed models according to a specified tuning algorithm. More specifically, the control scheme consists of a switching fuzzy controller which is based on some semi-fixed models, one adaptive model, and one reinitialized adaptive model. The semi-fixed models are changing their parameter estimations towards a direction where the performance index minimizes its value. In contrast to fixed model methods, the semi-fixed models offer a better initialization to the reinitialized adaptive model and thus improve the performance of the system requiring only a moderate increase in the computational burden. The asymptotic stability of the system and the asymptotic tracking of a stable reference model are ensured by using Lyapunov stability theory and some other tools such as a minimum time period which is left to elapse between switchings and a hysteresis in the switching rule. The effectiveness of the proposed method is demonstrated by some computer simulations which compare the effectiveness of three different methods.

References

1. Middleton, R., Goodwin, G., Hill, D., Mayne, D.: Design issues in adaptive control. IEEE Trans. Autom. Control **33**(1), 50–58 (1988)
2. Narendra, K., Balakrishnan, J.: Improving transient response of adaptive control systems using multiple models and switching. IEEE Trans. Autom. Control **39**(9), 1861–1866 (1994)
3. Narendra, K.S., Balakrishnan, J.: Adaptive control using multiple models. IEEE Trans. Autom. Control **42**, 171–187 (1997)
4. Ioannou, P.A., Kokotovic, P.V.: Instability analysis and improvement of robustness of adaptive control. Automatica **20**(5), 583–594 (1984)
5. Tao, G.: Adaptive Control Design and Analysis. Wiley, New York (2003)
6. Narendra, K., George, K.: Adaptive control of simple nonlinear systems using multiple models. In: Proceedings of the American Control Conference, vol. 3, pp. 1779–1784 (2002)
7. Narendra, K.S., Driollet, O.A., Feiler, M., George, K.: Adaptive control using multiple models, switching and tuning. Int. J. Adapt. Control Signal Process. **17**(2), 87–102 (2003)
8. Ye, X.: Nonlinear adaptive control using multiple identification models. Syst. Control Lett. **57**(7), 578–584 (2008)
9. Kuipers, M., Ioannou, P.: Multiple model adaptive control with mixing. IEEE Trans. Autom. Control **55**(8), 1822–1836 (2010)
10. Narendra, K., Han, Z.: Location of models in multiple-model based adaptive control for improved performance. In: American Control Conference (ACC), 30 July 2010, pp. 117–122
11. Narendra, K.S., Han, Z.: The changing face of adaptive control: the use of multiple models. Ann. Rev. Control **35**(1), 1–12 (2011)
12. Sofianos, N.A., Boutalis, Y.S., Christodoulou, M.A.: Feedback linearization adaptive fuzzy control for nonlinear systems: a multiple models approach. In: 19th Mediterranean Conference on Control and Automation (MED), June 2011, pp. 1453–1459
13. Sofianos, N.A., Boutalis, Y.S., Mertzios, B.G., Kosmidou, O.I.: Adaptive switching control of fuzzy dynamical systems based on hybrid T-S multiple models. In: IEEE 6th International Conference "Intelligent Systems" (IS), Sept 2012, pp. 397–402
14. Sofianos, N.A., Boutalis, Y.S.: Stable indirect adaptive switching control for fuzzy dynamical systems based on T-S multiple models. Int. J. Syst. Sci. **44**(8), 1546–1565 (2013)
15. Huang, L., Wang, S., Wang, K., You B.: Guaranteed cost reliable control for discrete systems based on LMIs. In: 7th International Forum on Strategic Technology (IFOST), 2012, pp. 1–6
16. Su, Z., Zhang, Q., Ai, J.: Practical and finite-time fuzzy adaptive control for nonlinear descriptor systems with uncertainties of unknown bound. Int. J. Syst. Sci. **44**(12), 2223–2233 (2013)

17. Hu, X., Wu, L., Hu, C., Wang, Z., Gao, H.: Dynamic output feedback control of a flexible air-breathing hypersonic vehicle via T-S fuzzy approach. Int. J. Syst. Sci., 1–17 (2013)
18. Takagi, T., Sugeno, M.: Fuzzy identification of systems and its applications to modeling and control. IEEE Trans. Syst. Man Cybern. **15**, 116–132 (1985)
19. Sofianos, N.A., Boutalis, Y.S.: Multiple models fuzzy control: a redistributed fixed models based approach. Int. J. Intell. Autom. Soft Comput. **20**(2), 229–243 (2014)
20. Ioannou, P., Sun, J.: Robust Adaptive Control. Prentice Hall (1996)
21. Liberzon, D.: Switching in Systems and Control. Birkhauser, Boston (2003)
22. Lian, K.Y., Liou, J.J.: Output tracking control for fuzzy systems via output feedback design. IEEE Trans. Fuzzy Syst. **14**(5), 628–639 (2006)

Optimization of Linear Objective Function Under min −Probabilistic Sum Fuzzy Linear Equations Constraint

Ketty Peeva

Abstract We present here linear optimization problem resolution, when the cost function is subject to fuzzy linear systems of equations as constraint.

1 Introduction

We study optimization of the linear objective function

$$Z = \sum_{j=1}^{n} c_j x_j, c_j \in \mathbb{R}, 0 \le x_j \le 1, 1 \le j \le n\,, \tag{1}$$

with traditional addition and multiplication, if $c = (c_1, \ldots, c_n)$ is the cost vector and (1) is subject to fuzzy linear system of equations as constraint

$$A \otimes X = B\,. \tag{2}$$

In (2) $A = (a_{ij})_{m \times n}$ stands for the matrix of coefficients, $X = (x_j)_{n \times 1}$ stands for the matrix of unknowns, $B = (b_i)_{m \times 1}$ is the right-hand side of the system and for each $i, 1 \le i \le m$, and for each $j, 1 \le j \le n$, we have a_{ij}, b_i, $x_j \in [0, 1]$. The composition written as $\otimes$ is min −probabilistic sum. The aim is to minimize or maximize (1) subject to constraint (2). The results for solving this linear optimization problem are provided by the inverse problem resolution for fuzzy linear system of equations (FLSE) with min −probabilistic sum composition as presented in [11] and next developed here for optimization.

In Sect. 2 we introduce basic notions. Sections 3 and 4 present method and algorithm for solving fuzzy linear systems of equations with min −probabilistic sum composition, following [11]. Suitable list manipulations lead to all maximal solutions.

K. Peeva (✉)
Faculty of Applied Mathematics and Informatics,
Technical University of Sofia, 8, Kl. Ohridski St. Sofia, 1000 Sofia, Bulgaria
e-mail: kgp@tu-sofia.bg

V. Sgurev et al. (eds.), *Recent Contributions in Intelligent Systems*,
Studies in Computational Intelligence 657,
DOI 10.1007/978-3-319-41438-6_10

When the system (2) is consistent, its solution set is determined. In Sect. 4 we propose method and algorithm for solving linear optimization problem (1) subject to constraint (2). Section 5 contains conclusions.

Terminology for algebra, orders and lattices are given according to [5, 9], for fuzzy sets and fuzzy relations—according to [2, 3, 7, 12], for computational complexity and algorithms is as in [4].

2 Basic Notions

Partial order relation on a partially ordered set (poset) P is denoted by the symbol $\leq$. By a *greatest element* of a poset P we mean an element $b \in P$ such that $x \leq b$ for all $x \in P$. The *least element* of P is defined dually. The (unique) least and greatest elements of P, when they exist, are called *universal bounds* of P and are denoted by 0 and 1, respectively.

Let $x, y \in [0, 1]$. The following operations will be used

(1) *Minimum*: $\min\{x, y\} = x \wedge y$.
(2) *Maximum*: $\max\{x, y\} = x \vee y$;
(3) *Probabilistic sum*: $x \oplus y = x + y - xy$.
Its residuum $\rightarrow$ is:

$$x \rightarrow y = \begin{cases} 0 \text{ if } x \geq y \\ \frac{y-x}{1-x} \text{ if } x < y \end{cases} . \tag{3}$$

The matrix $A = (a_{ij})_{m\times n}$, with $a_{ij} \in [0, 1]$ for each $i, j, 1 \leq i \leq m, 1 \leq j \leq n$, is called a *membership matrix* [7]. In what follows we write 'matrix' instead of 'membership matrix'.

Two finite matrices are called *conformable*, if the number of the columns in the first matrix equals the number of the rows in the second matrix. The matrices $A = (a_{ij})_{m\times p}$ and $B = (b_{ij})_{p\times n}$ are conformable and their product $C_{m\times n} = AB$, in this order, makes sense.

Definition 1 Let $A = (a_{ij})_{m\times p}$ and $B = (b_{ij})_{p\times n}$ be given matrices. The matrix

(i) $C = A \otimes B$, $C_{m\times n} = (c_{ij})$, is called the min −probabilistic sum product of A and B if

$$c_{ij} = \bigwedge_{k=1}^{p} (a_{ik} \oplus b_{kj}) \text{ when } 1 \leq i \leq m,\ 1 \leq j \leq n.$$

(ii) $C = A \rightarrow B$, $C_{m\times n} = (c_{ij})$, is called the max $-\rightarrow$ product of A and B if

$$c_{ij} = \bigvee_{k=1}^{p} (a_{ik} \rightarrow b_{kj}) \text{ when } 1 \leq i \leq m,\ 1 \leq j \leq n .$$

If there does not exist danger of confusion, we write AB for any of the matrix products in Definition 1.

Let $A = (a_{ij})_{m \times p}$ and $B = (b_{ij})_{p \times n}$ be given matrices. Computing their product AB is called ***direct problem resolution***.

If $A = (a_{ij})_{m \times p}$ and $C = (c_{ij})_{m \times n}$ are given matrices, computing an unknown matrix $B = (b_{ij})_{p \times n}$ such that $AB = C$ is called ***inverse problem resolution***. The algorithms for inverse problem resolution have exponential time complexity (see [1, 10]).

Let $\mathbb{B}$ denote the set of all matrices B, such that $A \otimes B = C$, when the matrices A and C are given; $B \in \mathbb{B}$ means that $A \otimes B = C$ is true.

Theorem 1 ([11]) *Let $A = (a_{ij})_{m \times p}$ and $C = (c_{ij})_{m \times n}$ be given matrices. Then for inverse problem resolution of $A \otimes B = C$ we have*

(i) The set $\mathbb{B}$ is not empty iff $A^t \rightarrow C \in \mathbb{B}$.
(ii) If $\mathbb{B} \neq \emptyset$ then $A^t \rightarrow C$ is the least element in $\mathbb{B}$.□

3 Fuzzy Linear Systems of Equations

We present method and algorithm for inverse problem resolution of fuzzy linear systems of equations when the composition is min−probabilistic sum (notation ⊗−FLSE) according to [11].

The ⊗−FLSE has the following long form description:

$$\left| \begin{array}{ccccc} (a_{11} \oplus x_1) \wedge \cdots \wedge (a_{1n} \oplus x_n) & = & b_1 \\ \cdots \quad \cdots \quad \cdots & & \cdots \cdots \\ (a_{m1} \oplus x_1) \wedge \cdots \wedge (a_{mn} \oplus x_n) & = & b_m \end{array} \right. \tag{4}$$

and its equivalent short form matrix description is

$$A \otimes X = B. \tag{5}$$

Here $a_{ij}, b_i, x_j \in [0, 1]$ for each $i = 1, \ldots, m$ and $j = 1, \ldots, n$, $A = (a_{ij})_{m \times n}$ stands for the matrix of coefficients, $X = (x_j)_{n \times 1}$ – for the matrix of unknowns, $B = (b_i)_{m \times 1}$ is the right-hand side of the system.

For $X = (x_j)_{n \times 1}$ and $Y = (y_j)_{n \times 1}$ the inequality $X \leq Y$ means $x_j \leq y_j$ for each $j = 1, \ldots, n$.

Definition 2 Let the FLSE $A \otimes X = B$ in n unknowns be given.

(i) The vector $X^0 = (x_j^0)_{n \times 1}$ with $x_j^0 \in [0, 1]$, when $1 \leq j \leq n$, is called a ***solution*** of $A \otimes X = B$ if $A \otimes X^0 = B$ holds.
(ii) The set of all solutions of $A \otimes X = B$ is called ***complete solution set*** and it is denoted by $\mathbb{X}$. If $\mathbb{X} \neq \emptyset$ then $A \otimes X = B$ is called ***consistent***, otherwise it is called ***inconsistent***.
(iii) A solution $\check{X} \in \mathbb{X}$ is called a ***lower*** or ***minimal solution*** of $A \otimes X = B$ if for any $X \in \mathbb{X}$ the relation $X \leq \check{X}$ implies $X = \check{X}$, where $\leq$ denotes the partial order,

induced in $\mathbb{X}$ by the order of [0, 1]. Dually, a solution $\hat{X} \in \mathbb{X}$ is called an ***upper*** or ***maximal*** solution of $A \otimes X = B$ if for any $X \in \mathbb{X}$ the relation $\hat{X} \leq X$ implies $X = \hat{X}$. When the upper solution is unique, it is called the ***greatest*** or ***maximum solution***. When the lower solution is unique, it is called the ***least*** or ***minimum solution***.

(iv) The n–tuple of intervals $(X_1, \dots, X_n)$ with $X_j \subseteq [0, 1]$ for each j, $1 \leq j \leq n$, is called an ***interval solution*** of the system $A \otimes X = B$ if any $X^0 = (x_j^0)_{n\times 1}$ belongs to $\mathbb{X}$ when $x_j^0 \in X_j$ for each j, $1 \leq j \leq n$.

(v) Any interval solution of (4), whose components are bounded by the least solution from the left and by an upper solution from the right, is called its ***maximal interval solution***.

3.1 Inverse Problem Resolution for ⊗–FLSE—Basic Results

We propose a way (with list operations) to compute the complete solution set of (5).

We assign to $A \otimes X = B$ a new matrix $A^* = (a_{ij}^*)$, where $a_{ij}^* = a_{ij} \rightarrow b_i$ is determined from A and B:

$$a_{ij}^* = a_{ij} \rightarrow b_i = \begin{cases} 0, & \text{if } a_{ij} \geq b_i \\ \frac{b_i - a_{ij}}{1 - a_{ij}}, & \text{if } a_{ij} < b_i \end{cases}. \tag{6}$$

The matrix $(A^* : B)$ with elements a_{ij}^*, determined by (6), is called ***augmented matrix*** of the system $A \otimes X = B$.

The system $A^* \otimes X = B$ is called ***associated*** to the system $A \otimes X = B$.

The systems $A \otimes X = B$ and its associated system $A^* \otimes X = B$ are equivalent. This permits to investigate the associated system instead of the original one. This reduces the size of the instant and makes easier to solve the original ⊗–FLSE.

We introduce the vector $\check{B} = A^t \rightarrow B = (\check{b}_j)$ with elements $\check{b}_j$, $j = 1, \dots, n$:

$$\check{b}_j = \max_{i=1}^{m} \{a_{ij}^*\}, \quad j = 1, \dots, n, \tag{7}$$

where a_{ij}^* is according to (6).

We denote by $A^*(j)$ the jth column of the matrix A^*.

Theorem 2 *Let the system $A \otimes X = B$ be given.*

(i) If $A^*(j)$ contains coefficient(s) $a_{ij}^* \neq 0$, then

$$\check{x}_j = \check{b}_j = \max_{i=1}^{m} \left\{ \frac{b_i - a_{ij}}{1 - a_{ij}} \right\} \quad \textit{implies:}$$

(a) $a_{ij} \oplus \check{x}_j = b_i$ when $a_{ij}^* = \check{x}_j$;

(b) $a_{ij} \oplus \check{x}_j > b_i$ *when* $a^*_{ij} \neq \check{x}_j$ *(In this case* $a^*_{ij} < \check{x}_j$*);*

(ii) If in $A^*(j)$ *all coefficients* $a^*_{ij} = 0$*, then* $\check{x}_j = \check{b}_j = 0$ *and*

(a) $a_{ij} \oplus \check{x}_j = b_i$ *when* $a_{ij} = b_i$ *and* $a^*_{ij} = 0 = \check{x}_j$*;*
(b) $a_{ij} \oplus \check{x}_j > b_i$ *when* $a_{ij} > b_i$ *and* $a^*_{ij} = 0 = \check{x}_j$*.*□

Remarks:

1. According to Theorem 2 ib: in $A^*(j)$ there may exist(s) $a^*_{ij} \neq 0$, that do(es) not contribute for solving the system.
2. According to Theorem 2 iia: nevertheless in $A^*(j)$ we have $a^*_{ij} = 0$ for $i = 1, \ldots, n$, there may exist zero coefficients that contribute for solving the system.
3. If in $A^*(j)$ all coefficients $a^*_{ij} = 0$ and corresponding $a_{ij} > b_i$ for $i = 1, \ldots, n$, then $a_{ij} \oplus \check{x}_j > b_i$ for each $i = 1, \ldots, n$. Hence there does not exist equation that could be satisfied by $a_{ij} \oplus \check{x}_j$ for each $i = 1, \ldots, n$.

3.2 Least Solution of ⊗−FLSE

We denote by $\check{X} = (\check{x}_j)$ the vector with components $\check{x}_j$, computed according to Theorem 2, i.e.,:

$$\check{x}_j = \check{b}_j = \begin{cases} \max\limits_{i=1}^{m} \left\{ \frac{b_i - a_{ij}}{1 - a_{ij}} \right\} & \text{if there exists } a^*_{ij} \neq 0 \\ 0, \text{ othervise} \end{cases}. \tag{8}$$

Corollary 1 *If the FLSE* $A \otimes X = B$ *is consistent, then the vector* $\check{X} = (\check{x}_j)$ *with components determined by (8) is its least solution.*□

Theorem 3 *If the system* $A \otimes X = B$ *is consistent, then the following expressions for its least solution* $\check{X}$ *are equivalent:*

(i) $\check{X} = A^t \rightarrow B$*;*
(ii) $\check{X} = (A^*)^t \rightarrow B$*.*□

In what follows the vector *IND* is used to establish consistency of the system by the eventual $\check{X}$: if all components in *IND* are *TRUE*, the system is consistent and $\check{X}$ contains its least solution, otherwise the system is inconsistent; $i = \overline{1, m}$ means $1 \leq i \leq m$ and $j = \overline{1, n}$ means $1 \leq j \leq n$.

Algorithm 1—least solution and consistency

Step 1. Initialize the vector $\check{X} = (\check{x}_j)$ with $\check{x}_j = 0$ for $j = \overline{1, n}$.
Step 2. Initialize a boolean vector *IND* with $IND_i = FALSE$ for $i = \overline{1, m}$.
Step 3. For each $j = \overline{1, n}$ and for each $i = \overline{1, m}$ determine $\check{x}_j$ according to (8), upgrade $\check{X}$ and correct IND_i to *TRUE* if $a^*_{ij} = \check{x}_j$ when $a_{ij} \leq b_i$.

Step 4. Check if all components of *IND* are set to *TRUE*.

a. If $IND_i = FALSE$ for some i, the system $A \otimes X = B$ is inconsistent. Go to Step 5.
b. If $IND_i = TRUE$ for all $i = \overline{1, m}$, the system $A \otimes X = B$ is consistent and its least solution is $\check{X}$.

Step 5. Exit.

Example 1 Find the least solution of the system by Algorithm 1

$$\begin{pmatrix} 1 & 0.95 & 0.6 & 1 \\ 1 & 0.7 & 0.8 & 0.9 \\ 0.2 & 0.76 & 1 & 1 \\ 0.7 & 1 & 0.2 & 0.85 \\ 1 & 0.88 & 0.58 & 0.9 \end{pmatrix} \otimes \begin{pmatrix} x_1 \\ x_2 \\ x_3 \\ x_4 \end{pmatrix} = \begin{pmatrix} 0.9 \\ 0.7 \\ 0.76 \\ 0.8 \\ 0.88 \end{pmatrix}. \tag{9}$$

The augmented matrix for (9) is

$$A^* = \begin{pmatrix} 0 & 0 & \frac{09.-0.6}{0.4} & 0 \\ 0 & 0 & 0 & 0 \\ \frac{0.76-0.2}{0.8} & 0 & 0 & 0 \\ \frac{0.8-0.7}{0.3} & 0 & \frac{0.8-0.2}{0.8} & 0 \\ 0 & 0 & \frac{0.88-0.58}{0.42} & 0 \end{pmatrix}, \tag{10}$$

i.e.,

$$A^* = \begin{pmatrix} 0 & 0 & 0.75 & 0 \\ 0 & 0 & 0 & 0 \\ 0.7 & 0 & 0 & 0 \\ \frac{1}{3} & 0 & 0.75 & 0 \\ 0 & 0 & \frac{5}{7} & 0 \end{pmatrix} \tag{11}$$

We denote by $A^*(j)$ the jth column of A^*.

1. Initialize $\check{X} = (0\ \ 0\ \ 0\ \ 0)'$.
2. Initialize $IND = \left(IND_i\right) = FALSE$ for each $i = \overline{1, 5}$.
3. For each $i = \overline{1, 5}$:

(A1) In $A^*(1)$, $\check{x}_1 = \max\{0.7, \frac{1}{3}\} = 0.7$. Upgrade $\check{x}_1 = 0.7$ and $\check{X} = (0.7\ \ 0\ \ 0\ \ 0)'$; Since $a^*_{31} = 0.7$, $IND_3 = TRUE$.

(A2) In $A^*(2)$, $\check{x}_2 = 0$. No upgrade for $\check{X}$. Since $a_{22} = b_2, a_{32} = b_3, a_{52} = b_5, IND_2 = TRUE, IND_5 = TRUE, IND_3$ is already *TRUE*.

(A3) In $A^*(3)$, $\check{x}_3 = \max\{0.75, \frac{1}{14}\} = 0.75$. Upgrade $\check{x}_3 = 0.75$ and $\check{X} = (0.7\ \ 0\ \ 0.75\ \ 0)'$. Since $a^*_{13} = a^*_{43} = 0.75$, $IND_4 = TRUE$ and $IND_1 = TRUE$.

All elements in *IND* are set to *TRUE* on this step. This means that the system is consistent.

(A4) In $A^*(4)$, $\check{x}_4 = 0$. No upgrade for $\check{X}$. Since $a_{ij} > b_i$ for each $i = \overline{1,5}$, no upgrade for *IND*.

4. All components of *IND* are set to *TRUE*. The system is consistent and its least solution is: $\check{X} = (0.7 \;\; 0 \;\; 0.75 \;\; 0)'$.

Corollary 2 *Let the FLSE $A \otimes X = B$ be consistent. Then:*

(i) If $\hat{X} = (\hat{x}_j)$ is its upper solution, then either $\hat{x}_j = 1$ or $\hat{x}_j = \check{x}_j$;

(ii) If $a_{ij} > b_i$ for each $i = 1, \ldots, m$, then $\check{x}_j = 0$ in the least solution and $\hat{x}_j = 1$ in any upper solution. □

3.3 Selected Coefficients in ⊗−FLSE

There exist coefficients that contribute for solving the system and such that do not contribute to solve it—see Theorem 2 and the remarks after it. This is the reason to propose a selection of all coefficients that do contribute for solving the system.

For consistent ⊗−FLSE (4) with the least solution $\check{X} = (\check{x}_j)$:

1. We find the value of the unknown $\check{x}_j$ in $\check{X} = (\check{x}_j)$ only from the jth column $A^*(j)$ of the matrix A^*.
2. For each i, $1 \leq i \leq m$, only coefficients a_{ij} with the property $a_{ij} \oplus \check{x}_j = b_i$ lead to a solution.

If the FLSE (4) is consistent, any coefficient a_{ij} with property $a_{ij} \oplus \check{x}_j = b_i$ is called ***selected coefficient***.

Corollary 3 *The FLSE $A \otimes X = B$ is consistent iff for each i, $1 \leq i \leq m$, there exists at least one selected coefficient, otherwise it is inconsistent.*

The time complexity function for establishing the consistency of FLSE $A \otimes X = B$ and for computing $\check{X}$ is $O(mn)$

If $|H_i|$ denotes the number of selected coefficients in the i−th equation of (4) then the number of its potential upper solutions does not exceed the estimation

$$PN1 = \prod_{i=1}^{m} |H_i|. \tag{12}$$

4 Algorithms for Solving $A \otimes X = B$

The complete solution set of $A \otimes X = B$ is the set of all its maximal interval solutions. Since there exists analytical expression for the least solution, attention is paid on computing the upper solutions [11].

4.1 Upper Solutions—List Manipulations

The hearth of the next approach is to find all selected coefficients a_{ij} in A—they contribute to satisfy the ith equation by the term $a_{ij} \oplus \check{x}_j = b_i$. Next, according to Corollaries 1, 2 we give to $\hat{x}_j$ either the value $\check{x}_j$ when the coefficient contributes to solve the system or 1 when it does not.

Obviously, there exist solutions that are neither lower, nor greatest, see [11]. In order to extract only upper solutions and to skip not extremal solutions, a method, based on *list manipulation* techniques, is presented in [11].

4.2 Removing Redundant Equations

We suppose that the system $A \otimes X = B$ is consistent and $\check{X}$ is its least solution. First, using $\check{X}$ we select and mark all coefficients that contribute to solve the system. Then we associate with each equation i, $1 \leq i \leq m$, a set E_i—it contains list of all indices j of the selected coefficients a_{ij} that contribute to satisfy the ith equation with $a_{ij} \oplus \check{x}_j = b_i$.

Definition 3 The set E_i, which elements are the indices $j \in \{1, \ldots, n\}$ in the ith equation of (4), such that $a_{ij} \oplus \check{x}_j = b_i$, is called a ***marking set*** for the i–th equation.

$\mathbb{E}$ denotes the set of all marking sets for the system (4).

Rather than work with equations, we use the marking sets—they capture all the properties of the equations with respect to solutions. In this manner we reduce the complexity of exhaustive search by making a more clever choice of the objects, over which the search is performed.

Definition 4 Let $E_l, E_k \in \mathbb{E}$ be two marking sets for (4). If $E_l \subseteq E_k$, then E_l is called a **dominant set** to E_k and E_k is called a **dominated set** by E_l.

The set of all non-dominated marking sets for (4) is denoted by $\hat{\mathbb{E}}$. Each E_i from $\hat{\mathbb{E}}$ corresponds to the ith row in A^* or ith equation in (4).

Each dominated set corresponds to redundant equation in (4). We implement Definition 4 to remove redundant equations from (4).

The number of upper solutions now does not exceed the estimation

$$PN2 = \prod_{i=1}^{m} | E_i |, \text{ where } E_i \in \hat{\mathbb{E}}. \tag{13}$$

If we compare (12) and (13), obviously $PN1 \geq PN2$.

Example 2 For the system

$$\begin{pmatrix} 1 & 0.8 & 0.95 & 1 & 0.6 & 1 \\ 0.9 & 0.4 & 0.7 & 1 & 0.8 & 0.8 \\ 0.2 & 0.52 & 0.76 & 0.52 & 1 & 1 \\ 0.7 & 1 & 1 & 0.6 & 0.2 & 0.85 \\ 1 & 1 & 0.88 & 0.8 & 0.52 & 0.9 \\ 0.5 & 0.7 & 1 & 0.9 & 0.4 & 1 \end{pmatrix} \otimes \begin{pmatrix} x_1 \\ x_2 \\ x_3 \\ x_4 \\ x_5 \\ x_6 \end{pmatrix}$$

$$= \begin{pmatrix} 0.9 \\ 0.7 \\ 0.76 \\ 0.8 \\ 0.88 \\ 0.85 \end{pmatrix}. \tag{14}$$

obtain the sets $\mathbb{E}$ and $\hat{\mathbb{E}}$ and all its upper solutions.

The system is consistent and

$$\check{X} = (0.7 \;\; 0.5 \;\; 0 \;\; 0.5 \;\; 0.75 \;\; 0)'.$$

The marking sets are

$$E_1 = \{2, 5\}, E_2 = \{2, 3\}, E_3 = \{1, 2, 3, 4\},$$

$$E_4 = \{4, 5\}, E_5 = \{3, 5\}, E_6 = \{1, 2, 5\},$$

and hence

$$\mathbb{E} = \{\{2, 5\}, \{2, 3\}, \{1, 2, 3, 4\}, \{4, 5\}, \{3, 5\}, \{1, 2, 5\}\}, \tag{15}$$

leading to $PN1 = 2.2.4.2.2.3 = 172$ potential solutions.

In (15): $E_1 \subset E_6$, $E_2 \subset E_3$ and thus

$$\hat{\mathbb{E}} = \{\{2, 5\}, \{2, 3\}, \{4, 5\}, \{3, 5\}\}, \tag{16}$$

We expand $\hat{\mathbb{E}}$ step by step:

1. Expand the first set in the list, namely $E_1 = \{2, 5\}$.
   1.1 Begin with the first element $\{2\}$: since $\{2\} \subset \{2, 3\}$, we remove $\{2, 3\}$ from the list (16) and investigate the path determined by the sequence $\{2\}$, $\{4, 5\}$, $\{3, 5\}$.
   1.1.1 Expand the path determined by the first element 4 in $\{4, 5\}$. Since 4 does not belong to the set $\{3, 5\}$, there exist two paths, namely

{2}, {3}, {4} and {2}, {4}, {5}.

1.1.2 Backtracking to the second element 5 in {4, 5} gives that {3, 5} should be removed because {5} ⊂ {3, 5} and this creates the only path {2}, {5}. Note that this path covers a path from previous step 1.1.1, namely ({2}, {4}, {5})

Hence the irreducible paths from this branch are:

a. {2}, {3}, {4};
b. {2}, {5}.

1.2 Backtracking to the second 5 element in {2, 5} gives that {5} ⊂ {4, 5}, {5} ⊂ {3, 5} and we remove the sets {4, 5} and {3, 5} from the list (16). Then {{5}, {2, 3}} gives two paths

a. {2}, {5} (already known).
b. {3}, {5}.

Finally, we obtain three different paths leading to three upper solutions:

a. {2}, {3}, {4} leading to $\hat{X}_1 = (1\ \ 0.5\ \ 0\ \ 0.5\ \ 1\ \ 1)'$;
b. {2}, {5} leading to $\hat{X}_2 = (1\ \ 0.5\ \ 1\ \ 1\ \ 0.75\ \ 1)'$.;
c. {3}, {5} leading to $\hat{X}_3 = (1\ \ 1\ \ 0\ \ 1\ \ 0.75\ \ 1)'$.

$$sols = \begin{pmatrix} 1 & 0.5 & 0 & 0.5 & 1 & 1 \\ 1 & 0.5 & 1 & 1 & 0.75 & 1 \\ 1 & 1 & 0 & 1 & 0.75 & 1 \end{pmatrix} \tag{17}$$

According to (13) and (16) $PN2 = 2.2.2.2 = 16$.

In this manner with this very simple operation we reduce the size of the instant: according to (12) we should investigate 172 paths for upper solutions, while (13) requires to investigate only 16 paths.

"Backtracking" based algorithm using that principle is presented next. Backtracking reduces the size of the instant.

Let $\hat{X} = (\hat{x}_j)$ denote an upper solution of (4).

Example 3 Find all upper solutions for the FLSE (14).

Algorithm 2—extracting the upper solutions from $\hat{\mathbb{E}}$

Step 1. Form the sets $\mathbb{E}$ and $\hat{\mathbb{E}}$.
Step 2. Initialize solution vector $\hat{X}_0(j) = 1,\ j = \overline{1, n}$.
Step 3. Initialize a vector *rows*(i), $i = \overline{1, m}$ which holds all consecutive row numbers—the indices i for each $E_i \in \hat{\mathbb{E}}$.
Step 4. Initialize i with the first element in *rows*.
Step 5. Initialize *sols* to be the empty set of vectors, which is supposed to be the set of all minimal solutions for the current problem.
Step 6. Check if *rows* = ∅. If so, add $\hat{X}_{ij}$ to *sols* and go to step 8.

Step 7. For each i in *rows* expand the set E_i: for each $j \in E_i$ create a successor $\hat{X}_{ij}$ of $\hat{X}_0$ and update its jth element to be equal to $\check{x}_j$. Remove i from *rows*. Go to step 6 with copied in this step *rows* and $\hat{X}_{ij}$.
Step 8. Exit.

Corollary 4 *For any consistent* $\otimes$*−FLSE the upper solutions are computable and the set of all its upper solutions is finite.* □

Algorithm 3—solving $A \otimes X = B$

Step 1. Input the matrices A and B.
Step 2. Obtain least solution for the system and check it for consistency (Algorithm 1).
Step 3. If the system is inconsistent go to step 5.
Step 4. Obtain all upper solutions (Algorithm 2).
Step 5. Exit.

5 Linear Optimization Problem—The Algorithm

Our aim is to solve the optimization problem, when the linear objective function (1) is subject to the constraints (2). We first decompose the linear objective function Z in two functions Z' and Z'' by separating the non-negative and negative coefficients (as it is proposed in [8] for instance). Using the extremal solutions for constraint and the above two functions, we solve the optimization problem, as described below.

The linear objective function

$$Z = \sum_{j=1}^{n} c_j x_j, \quad c_j \in \mathbb{R}, \; 0 \leq x_j \leq 1, \; 1 \leq j \leq n, \tag{18}$$

determines a cost vector $Z = (c_1, c_2, \ldots, c_n)$. We decompose Z into two vectors with suitable components

$$Z' = (c'_1, c'_2, \ldots, c'_n) \text{ and } Z'' = (c''_1, c''_2, \ldots, c''_n),$$

such that the objective value is

$$Z = Z' + Z''$$

and cost vector components are

$$c_j = c'_j + c''_j, \quad \text{for each } j = 1, \ldots, n,$$

where

$$c'_j = \begin{cases} c_j, & \text{if } c_j > 0, \\ 0, & \text{if } c_j \le 0 \end{cases}, \tag{19}$$

$$c''_j = \begin{cases} 0, & \text{if } c_j \ge 0, \\ c_j, & \text{if } c_j < 0 \end{cases}. \tag{20}$$

Hence the components of Z' are nonnegative, the components of Z'' are non-positive.

We study how to maximize (or minimize, respectively) the linear objective function (18), subject to the constraint (2).

5.1 Maximize the Linear Objective Function, Subject to Constraint (2)

The original problem: to maximize Z subject to constraint (2) splits into two problems, namely to maximize both

$$Z' = \sum_{j=1}^{n} c'_j x_j \tag{21}$$

and

$$Z'' = \sum_{j=1}^{n} c''_j x_j \tag{22}$$

with constraint (2), i.e., for the problem (18) Z takes its maximum when both Z' and Z'' take maximum.

Since the components c'_j, $1 \le j \le n$, in Z' are nonnegative (see 19), Z' takes its maximum among the maximal solutions of (2). Hence for the problem (21) the maximal solutionis among the maximal solutions $\hat{X} = (\hat{x}_1, \ldots, \hat{x}_n)$ of the system (2).

Since the components c''_j, $1 \le j \le n$, in Z'' are non-positive (see 20), Z'' takes its maximum for the least solution $\check{X} = (\check{x}_1, \ldots, \check{x}_n)$ of (2).

The maximal solution of the problem (18) with constraint (2) is

$$X^* = (x^*_1, \ldots, x^*_n),$$

where

$$x^*_i = \begin{cases} \check{x}_i, & \text{if } c_i < 0 \\ \hat{x}_i, & \text{if } c_i > 0 \\ 0, & \text{if } c_i = 0 \end{cases} \tag{23}$$

and the optimal value is

$$Z^* = \sum_{j=1}^{n} c_j x_j^* = \sum_{j=1}^{n} c_j' \hat{x}_j + c_j'' \check{x}_j. \tag{24}$$

Example 4 Maximize the following linear objective function

$$Z = -3x_1 + 0.5x_2 - 2x_4 + x_5 + 2x_6$$

under the constrains (14).

The cost vector is

$$c = \left(-3\ 0.5\ 0\ -2\ 1\ 2 \right)$$

and hence according to (19) and (20) we obtain:

$$c' = \left(0\ 0.5\ 0\ 0\ 1\ 2 \right),$$

$$c'' = \left(-3\ 0\ 0\ -2\ 0\ 0 \right).$$

As obtained in Example 2, the FLSE has the following extremal solutions:

$$\check{X} = (0.7\ \ 0.5\ \ 0\ \ 0.5\ \ 0.75\ \ 0)'$$

and

$$\hat{X}_1 = (1\ \ 0.5\ \ 0\ \ 0.5\ \ 1\ \ 1)';$$

$$\hat{X}_2 = (1\ \ 0.5\ \ 1\ \ 1\ \ 0.75\ \ 1)';$$

$$\hat{X}_3 = (1\ \ 1\ \ 0\ \ 1\ \ 0.75\ \ 1)'.$$

First, using $\hat{X}$ and c' we obtain for $Z'_{\max}$

$$Z'_{\max} = 0, 5\hat{x}_2 + 1\hat{x}_5 + 2.\hat{x}_6,$$

with

$$Z'_{\max}(1) = 0, 25 + 1 + 2 = 3, 25,$$

$$Z'_{\max}(2) = 0, 25 + 0, 75 + 2 = 3,$$

$$Z'_{\max}(3) = 0, 5 + 0, 75 + 2.1 = 3, 25.$$

Then, for $\check{X} = (0.7\ \ 0,5\ \ 0\ \ 0.5\ \ 0.75\ \ 0)'$ and c'' we obtain

$$Z''_{\max} = -3\check{x}_1 - 2\check{x}_4 = -3.0,7 - 2.0,5 = -3,1$$

Next, implementing (24) we obtain:

$$Z^* = \sum_{j=1}^{n} c_j x_j^* = \sum_{j=1}^{n} c'_j \check{x}_j + c''_j \hat{x}_j, \tag{25}$$

leading to three candidates:

$$Z^*(1) = -3,1 + 3,25 = 0,15$$

$$Z^*(2) = -3,1 + 3 = -0,1$$

$$Z^*(3) = -3,1 + 3,25 = 0,15.$$

Both of the maximal solutions of the constraint are equal and together with the least solution they lead to the same optimal solution of the problem $Z^*(1) = Z^*(3) = 0,15$ with

$$X_1^* = (\,0.7\ 0.5\ 0\ 0.5\ 1\ 1\,)$$

$$X_2^* = (\,0.7\ \ 1\ 0\ 0.5\ 0.75\ 1\,).$$

The optimal value of this optimization problem is 0, 15.

5.2 *Minimize the Linear Objective Function, Subject to Constraint (2)*

If the aim is to minimize the linear objective function (18), we again split it, but now for Z'' the optimal solution is among the maximal solutions of the system (2), for Z' the optimal solution is $\check{X}$. In this case the optimal solution of the problem is

$$X^* = (x_1^*, \ldots, x_n^*),$$

where

$$x_j^* = \begin{cases} \hat{x}_j, & \text{if } c_j < 0 \\ \check{x}_j, & \text{if } c_j > 0 \\ 0, & \text{if } c_j = 0 \end{cases}. \tag{26}$$

and the optimal value is

$$Z^* = \sum_{j=1}^{n} c_j x_j^* = \sum_{j=1}^{n} c_j' \check{x}_j + c_j'' \hat{x}_j. \tag{27}$$

5.3 Algorithm for Finding Optimal Solutions

1. Enter the matrices $A_{m\times n}$, $B_{m\times 1}$ and the cost vector $C_{1\times n}$.
2. Establish consistency of the system (2). If the system is inconsistent go to step 8.
3. Compute $\check{X}$ and all maximal solutions of (2).
4. If finding $Z_{\min}$ go to Step 6.
5. For finding $Z_{\max}$ compute $x_j^*, j = 1, \ldots, n$ according to (23). Go to Step 7.
6. For finding $Z_{\min}$ compute $x_j^*, j = 1, \ldots, n$ according to (26).
7. Compute the optimal value according to (24) (for maximizing) or (27) (for minimizing).
8. End.

References

1. Chen, L., Wang, P.: Fuzzy relational equations (I): the general and specialized solving algorithms. Soft Comput. **6**, 428–435 (2002)
2. De Baets, B.: Analytical solution methods for fuzzy relational equations. In: Dubois, D., Prade, H. (eds.) Handbooks of Fuzzy Sets Series: Fundamentals of Fuzzy Sets, vol. 1, pp. 291–340. Kluwer Academic Publishers (2000)
3. Di Nola, A., Pedrycz, W., Sessa, S., Sanchez, E.: Fuzzy Relation Equations and Their Application to Knowledge Engineering. Kluwer Academic Press, Dordrecht (1989)
4. Garey, M.R., Johnson, D.S.: Computers and Intractability. A Guide to the Theory of NP-Completeness, Freeman, San Francisco, CA (1979)
5. Grätzer, G.: General Lattice Theory. Akademie-Verlag, Berlin (1978)
6. Guu, S.M., Wu, Y.-K.: Minimizing a linear objective function with fuzzy relation equation constraints. Fuzzy Optim. Decis. Making **4**(1), 347–360 (2002)
7. Klir, G.J., Clair, U.H.S., Yuan, B.: Fuzzy Set Theory Foundations and Applications. Prentice Hall PRT (1977)
8. Loetamonphong, J., Fang, S.-C.: An efficient solution procedure for fuzzy relational equations with max-product composition. IEEE Trans. Fuzzy Syst. **7**(4), 441–445 (1999)
9. MacLane, S., Birkhoff, G.: Algebra. Macmillan, New York (1979)
10. Peeva, K.: Resolution of Fuzzy relational equations—method, algorithm and software with applications, information sciences. Special Issue (2011). doi:10.1016/j.ins.2011.04.011
11. Peeva, K.: Inverse Problem Resolution for min-probabilistic sum Fuzzy relational equations—method and algorithm. In: 2012 VIth International IEEE Conference "Intelligent Systems", vol. 1, pp. 489– 494. Sofia 6–8 Sept. (2012). ISBN 978-1-4673-2277-5
12. Peeva, K., Kyosev, Y.: Fuzzy relational calculus-theory, applications and software (with CD-ROM). In the Series Advances in Fuzzy Systems—Applications and Theory, vol. 22. World Scientific Publishing Company (2004)

Intuitionistic Fuzzy Logic Implementation to Assess Purposeful Model Parameters Genesis

Tania Pencheva and Maria Angelova

Abstract In this investigation, intuitionistic fuzzy logic is implemented to derive intuitionistic fuzzy estimations of model parameters of yeast fed-batch cultivation. Two kinds of simple genetic algorithms with operators sequences selection-crossover-mutation and mutation-crossover-selection are here considered, both applied for the purposes of parameter identification of *S. cerevisiae* fed-batch cultivation. Intuitionistic fuzzy logic overbuilds the results achieved by the application of recently developed purposeful model parameters genesis procedure in order to keep promising results obtained. Behavior of applied algorithms has also been examined at different values of the genetic algorithms parameter generation gap, proven as the most sensitive parameter toward convergence time. Results obtained after the implementation of intuitionistic fuzzy logic for the assessment of algorithms performances have been compared and based on the evaluations in each case the most reliable algorithm has been distinguished.

Keywords Intuitionistic fuzzy logic · Genetic algorithm · Parameter identification · Fermentation process · *Saccharomyces cerevisiae*

1 Introduction

Genetic algorithms [11], inspired by Darwin's theory of "survival of the fittest", are a stochastic global optimization technique with applications in different areas [11, 15]. Some properties such as hard problems solving, noise tolerance, easiness to interface and hybridize, make GA a suitable and quite workable technique especially for incompletely determined tasks, and in particular—for parameter identi-

T. Pencheva (✉) · M. Angelova
Institute of Biophysics and Biomedical Engineering, Bulgarian Academy of Sciences, Sofia, Bulgaria
e-mail: tania.pencheva@biomed.bas.bg

M. Angelova
e-mail: maria.angelova@biomed.bas.bg

V. Sgurev et al. (eds.), *Recent Contributions in Intelligent Systems*,
Studies in Computational Intelligence 657,
DOI 10.1007/978-3-319-41438-6_11

fication of fermentation process models [1–7, 12–15]. Fermentation processes (FP) are well known as complex, dynamic systems with interdependent and time-varying process variables. These peculiarities make their modelling specific and difficult to be solved task. Inability of conventional optimization methods such as Nelder-Mead's minimization, sequential quadratic programming, quasi-Newton algorithms (i.e., Broyden, Fletcher, Goldfarb, and Shanno), etc., to reach a satisfactory solution for model parameter identification of FP [14] provokes the idea genetic algorithms (GA) to be tested. Promising results obtained through GA application, evaluated by the model accuracy achieved and the convergence time needed, encourage their future investigation.

For the purposes of current investigation, standard simple genetic algorithm (SGA), originally presented in [11] with the sequential execution of the main genetic operators selection-crossover-mutation, is here denoted with the abbreviation SGA-SCM. Starting algorithm, chromosomes (a coded parameter set) representing better possible solutions according to their own objective function values are chosen from the population by means of selection. After that, crossover takes place in order to form new offspring. Mutation is then applied with a determinate probability, aiming to prevent all solutions in the population from falling into a local optimum of the problem. Many modifications of SGA-SCM have been elaborated with the purpose of improvement of model accuracy and algorithm convergence time [5]. These modifications differ from one to another in the sequence of execution of GA operators—selection, crossover, and mutation. As such, SGA-MCS (coming from mutation-crossover-selection) is proposed and thoroughly investigated in [5]. In it, selection operator has been applied after crossover and mutation, in order to avoid the loss of any eventually reached "good" solution as a result of either crossover or mutation, or both. In this case, SGA-MCS calculates the objective function for the offspring after the reproduction and the best fitted individuals are selected to replace the parents.

According to [11], SGA-MCS working principle is shown as follows:

1. **[Start]**
   Generate random population of n chromosomes
2. **[Object function]**
   Evaluate the object function of each chromosome x in the populations
3. **[Fitness function]**
   Evaluate the fitness function of each chromosome n in the populations
4. **[New population]**
   Create a new population by repeating following steps:

   4.1. **[Mutation]**
   Mutate new offspring at each locus with a mutation probability

   4.2. **[Crossover]**
   Cross over the parents to form new offspring with a crossover probability

4.3. **[Selection]**
Select parent chromosomes from the population according to their fitness function

5. [**Accepting**]
Place new offspring in a new population
6. **[Replace]**
Use new generated population for a further run of the algorithm
7 **[Test]**
If the end condition is satisfied, stop and return the best solution in current population, else move to **Loop** step
8 **[Loop]**
Go to **Fitness step**.

Aiming to obtain reliable results in parameter identification of a fermentation process model when applying GA, a great number of algorithm runs have to be executed due to their stochastic nature. First, the genetic algorithm searches for model parameters estimations in wide but reasonably chosen boundaries. When results from many algorithms executions were accumulated and analyzed, they showed that the values of model parameters can be assembled and predefined boundaries could be straitened. Thus the development of a procedure for purposeful model parameters genesis was forced [7]. Its implementation results in the determination of more appropriate boundaries for variation of model parameters values, aiming to decrease convergence time while at least saving model accuracy.

The purpose of this study is intuitionistic fuzzy logic to be implemented to assess the performance quality of two modifications of simple genetic algorithms. Aiming to save decreased convergence time while keeping or even improving model accuracy, intuitionistic fuzzy estimations overbuild the results achieved by purposeful model parameters genesis application. Intuitionistic fuzzy logic is going to be applied for the purposes of parameter identification of *S. cerevisiae* fed-batch cultivation in the order

- to assess the performance of SGA at different values of generation gap (GGAP), proven as the most sensitive genetic algorithm parameter;
- to assess the performance of two modifications of SGA, namely SGA-SCM and SGA-MCS.

2 Background

A. Model of S. cerevisiae fed-batch cultivation

Experimental data of *S. cerevisiae* fed-batch cultivation is obtained in the *Institute of Technical Chemistry—University of Hannover, Germany* [14]. The cultivation of the yeast *S. cerevisiae* is performed in a 1.5 l reactor using a Schatzmann medium. Glucose in feeding solution is 50 g/l. The temperature was controlled at 30 °C, the pH at 5.7. The stirrer speed was set to 500 rpm. Biomass and ethanol were measured off-line, while substrate (glucose) and dissolved oxygen were measured online.

Mathematical model of *S. cerevisiae* fed-batch cultivation is commonly described as follows, according to the mass balance: [14]:

$$\frac{dX}{dt} = \mu X - \frac{F}{V}X \tag{1}$$

$$\frac{dS}{dt} = -q_S X + \frac{F}{V}(S_{in} - S) \tag{2}$$

$$\frac{dE}{dt} = q_E X - \frac{F}{V}E \tag{3}$$

$$\frac{dO_2}{dt} = -q_{O_2} X + k_L^{O_2} a(O_2^* - O_2) \tag{4}$$

$$\frac{dV}{dt} = F \tag{5}$$

where X is the concentration of biomass, (g/l);

S	concentration of substrate (glucose), (g/l);
E	concentration of ethanol, (g/l);
O_2	concentration of oxygen, (%);
O_2^*	dissolved oxygen saturation concentration, (%);
F	feeding rate, (l/h);
V	volume of bioreactor, (l);
$k_L^{O_2} a$	volumetric oxygen transfer coefficient, (1/h);
S_{in}	initial glucose concentration in the feeding solution, (g/l);
μ, q_S, q_E, q_{O_2}	specific growth/utilization rates of biomass, substrate, ethanol, and dissolved oxygen, (1/h).

All functions are continuous and differentiable.

The fed-batch cultivation of *S. cerevisiae* considered here is characterized by keeping glucose concentration equal to or below its critical level ($S_{crit} = 0.05$ g/l), sufficient dissolved oxygen $O_2 \geq O_{2crit}$ ($O_{2crit} = 18$ %), and availability of ethanol in the broth. This state corresponds to the so-called mixed oxidative state (FS II) according to functional state modeling approach [14]. Hence, specific rates in the model (1)–(5) are as follows:

$$\mu = \mu_{2S} \frac{S}{S + k_S} + \mu_{2E} \frac{E}{E + k_E}, \tag{6}$$

$$q_S = \frac{\mu_{2S}}{Y_{SX}} \frac{S}{S + k_S}, \tag{7}$$

$$q_E = -\frac{\mu_{2E}}{Y_{EX}} \frac{E}{E + k_E}, \tag{8}$$

$$q_{O_2} = q_E Y_{OE} + q_S Y_{OS} \tag{9}$$

where μ_{2S}, μ_{2E} are the maximum growth rates of substrate and ethanol, (1/h);
k_S, k_E saturation constants of substrate and ethanol, (g/l);
Y_{ij} yield coefficients, (g/g).
All model parameters fulfill the nonzero division requirement.

As an optimization criterion, mean square deviation between the model output and the experimental data obtained during cultivation has been used

$$J_Y = \sum (Y - Y^*)^2 \rightarrow \min, \tag{10}$$

where Y is the experimental data;

Y^*—model predicted data;
$Y = [X, S, E, O_2]$.

B. Procedure for purposeful model parameters genesis

The procedure for purposeful model parameter genesis (PMPG), originally developed for SGA [7], consists of six steps shortly outlined below. First, a number of genetic algorithm runs have to be performed in as here denoted "broad" range and the minimum and maximum values of the objective function to be determined. Then, following the scheme presented in [7], top level (TL), middle level (ML), and low level (LL) of performance with corresponding low boundary (LB) and up boundary (UB) are constructed. At the next step, minimum, maximum, and average value for each parameter at each level is set. The last step is a determination of new intervals of model parameters variations—as here denoted "narrow" range, which is user-defined decision. Then, the genetic algorithm is again performed with boundaries determined in the last step.

This stepwise procedure passes through all the six steps described above, not omitting any of them and without cycles. Originally developed for standard SGA (with a sequence selection-crossover-mutation), the procedure has been implemented successfully further to some of the developed modifications, among them for SGA-MCS [3], as well as to multi-population genetic algorithm.

Following model (1)–(9) of *S. cerevisiae* fed-batch cultivation, nine model parameters have to be estimated altogether. The procedure for purposeful model parameter genesis has been applied to SGA_SCM and SGA_MCS for the purposes of parameter identification of model parameters. Parameter identification has been performed using *Genetic Algorithm Toolbox* [10] in *Matlab 7* environment. Based on the previous authors' investigation [2], the values of genetic algorithms parameters and the type of genetic operators have been accepted as presented below

Genetic algorithm parameters

• number of variables (NVAR)	9
• precision of binary representation (PRECI)	20
• number of individuals (NIND)	20
• maximum number of generations (MAXGEN)	100
• generation gap (GGAP)	0.1–0.9
• crossover rate (XOVR)	0.85
• mutation rates (MUTR)	0.1

Type of genetic operators:

• Encoding	binary
• Reinsertion	fitness-based
• Crossover	double point
• Mutation	bit inversion
• Selection	roulette wheel selection
• Fitness function	linear ranking

GA is terminated when a certain number of generations is performed, in this case 100. Scalar relative error tolerance *RelTol* is set to $1e^{-4}$, while the vector of absolute error tolerances (all components) *AbsTol*—to $1e^{-5}$. All the computations are performed using a PC Intel Pentium 4 (2.4 GHz) platform running Windows XP.

C. Intuitionistic fuzzy estimations

In intuitionistic fuzzy logic (IFL) [8, 9] if p is a variable, then its truth value is represented by the ordered couple

$$V(p) = \langle M(p), N(p) \rangle, \tag{11}$$

so that $M(p)$, $N(p)$, $M(p) + N(p) \in [0, 1]$, where $M(p)$ and $N(p)$ are degrees of validity and of nonvalidity of p. These values can be obtained applying different formula depending on the problem considered. In [8, 9], the relation $\leq$ between the intuitionistic fuzzy pairs $\langle a, b \rangle$ and $\langle c, d \rangle$ is defined by

$$\langle a, b \rangle \leq \langle c, d \rangle \text{ if and only if } a \leq c \text{ and } b \geq d.$$

In the frame of intuitionistic fuzzy logic [9] different (standard) logic operations are introduced: *conjunction*, *disjunction*, more than 40 different *negations*, more than 180 different *implications*. Some types of modal logic operators are introduced. They include as particular case the standard modal logic operators "*necessity*" and "*possibility*". Now there is also temporal intuitionistic fuzzy logic, elements of which will be included in the future research. As a component of intuitionistic fuzzy logic, an intuitionistic fuzzy predicate calculus is developed.

For the purpose of this investigation, the degrees of validity/nonvalidity can be obtained, e.g., by the following formula:

$$M(p) = \frac{m}{u}, \quad N(p) = 1 - \frac{n}{u}, \tag{12}$$

where m is the lower boundary of the "narrow" range; u—the upper boundary of the "broad" range; n—the upper boundary of the "narrow" range.

If there is a database collected having elements with the form $<p, M(p), N(p)>$, different new values for the variables can be obtained. In case of two records in the database, the new values might be as follows:

- *optimistic prognosis*

$$V_{\text{opt}}(p) = \ <\max(M_1(p), M_2(p)), \min(N_1(p), N_2(p))>, \tag{13}$$

- *average prognosis*

$$V_{\text{aver}}(p) = <\left(\frac{M_1(p) + M_2(p)}{2}, \frac{N_1(p) + N_2(p)}{2}\right)>, \tag{14}$$

- *pessimistic prognosis*

$$V_{pes}(p) = \ <\min(M_1(p), M_2(p)), \max(N_1(p), N_2(p))> \tag{15}$$

Therefore, for each p

$$V_{\text{pes}}(p) \leq V_{\text{aver}}(p) \leq V_{\text{opt}}(p).$$

In case of three records in the database, the following new values can be obtained:

- *strong optimistic prognosis*

$$\begin{aligned} V_{\text{strong\_opt}}(p) = \ & <M_1(p) + M_2(p) + M_3(p) - \\ & - M_1(p)M_2(p) - M_1(p)M_3(p) - M_2(p)M_3(p) + \\ & + M_1(p)M_2(p)M_3(p), N_1(p)N_2(p)N_3(p)> \end{aligned} \tag{16}$$

- *optimistic prognosis*

$$V_{\text{opt}}(p) = \ <\max(M_1(p), M_2(p), M_3(p)), \min(N_1(p), N_2(p), N_3(p))>, \tag{17}$$

- *average prognosis*

$$V_{\text{aver}}(p) = <\frac{M_1(p) + M_2(p) + M_3(p)}{3}, \frac{N_1(p) + N_2(p) + N_3(p)}{3})>, \tag{18}$$

- *pessimistic prognosis*

$$V_{\text{pes}}(p) = \ <\min(M_1(p), M_2(p), M_3(p)), \max(N_1(p), N_2(p), N_3(p))>, \quad (19)$$

- *strong pessimistic prognosis*

$$V_{\text{strong\_pes}}(p) = \ <M_1(p)M_2(p)M_3(p), N_1(p) + N_2(p) + N_3(p) - N_1(p)N_2(p) - N_1(p)N_3(p) - N_2(p)N_3(p) + N_1(p)N_2(p)N_3(p)> \quad (20)$$

Therefore, for each p

$$V_{\text{strong\_pes}}(p) \leq V_{\text{pes}}(p) \leq V_{\text{aver}}(p) \leq V_{\text{opt}}(p) \leq V_{\text{strong\_opt}}(p).$$

D. Procedure for genetic algorithms quality assessment applying IFL

The implementation of IFL for assessment of genetic algorithms performance quality requires construction of degrees of validity and nonvalidity in two different intervals of model parameters variation: so-called “broad” range as known from the literature and so-called “narrow” range which is user-defined. In [13] authors have proposed a procedure for assessment of algorithm quality performance implementing IFL. The procedure starts with the performance of a number of runs of each of the algorithms, object of the investigation, in both “broad” and “narrow” ranges of model parameters. Then the average values of the objective function, algorithms convergence time and each of the model parameters for each one of the ranges and each one of the investigated algorithms are obtained. According to (12), degrees of validity/nonvalidity for each of the investigated algorithms are determined. Then, in case of two objects, the ranges for *optimistic*, *average,* and *pessimistic* prognosis are calculated for each one of the model parameters according to (13)–(15), while in case of three objects, the ranges for *strong optimistic*, *optimistic*, *average*, *pessimistic,* and *strong pessimistic* prognosis are calculated according to (16)–(20). Next, for each of the algorithms, in both considered ranges, each of the model parameters is assigned a value based on the determined in such a way prognoses. Finally, on the basis of these assigns, the quality of each one of the considered algorithms is assessed.

3 Intuitionistic Fuzzy Estimations of Model Parameters of *S. cerevisiae* Fed-Batch Cultivation

A. *IFL applied to assess the performance of SGA at different values of GGAP*

According to previous authors’ investigations [2], generation gap has been proven as the most sensitive genetic algorithm parameter toward convergence time. In this section, the performance quality of two kinds of SGA considered here is going to be

Table 1 SGA performance at different values of GGAP

	Objective function			Levels of performance			Average convergence time	
		SCM[a]	MCS[a]		SCM[a]	MCS[a]	SCM[a]	MCS[a]
GGAP = 0.9	minJ	0.0221	0.0222	TL_LB	0.0221	0.0222	81.67	71.16
				TL_UB	~0.0222	~0.0227		
	avrgJ	0.0222	0.0225	ML_LB	0.0222	0.0227		
				ML_UB	0.0222	~0.0231		
	maxJ	0.0223	0.0236	LL_LB	0.0222	0.0231		
				LL_UB	0.0223	0.0236		
GGAP = 0.5	minJ	0.0222	0.0221	TL_LB	0.0222	0.0221	46.99	40.53
				TL_UB	~0.0223	~0.0223		
	avrg	0.0223	0.0224	ML_LB	0.0223	0.0223		
				ML_UB	~0.0225	~0.0226		
	maxJ	0.0226	0.0228	LL_LB	0.0225	0.0226		
				LL_UB	0.0226	0.0228		
GGAP = 0.1	minJ	0.0224	0.0223	TL_LB	0.0224	0.0223	25.98	25.89
				TL_UB	~0.0225	~0.0225		
	avrgJ	0.0225	0.0225	ML_LB	0.0225	0.0225		
				ML_UB	0.0226	~0.0227		
	maxJ	0.0228	0.0230	LL_LB	~0.0226	0.0227		
				LL_UB	0.0228	0.0230		

[a]Short denotations used: SCM instead of SGA-SCM and MCS instead of SGA_MCS

assessed at different values of GGAP. In [4] the influence of GGAP has been thoroughly investigated for SGA-SCM by applying three different values of GGAP: GGAP = 0.9, GGAP = 0.5, and GGAP = 0.1 as representatives of two "extreme" cases and the middle one. Results from this investigation are shortly repeated here in order to be compared to SGA_MCS, for which investigation begun in [3] but here it is thoroughly expanded. Table 1 presents the results from the first step of the procedure for purposeful model parameter genesis application. Thirty runs, which are assumed as a representative from a statistical point of view, have been performed in order to obtain reliable results.

For each of the levels, constructed in such a way, the minimum, maximum, and average values of each model parameter have been determined. Table 2 presents these values only for the top levels, according to Table 1.

Table 3 presents previously used "broad" boundaries for each model parameter as well as new boundaries proposed based on the procedure for purposeful model parameter genesis when applying corresponding SGA. In this investigation, the "narrow" range is constructed in a way that the new minimum is lower but close to the minimum of the top level, and the new maximum is higher but close to the maximum of the top level. Additionally, Table 3 consists of intuitionistic fuzzy estimations, obtained based on (12) as described in the *Background section C.*

Table 2 Model parameters values for the top levels

			μ_{2S}	μ_{2E}	k_S	k_E	Y_{SX}	Y_{EX}	$k_L^{O_2}a$	Y_{OS}	Y_{OE}
GGAP = 0.9	SCM[a]	min	0.94	0.14	0.13	0.80	0.39	1.81	40.42	333.03	35.73
		max	0.99	0.15	0.14	0.80	0.40	2.00	95.53	785.10	96.73
		avrg	0.97	0.14	0.13	0.80	0.39	1.92	63.24	515.78	61.78
	MCS[a]	min	0.90	0.09	0.10	0.71	0.39	1.13	59.56	507.42	9.94
		max	0.95	0.13	0.13	0.80	0.44	1.70	138.70	979.42	292.97
		avrg	0.92	0.10	0.11	0.78	0.42	1.38	101.53	787.85	189.82
GGAP = 0.5	SCM[a]	min	0.91	0.11	0.11	0.79	0.39	1.47	99.59	768.66	102.84
		max	1.00	0.15	0.14	0.80	0.40	2.04	126.78	983.37	261.13
		avrg	0.95	0.14	0.13	0.80	0.40	1.84	108.41	853.07	216.27
	MCS[a]	min	0.90	0.10	0.10	0.78	0.40	1.28	28.53	225.11	116.93
		max	0.95	0.14	0.13	0.80	0.42	1.98	127.80	997.50	252.26
		avrg	0.92	0.12	0.12	0.80	0.40	1.58	90.23	708.44	203.38
GGAP = 0.1	SCM[a]	min	0.91	0.08	0.10	0.71	0.39	1.06	30.53	244.17	78.47
		max	0.99	0.14	0.15	0.80	0.44	1.84	128.75	976.78	220.70
		avrg	0.94	0.11	0.13	0.77	0.41	1.36	79.55	639.94	171.59
	MCS[a]	min	0.91	0.09	0.11	0.77	0.39	1.19	51.37	364.45	11.68
		max	0.93	0.13	0.13	0.79	0.47	1.63	107.22	696.21	239.89
		avrg	0.92	0.11	0.12	0.77	0.43	1.43	70.33	526.30	140.25

[a]Short denotations used: SCM instead of SGA-SCM and MCS instead of SGA_MCS

Table 3 Model parameters boundaries for SGA-SCM and SGA-MCS

Previously used			μ_{2S}	μ_{2E}	k_S	k_E	Y_{SX}	Y_{EX}	$k_L^{O_2}a$	Y_{OS}	Y_{OE}
		LB	0.90	0.05	0.08	0.50	0.30	1.00	0.001	0.001	0.001
		UB	1.00	0.15	0.15	0.80	10.00	10.00	1000.00	1000.00	300.00
SGA-SCM at GGAP = 0.9	Advisable after procedure application	LB	0.9	0.13	0.12	0.7	0.38	1.8	40	300	35
		UB	1	0.15	0.15	0.8	0.4	2.1	100	800	100
	Degrees of validity of p	$M_1(q)$	0.90	0.87	0.80	0.88	0.13	0.18	0.04	0.30	0.12
	Degree of non-validity of p	$N_1(q)$	0.00	0.00	0.00	0.00	0.87	0.79	0.09	0.20	0.67
SGA-SCM at GGAP = 0.5	Advisable after procedure application	LB	0.9	0.1	0.1	0.7	0.39	1.4	90	760	100
		UB	1	0.15	0.15	0.8	0.45	2	130	990	270
	Degrees of validity of p	$M_2(p)$	0.90	0.67	0.67	0.88	0.13	0.14	0.09	0.76	0.33
	Degree of non-validity of p	$N_2(p)$	0.00	0.00	0.00	0.00	0.85	0.80	0.87	0.01	0.10
SGA-SCM at GGAP = 0.1	Advisable after procedure application	LB	0.9	0.07	0.1	0.7	0.39	1	30	240	75
		UB	1	0.15	0.15	0.8	0.45	1.9	130	980	240
	Degrees of validity of p	$M_3(q)$	0.90	0.47	0.67	0.88	0.13	0.10	0.03	0.24	0.25
	Degree of non-validity of p	$N_3(q)$	0.00	0.00	0.00	0.00	0.85	0.81	0.87	0.02	0.20
SGA-MCS at GGAP = 0.9	Advisable after procedure application	LB	0.90	0.09	0.10	0.70	0.30	1.00	50.00	500.00	5.00
		UB	0.95	0.13	0.13	0.80	0.50	2.00	140.00	1000.00	300.00
	Degrees of validity of p	$M_1(p)$	0.90	0.60	0.67	0.88	0.03	0.10	0.05	0.50	0.00
	Degree of non-validity of p	$N_1(p)$	0.05	0.13	0.13	0.00	0.95	0.80	0.86	0.00	0.00
SGA-MCS at GGAP = 0.5	Advisable after procedure application	LB	0.90	0.10	0.10	0.70	0.30	1.00	20.00	200.00	110.00
		UB	0.95	0.14	0.13	0.80	0.50	2.00	130.00	1000.00	260.00
	Degrees of validity of p	$M_2(q)$	0.90	0.67	0.67	0.88	0.03	0.10	0.02	0.20	0.37
	Degree of non-validity of p	$N_2(q)$	0.05	0.07	0.13	0.00	0.95	0.80	0.87	0.00	0.13
SGA-MCS at GGAP = 0.1	Advisable after procedure application	LB	0.91	0.09	0.11	0.70	0.30	1.00	50.00	300.00	10.00
		UB	0.93	0.13	0.13	0.80	0.50	2.00	110.00	700.00	250.00
	Degrees of validity of p	$M_3(p)$	0.91	0.60	0.73	0.88	0.03	0.10	0.05	0.30	0.03
	Degree of non-validity of p	$N_3(p)$	0.07	0.13	0.13	0.00	0.95	0.80	0.89	0.30	0.17

Table 4 presents the boundaries (low LB and up UB) for the *strong optimistic*, *optimistic*, *average*, *pessimistic,* and *strong pessimistic* prognoses for the performances of both SGA, obtained based on intuitionistic fuzzy estimations (12) and formula (16)–(20).

Investigated SGA has again been applied for the parameter identification of *S. cerevisiae* fed-batch cultivation involving newly proposed according to Table 3 boundaries at GGAP = 0.9, GGAP = 0.5, and GGAP = 0.1. Again thirty runs of the algorithm have been performed in order to obtain reliable results.

Table 5 presents the average values of the objective function, convergence time, and model parameters when both SGA considered here have been executed at three investigated values of GGAP before and after the application of the purposeful model parameter genesis.

The applied procedure for PMPG leads to the reduction of the convergence time in both SGA at any of investigated values of GGAP. Meanwhile, the procedure ensures saving and even slightly improving of model accuracy. When SGA_SCM has been applied, between 29 % and 44 % of the convergence time has been saved for different GGAP values. But if one compares the "fastest" case—at GGAP = 0.1 after PMPG to the slowest one—at GGAP = 0.9 before PMPG, about 77 % time saving has been realized keeping the model accuracy at the highest achieved value. When applying SGA_MCS, the algorithm calculation time decreases from 25 to 29 % for different GGAP values without loss of model accuracy. Similarly, GGAP = 0.1 after PMPG instead of 0.9 before PMPG leads to saving of 73 % of calculation time. In such a way, good effectiveness of proposed procedure for purposeful model parameter genesis has been demonstrated for both applied SGA.

Table 6 lists the estimations assigned to the each of the parameters concerning Table 4 for the three values of GGAP in "broad" and "narrow" range, i.e., before and after PMPG application, for both SGA considered here.

As presented in previous authors' investigation [4], there are only two *strong pessimistic* prognoses when applying SGA_SCM—one at GGAP = 0.9 before PMPG and one at GGAP = 0.5 but after PMPG, while there are no *pessimistic* ones. In five out of six cases there are three *strong optimistic* prognoses, except of GGAP = 0.1 before PMPG. The number of *optimistic* and *average* prognosis is relatively equal, with a slight dominance of the *average* ones. The analysis shows that there are three absolutely equal performances—one of each value of GGAP: for GGAP = 0.9 and GGAP = 0.1 after PMPG, and GGAP = 0.5 before PMPG. In all the cases, the value of the objective function is very close to the lowest one that means they are with the highest achieved degree of accuracy. But if one compares the time, the SGA_SCM with GGAP = 0.1 after PMPG is about 58 % faster than SGA_SCM with GGAP = 0.9 after PMPG and about 39 % faster than GGAP = 0.5 before PMPG. Thus, based on the IFL estimations of the model parameters and further constructed prognoses, SGA_SCM with GGAP = 0.1 and after the procedure for purposeful model parameter genesis has been distinguished as more reliable algorithm, if one would like to obtain results with a high level of relevance and for less computational time.

Table 4 Prognoses for SGA-SCM and SGA-MCS performances

	μ_{2S}		μ_{2E}		k_S		k_E		Y_{SX}		Y_{EX}		$k_L^{O_2}a$		Y_{OS}		Y_{OE}	
	LB	UB	LB	UB	LB	UB	LB	UB	LB	UB	LB	UB	LB	UB	LB	UB	LB	UB
SCM[a] V_{str_opt}	1.00	1.00	0.15	0.15	0.15	0.15	0.80	0.80	1.02	1.12	3.65	4.88	167.50	296.00	872.32	999.96	152.61	931.88
SCM[a] V_{opt}	0.90	1.00	0.13	0.15	0.12	0.15	0.70	0.80	0.39	0.45	1.80	2.10	99.00	270.00	760.00	990.00	90.00	910.00
SCM[a] V_{aver}	0.90	1.00	0.10	0.15	0.11	0.15	0.70	0.80	0.39	0.42	1.40	2.00	70.00	204.00	433.33	920.00	53.33	390.00
SCM[a] V_{pes}	0.90	1.00	0.07	0.15	0.10	0.15	0.70	0.80	0.39	0.39	1.00	1.90	36.00	99.00	240.00	800.00	30.00	130.00
SCM[a] V_{str_pes}	0.73	1.00	0.04	0.15	0.05	0.15	0.54	0.80	0.00	0.00	0.03	0.10	2.92	72.00	54.72	780.00	0.11	20.00
MCS[a] V_{str_opt}	1.00	1.00	0.14	0.15	0.15	0.15	0.80	0.80	0.87	1.43	2.71	4.88	115.55	334.10	720.00	1000.00	116.95	300.00
MCS[a] V_{opt}	0.91	0.95	0.10	0.14	0.11	0.13	0.70	0.80	0.30	0.50	1.00	2.00	50.00	140.00	500.00	1000.00	110.00	300.00
MCS[a] V_{aver}	0.90	0.94	0.09	0.13	0.10	0.13	0.70	0.80	0.30	0.50	1.00	2.00	40.00	126.67	333.33	900.00	40.33	270.00
MCS[a] V_{pes}	0.90	0.93	0.09	0.13	0.10	0.13	0.70	0.80	0.30	0.50	1.00	2.00	20.00	110.00	200.00	700.00	1.00	250.00
MCS[a] V_{str_pes}	0.74	0.84	0.04	0.11	0.05	0.10	0.54	0.80	0.00	0.00	0.01	0.08	0.05	2.00	30.00	700.00	0.01	216.67

[a]Short denotations used: SCM instead of SGA-SCM and MCS instead of SGA_MCS

Table 5 Results from model parameter identification

	SGA_SCM						SGA_MCS					
	GGAP = 0.9		GGAP = 0.5		GGAP = 0.1		GGAP = 0.9		GGAP = 0.5		GGAP = 0.1	
Parameter	Before PMPG	After PMPG	Before PMPG	After PMPG	Before PMPG	After PMPG	Before PMPG	After PMPG	Before PMPG	After PMPG	Before PMPG	After PMPG
J	0.0222	0.221	0.0223	0.0221	0.0225	0.0222	0.0226	0.0222	0.0224	0.0221	0.0225	0.0222
time, s	81.67	43.82	46.99	30.48	25.98	18.45	73.17	52.34	40.53	28.61	25.88	19.66
μ_{2S}	0.94	0.95	0.94	0.95	0.96	0.95	0.95	0.91	0.92	0.91	0.92	0.91
μ_{2E}	0.12	0.15	0.12	0.14	0.11	0.12	0.08	0.11	0.12	0.11	0.11	0.12
k_S	0.12	0.13	0.13	0.13	0.13	0.12	0.10	0.12	0.12	0.11	0.12	0.12
k_E	0.80	0.80	0.75	0.80	0.75	0.79	0.80	0.80	0.79	0.8	0.77	0.79
Y_{SX}	0.41	0.40	0.40	0.40	0.41	0.40	0.47	0.42	0.4	0.42	0.43	0.41
Y_{EX}	1.44	2	1.58	1.92	1.48	1.67	1.10	1.44	1.58	1.51	1.45	1.57
$k_L^{O_2}a$	70.28	70.32	109.27	108.31	79.31	94.03	72.21	74.19	90.23	69.23	70.33	81.71
Y_{OS}	544.37	553.71	858.75	864.15	627.78	729.57	556.95	575.74	708.44	550.75	526.30	650.34
Y_{OE}	102.31	59.01	156.92	155.49	187.09	192.43	223.53	282.79	203.38	194.26	140.25	138.08

Table 6 Models parameter estimations before and after PMPG

	SGA_SCM						SGA_MCS					
	GGAP = 0.9		GGAP = 0.5		GGAP = 0.1		GGAP = 0.9		GGAP = 0.5		GGAP = 0.1	
	Before PMPG	After PMPG	Before PMPG	After PMPG	Before PMPG	After PMPG	Before PMPG	After PMPG	Before PMPG	After PMPG	Before PMPG	After PMPG
str_opt	3	3	3	3	1	3	2	2	1	2	1	1
opt	1	3	3	2	4	3	5	7	8	6	8	8
aver	4	3	3	3	4	3	1	0	0	1	0	0
pes	0	0	0	0	0	0	0	0	0	0	0	0
str_pes	1	0	0	1	0	0	1	0	0	0	0	0

When applying SGA_MCS which is more in the focus of present investigation, there is only one *strong pessimistic* prognosis (in the case of GGAP = 0.9 before PMPG) compared to two in the case of SGA-SCM. In three of the cases there are two *strong optimistic* prognoses while another three are with one *strong optimistic* prognosis. Between the cases with two *strong optimistic* prognoses, slightly better performance has been achieved at GGAP = 0.9 after PMPG with seven compared to six (at GGAP = 0.5 after PMPG) and five (at GGAP = 0.9 before PMPG) *optimistic* prognoses. Moreover, SGA_MCS at GGAP = 0.9 before PMPG has been evaluated with one *average* and one *strong pessimistic* prognosis compared to none of them in the other two cases. In all these cases, the value of the objective function has been slightly improved after PMPG application, very close to the lowest one achieved. Another three cases—at GGAP = 0.5 before PMPG, at GGAP = 0.1 before and after PMPG, show absolutely equal performance—with only one *strong optimistic* prognoses, but followed by eight *optimistic* ones. But if one compares the convergence time between, from one side, distinguished as a leader based on the IFL estimations SGA_MCS at GGAP = 0.9 after PMPG, and—on the other side the fastest one SGA_MCS at GGAP = 0.1 after PMPG, the last one is about 2.77 times faster than the first one. Thus, based on the IFL estimations of the model parameters and further constructed prognoses, as an acceptable compromise SGA_MCS at GGAP = 0.1 after PMPG might be distinguished not as the most reliable one, but the second one, while this is the fastest one.

Due to the similarity of the results but focus pointed on the SGA_MCS, Fig. 1 presents results from experimental data and model prediction, respectively, for biomass, ethanol, substrate, and dissolved oxygen when applying SGA_MCS at GGAP = 0.1 in "narrow" range (that means after the PMPG application).

B. *IFL applied to assess the performance of SGA-SMC and SGA-MCS*

Following investigation is a logical consequence from previous subsection. The assessment of the performance quality of both SGA considered here is going to be done for the best results obtained at different GGAP values. As such, SGA_SCM and SGA_MCS are to be compared at GGAP = 0.1 after PMPG. The application of the procedure for quality assessment steps on the results is presented in previous section. Table 7 presents previously used "broad" range for each model as well as new boundaries proposed. In this case, the new boundaries are again advised based on the minimum and maximum values of the investigated parameters in the top level. Additionally, Table 7 consists of intuitionistic fuzzy estimations, obtained based on (12) as described above.

Table 8 presents the boundaries (low LB and up UB) for the *optimistic*, *average,* and *pessimistic* prognoses for the performances of both SGA, obtained based on intuitionistic fuzzy estimations (12) and formula (13)-(15).

Investigated here, two kinds of SGA have been again applied for parameter identification of *S. cerevisiae* fed-batch cultivation involving newly proposed boundaries at GGAP = 0.1, according to Table 7. Again thirty runs of the algorithms have been performed in order to obtain.

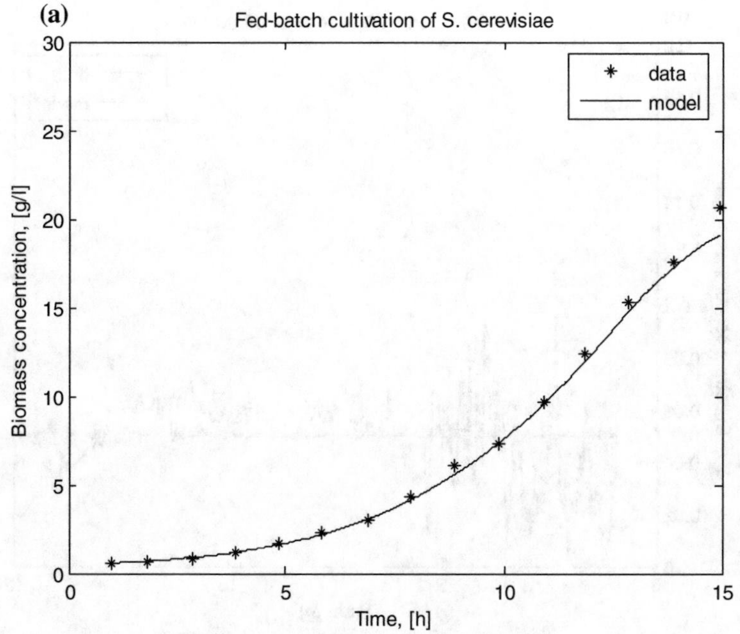

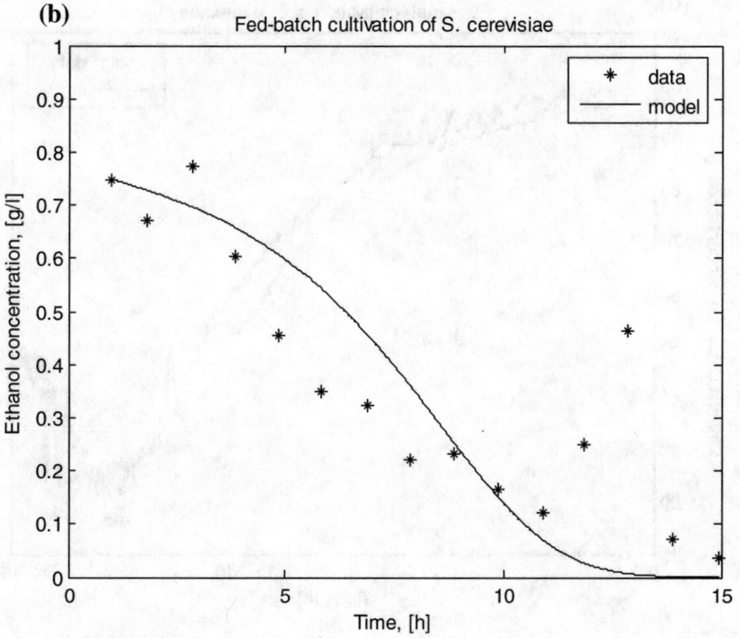

Fig. 1 Model prediction compared to experimental data when SGA_MCS at GGAP = 0.1 in "narrow" range has been applied. **a** Biomass concentration. **b** Ethanol concentration. **c** Substrate concentration. **d** Dissolved oxygen concentration

(c)

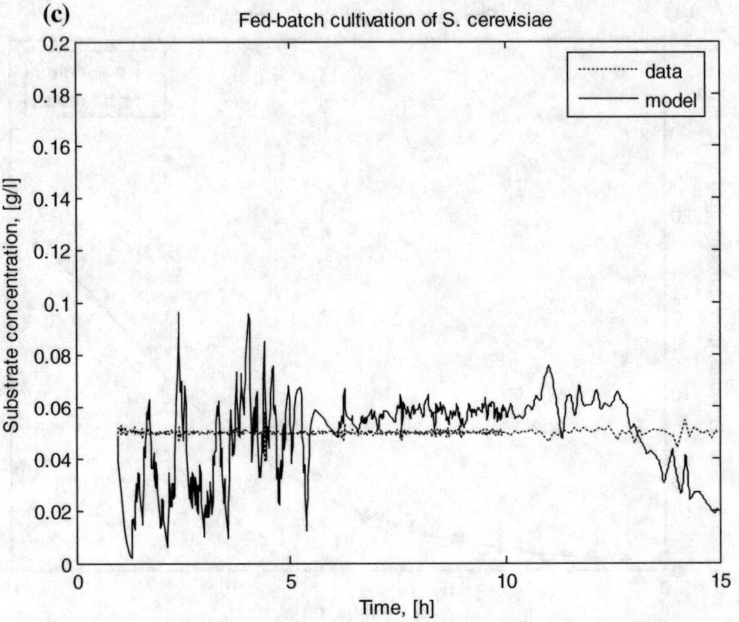

(d)

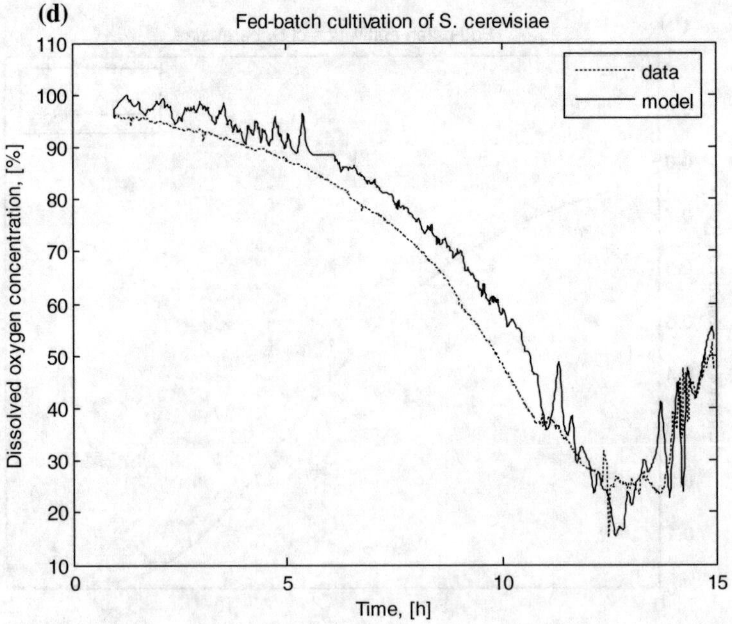

Fig. 1 (continued)

Table 7 Model parameters boundaries for SGA-SCM and SGA-MCS at GGAP = 0.1

Previously used			μ_{2S}	μ_{2E}	k_S	k_E	Y_{SX}	Y_{EX}	$k_L^{O_2}a$	Y_{OS}	Y_{OE}
		LB	0.9	0.05	0.08	0.5	0.3	1	0.001	0.001	0.001
		UB	1	0.15	0.15	0.8	10	10	1000	1000	300
SGA-SCM	Advisable after procedure application	LB	0.9	0.07	0.1	0.7	0.39	1	30	240	75
		UB	1	0.15	0.15	0.8	0.45	1.9	130	980	240
	Degrees of validity of p	$M_1(q)$	0.9	0.47	0.67	0.88	0.13	0.1	0.03	0.24	0.25
	Degree of non-validity of p	$N_1(q)$	0	0	0	0	0.85	0.81	0.87	0.02	0.2
SGA-MCS	Advisable after procedure application	LB	0.91	0.09	0.11	0.70	0.30	1.00	50.00	300.00	10.00
		UB	0.93	0.13	0.13	0.80	0.50	2.00	110.00	700.00	250.00
	Degrees of validity of p	$M_2(p)$	0.90	0.67	0.67	0.88	0.03	0.10	0.07	0.50	0.33
	Degree of non-validity of p	$N_2(p)$	0.07	0.13	0.13	0.00	0.95	0.80	0.90	0.20	0.33

Table 8 Prognoses for SGA-SCM and SGA-MCS Performances

	μ_{2S}		μ_{2E}		k_S		k_E		Y_{SX}		Y_{EX}		$k_L^{O_2}a$		Y_{OS}		Y_{OE}	
	LB	UB	LB	UB	LB	UB	LB	UB	LB	UB	LB	UB	LB	UB	LB	UB	LB	UB
V_{opt}	0.91	1.00	0.09	0.15	0.11	0.15	0.70	0.80	0.39	0.50	1.00	2.00	50.00	130.00	300.00	980.00	75.00	250.00
V_{aver}	0.91	0.97	0.08	0.14	0.11	0.14	0.70	0.80	0.35	0.48	1.00	1.95	40.00	120.00	270.00	840.00	42.50	245.00
V_{pes}	0.90	0.93	0.07	0.13	0.10	0.13	0.70	0.80	0.30	0.45	1.00	1.90	30.00	110.00	240.00	700.00	10.00	240.00

Table 9 Results from model parameter identification

	SGA_SCM		SGA_MCS	
	Before PMPG	After PMPG	Before PMPG	After PMPG
J	0.0225	0.0222	0.0225	0.0222
time, s	25.98	18.45	25.88	19.66
μ_{2S}	0.96	0.95	0.92	0.91
μ_{2E}	0.11	0.12	0.11	0.12
k_S	0.13	0.12	0.12	0.12
k_E	0.75	0.79	0.77	0.79
Y_{SX}	0.41	0.4	0.43	0.41
Y_{EX}	1.48	1.67	1.45	1.57
$k_L^{O_2}a$	79.31	94.03	70.33	81.71
Y_{OS}	627.78	729.57	526.3	650.34
Y_{OE}	187.09	192.43	140.25	138.08

Table 10 Models parameter estimations before and after PMPG

	SGA_SCM		SGA_MCS	
	GGAP = 0.1		GGAP = 0.1	
	Before PMPG	After PMPG	Before PMPG	After PMPG
opt	9	9	9	9
aver	0	0	0	0
pes	0	0	0	0

Table 9 presents the average values of the objective function, computation time, and model parameters when SGA-SCM and SGA-MCS have been executed at GGAP = 0.1 both in "broad" and "narrow" ranges, corresponding to before and after PMPG application.

As expected and shown in the previous subsection as well, the application of the purposeful model genesis procedure for both SGA considered here leads to expecting decrease of the convergence time. Meanwhile, even slight improvement of the model accuracy has been observed. In comparison to the results before the procedure application, up to 24 % reduction of the computation time of SGA-MCS without loss of model accuracy has been achieved, while for SGA-SCM even 29 % of computation time has been saved, thus showing good effectiveness of PMPG.

Table 10 lists the estimations assigned to each of the estimated parameters concerning Table 8 for the considered SGA_SCM nd SGA_MSC in "broad" and "narrow" ranges.

It is interesting to note that obtained results are absolutely equal for both SGA considered here, before and after the PMPG application. Specific at a first glance, results might be reasonably explained and had been expected in some way. The application of the procedure for assessment of the algorithm performance quality has been demonstrated here for "leaders of the leaders". Both algorithms have been

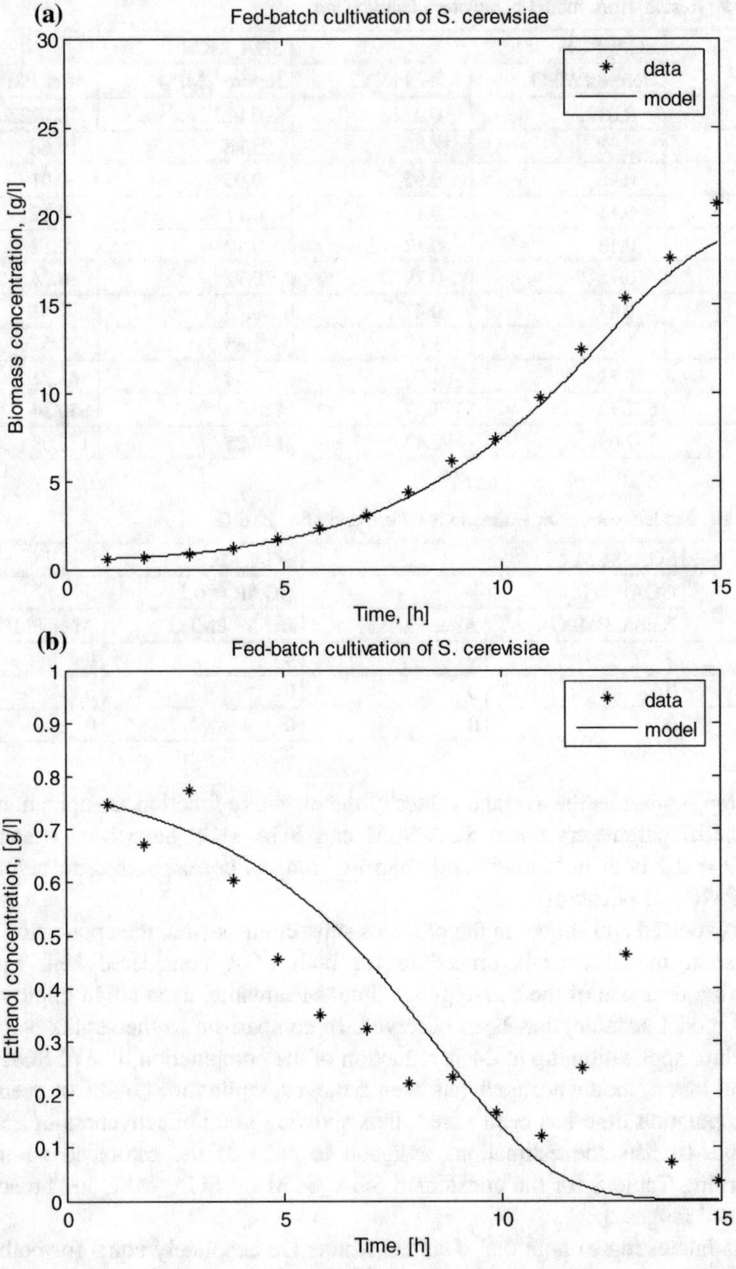

Fig. 2 Model prediction compared to experimental data when SGA_SCM at GGAP = 0.1 in "narrow" range has been applied. **a** Biomass concentration. **b** Ethanol concentration. **c** Substrate concentration. **d** Dissolved oxygen concentration

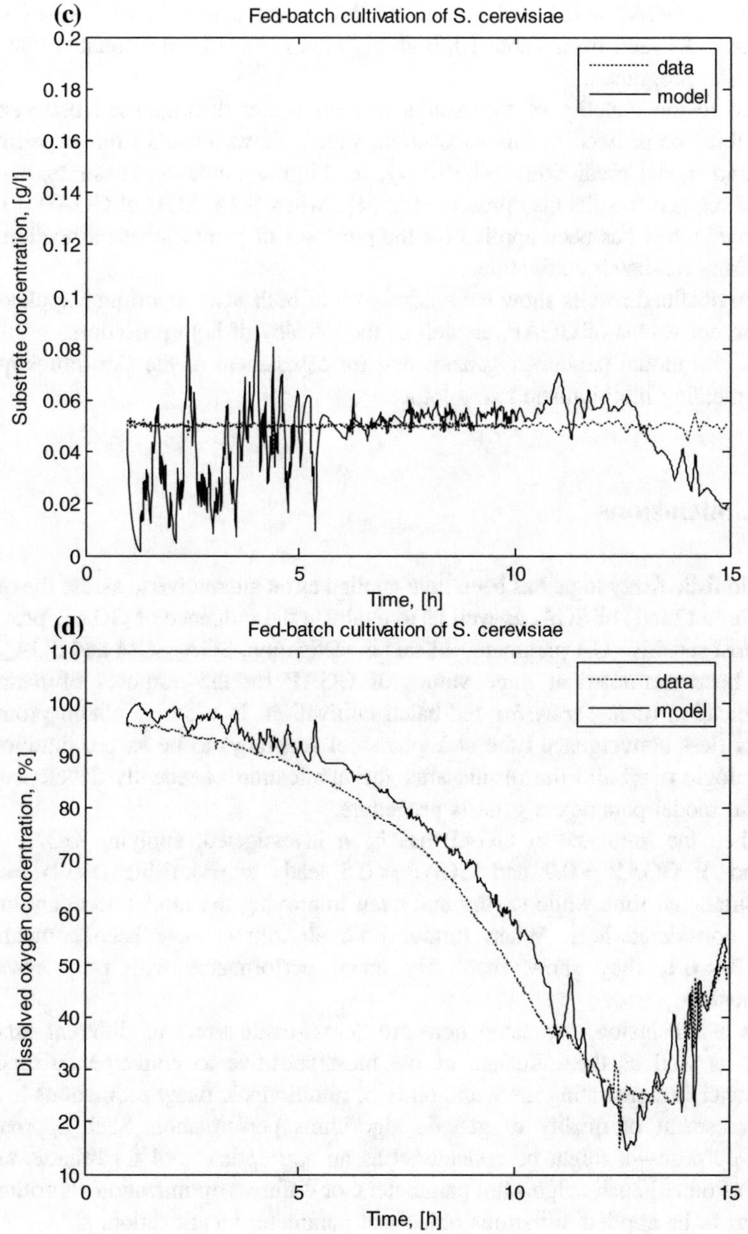

Fig. 2 (continued)

applied at GGAP = 0.1 and proved as the fastest case without lose of model accuracy. As seen from Table 10, both algorithms have been evaluated with only *optimistic* prognoses.

Due to the equality of the results and no leader distinguished between two algorithms considered in this subsection, Fig. 2 shows results from experimental data and model prediction, respectively, for biomass, ethanol, substrate, and dissolved oxygen (results also presented in [4]), when SGA_SCM at GGAP = 0.1 in "narrow" range has been applied for the purposes of parameter identification of *S. cerevisiae* fed-batch cultivation.

The obtained results show the workability of both SGA algorithms applied here at different values of GGAP, as well as the efficacy of both procedures used—for purposeful model parameter genesis and for assessment of the algorithms quality implementing intuitionistic fuzzy logic.

4 Conclusions

Intuitionistic fuzzy logic has been here applied as an alternative to assess the quality of different kinds of SGA, as well as to evaluate the influence of GGAP, proven as the most sensitive GA parameter. In this investigation, SGA_SCM and SGA_MCS have been examined at three values of GGAP for the purposes of parameter identification of *S. cerevisiae* fed-batch cultivation. In order to obtain promising results (less convergence time at kept model accuracy) to be saved, intuitionistic fuzzy logic overbuild the results after the application of recently developed purposeful model parameters genesis procedure.

When the influence of GGAP has been investigated, applying GGAP = 0.1 instead of GGAP = 0.9 and GGAP = 0.5 leads to promising results for less computational time while saving and even improving the model accuracy in both SGA considered here. When further both algorithms have been compared at GGAP = 0.1, they show absolutely equal performance with only *optimistic* prognoses.

As a conclusion, presented here are "cross-evaluation" of different kinds of SGA, as well as the influence of the most sensitive to convergence time GA parameter demonstrating the workability of intuitionistic fuzzy estimations to assist in assessment of quality of genetic algorithms performance. Such approach is multi-purpose—it might be considered as an appropriate tool for reliable assessment of other genetic algorithm parameters, or different optimization algorithms, as well as to be applied to various objects of parameter identification.

References

1. Adeyemo, J., Enitian, A.: Optimization of fermentation processes using evolutionary algorithms—a review. Sci. Res. Ess. **6**(7), 1464–1472 (2011)
2. Angelova, M., Pencheva, T.: Tuning genetic algorithm parameters to improve convergence time. Int. J. Chem. Eng., article ID 646917 (2011). http://www.hindawi.com/journals/ijce/2011/646917/cta/
3. Angelova, M., Atanassov, K., Pencheva, T.: Intuitionistic fuzzy estimations of purposeful model parameters genesis. In: IEEE 6th International Conference "Intelligent Systems", Sofia, Bulgaria, 6–8 Sept 2012, pp. 206–211 (2012)
4. Angelova, M., Atanassov, K., Pencheva, T.: Intuitionistic fuzzy logic based quality assessment of simple genetic algorithm, vol. 2. In: Proceedings of the 16th International Conference on System Theory, Control and Computing (ICSTCC), 12–14 Oct 2012, Sinaia, Romania, Elecrtonic edition (2012)
5. Angelova, M., Tzonkov, S., Pencheva, T.: Genetic algorithms based parameter identification of yeast fed-batch cultivation. Lecture Notes in Computer Science, vol. 6046, pp. 224–231 (2011)
6. Angelova, M., Pencheva, T.: Algorithms improving convergence time in parameter identification of fed-batch cultivation. Comptes Rendus de l'Académie Bulgare des Sciences **65**(3), 299–306 (2012)
7. Angelova, M., Pencheva, T.: Purposeful model parameters genesis in simple genetic algorithms. Comput. Math Appl. **64**, 221–228 (2012)
8. Atanassov, K.: Intuitionistic Fuzzy Sets. Springer, Heidelberg (1999)
9. Atanassov, K.: On intuitionistic Fuzzy Sets Theory. Springer, Berlin (2012)
10. Chipperfield, A.J., Fleming, P., Pohlheim, H., Fonseca, C.M.: Genetic algorithm toolbox for use with MATLAB, User's guide, version 1.2. Department of Automatic Control and System Engineering, University of Sheffield, UK (1994)
11. Goldberg, D.: Genetic Algorithms in Search, Optimization and Machine Learning. Addison-Wiley Publishing Company, Massachusetts (1989)
12. Jones, K.: Comparison of genetic algorithms and particle swarm optimization for fermentation feed profile determination. In: Proceedings of the CompSysTech'2006, 15–16 June 2006, Veliko Tarnovo, Bulgaria, IIIB.8-1–IIIB.8-7 (2006)
13. Pencheva, T., Angelova, M., Atanassov, K.: Intuitionistic fuzzy logic implementation to assess genetic algorithms quality. Comput. Math. Appl. (2012)
14. Pencheva, T., Roeva, O., Hristozov, I.: Functional state approach to fermentation processes modelling. In: Tzonkov, St., Hitzmann, B. (eds.) Prof. Marin Drinov Academic Publishing House, Sofia (2006)
15. Roeva, O. (ed.): Real-world application of genetic algorithms. InTech (2012)

Dynamic Representation and Interpretation in a Multiagent 3D Tutoring System

Patrick Person, Thierry Galinho, Hadhoum Boukachour, Florence Lecroq and Jean Grieu

Abstract In this paper we present an intelligent tutoring system which aims at decreasing students' dropout rate by offering the possibility of a personalized follow-up. We address the specific problem of the evolution of the large amount of data to be processed and interpreted in an intelligent tutoring system. In this regard we detail the architecture and experimental results of our decision support system used as the core of the intelligent tutor—which could be applied to a variety of teaching fields. The first part presents an overview of the characteristics of intelligent tutors, the chosen data organization—composed of a composite factual semantic feature descriptive representation associated to a multiagent system—and two examples used to illustrate the architecture of our prototype. The second and last part describes all the components of the prototype: student interface, dynamic representation layer, characterization, and interpretation layers. First, for the student interface, the system shows our 3D virtual campus named GE3D to be connected to the intelligent tutor. Then we explain how the agents of the first layer represent the evolution of the situation being analyzed. Next, we specify the use of the characterization layer to cluster the agents of representation layer and to compute compound parameters. Finally, we expose how—using compound parameters—the third layer can measure similarity between current target case and past cases to constitute an interpretation of cases according to a case-based reasoning paradigm.

P. Person · T. Galinho · H. Boukachour · F. Lecroq · J. Grieu (✉)
LITIS University of Le Havre, 25 Rue Philippe Lebon, BP 1123, 76063 Le Havre, France
e-mail: Jean.Grieu@univ-lehavre.fr

P. Person
e-mail: Patrick.Person@univ-lehavre.fr

T. Galinho
e-mail: Thierry.Galinho@univ-lehavre.fr

H. Boukachour
e-mail: Hadhoum.Boukachour@univ-lehavre.fr

F. Lecroq
e-mail: Florence.Lecroq@univ-lehavre.fr

V. Sgurev et al. (eds.), *Recent Contributions in Intelligent Systems*,
Studies in Computational Intelligence 657,
DOI 10.1007/978-3-319-41438-6_12

1 Introduction

Intelligent tutoring systems (ITS) might be one adequate answer toward helping decrease students' dropout rate by offering personalized follow-up either in blended or distance learning courses [20]. Quoting Hafner [8], an ITS: "is educational software containing an artificial intelligence component. The software tracks students' work, tailoring feedback, and hints along the way. By collecting information on a particular student's performance, the software can make inferences about strengths and weaknesses, and can suggest additional work." On the one hand, thanks to information feedback given by the decision support system we have conceived, our system lets the ITS deal with students who learn easily. On the other hand, the ITS spots students with most difficulties and calls on a human tutor for help. Thus the latter has more time to focus on students more likely to abandon their studies.

In the first part, we give an explanation of the dynamic aspect of the problem dealt with by an ITS. Considering the characteristic mentioned above we chose how to organize and process data to design an appropriate multiagent decision support system accordingly. Then, the architecture of the decision system is introduced and illustrated by two applications: first, a game to present technical details and second, the intelligent tutoring system itself. Finally, the design and realization of the ITS are described:

- the 3D virtual campus interface;
- the representation of successive dynamic situations;
- the clustering characterization and its associated computed parameters applied to dynamic representations;
- the specification of target case and past cases by using computed parameters.

2 Data Structures, Algorithms, and Dynamic Problems

Following Denning [5]: "the fundamental question underlying all of computing is, *what can be (efficiently) automated*?" Three broad categories of problems are identified: first, problems that are impossible to solve with a computer, such as predicting the next Lotto numbers drawn at random. Second problems that are easy to solve with a computer, such as searching whether a given integer exists in a billion integers. In this range of problems, there is only one possible answer or one optimal solution. And third, problems where it is possible to use a computer to achieve a correct output, even if not the best one. This is the typical situation where artificial intelligence is applied to find solutions to complex problems. For the second and third categories of problems—simple or complex—a correct answer depends on the quality of information used as input, as well as the data structures and data organization chosen according to the characteristics of the problems at hand. So, we will detail the input, structure, and organization of data for their relationships of interdependence.

Knuth [9] defines a data structure as "A table of data including structural relationships" and also includes means by which one may access and manipulate the data structures as data organization—which is "a way to represent information in a data structure, together with algorithms that access and/or modify the structure" [15].

Currently, intelligent systems have scarcely been applied to the educational field. The notion of an intelligent tutoring system presents complex problems—key among them, the number of dynamic elements needing to be taken into account, to interpret and model toward providing answers to students [3]. Consequently in line with Wooldrigde [22], considering the dynamic characteristics of the problem, we chose a multiagent system (MAS) as processing part of data organization.

Our data structures—used as inputs by agents—are descriptive representations of knowledge with few attribute-value couples grouped in composite factual semantic features (CFSFs). In linguistics, a semantic feature is a meaningful component used in text analysis. Here, it is used in another context. For us, a composite factual semantic feature is still a meaningful component but in addition, a CFSF is also an observable fact which is composite because it is composed of a few items. An item is a property-value pair. The generic model of CFSF is:

< !ELEMENT CFSF (Type, Name, Date, Item$^+$)>

To illustrate composite factual semantic features in a general context we give the following example chosen to be as simple as possible. The river Seine is 776 km long rising in the Langres plateau and flowing through Paris. On the March 21, 2012, the recorded flow of river Seine in Paris was 227 cubic meters per second and it was 1.1 m high [18]. Here is the CFSFs descriptive representation of this example:

```
<fact1>
        <type>   river           </type>
        <name>   Seine           </name>
        <date>   21 March 2012   </date>
        <long>   776             </long>
        <spring> Langres         </spring>
</fact1>
<fact2>
        <type>   river           </type>
        <name>   Seine           </name>
        <date>   21 March 2012   </date>
        <flow>   227             </flow>
        <height> 1.1             </height>
</fact2>
```

Using this same example we will expose the three classes of problems according to data variation analysis, as shown in Fig. 1. Unless dealing with a long geologic timescale, <fact1> will never change over time (case A, Fig. 1). Cases B and C in Fig. 1 could both exploit <fact2>. In case B, a fact has successive values over time, but knowing the last one is enough to answer the problem. Is there enough or too

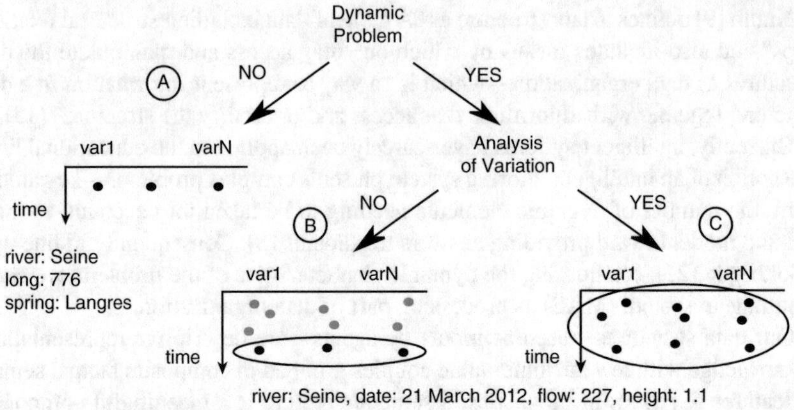

Fig. 1 Classes of problems according to data variation analysis

Fig. 2 Global architecture of the decision support system

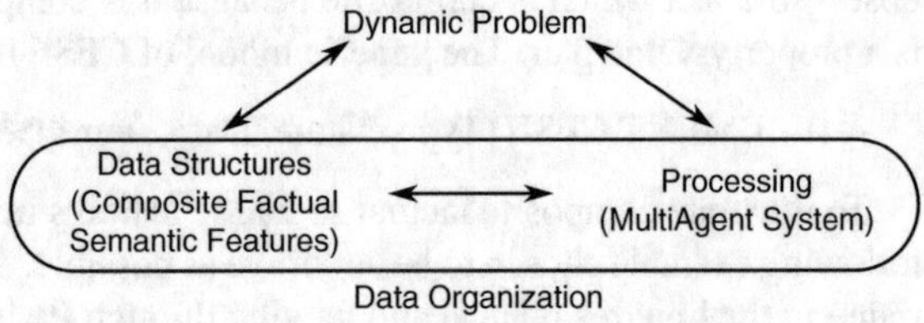

much water for a boat to go under the bridge facing you? Only knowing <fact2> is enough to decide to go under the bridge with a boat in Paris on 21st of March. In case C, the system needs successive values of the same fact to provide an answer. To predict a flood, you will collect <fact2$_i$> hour by hour to analyze the successive dynamic evolution.

Our decision support system will not fit for classes of A and B problems because the lack of variation of data implies no evolution to represent inside the MAS. However, it is designed specifically to handle type C dynamic problems with evolving data. Figure 2 summarizes the global architecture of our multiagent decision support system (DSS).

3 Architecture and Applications

In order to expose the architecture of our decision support system we now consider two applications. First, the example of the game of Risk which allows us to present the technical details then we expose the ITS whose core is the DSS.

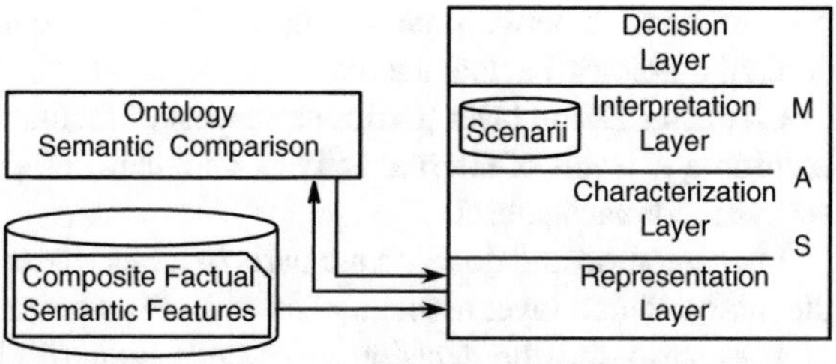

Fig. 3 Architecture of the intelligent system

3.1 Game of Risk

We have adapted our DSS to the game of Risk as it is not a toy problem. Indeed "Risk is far more complex than most traditional board games" [21]. Risk is a game of strategy for 2–6 players. The game board is a map of the world divided into 42 territories. A player wins by conquering all territories. In turn, after an initial placement of armies, each player receives and places new armies and may attack adjacent territories. An attack is one or more battles which are fought with dice. Rules and strategies are detailed in [13].

Figure 3 displays the main components of the system. From a set of files recorded during test phases, a file is chosen as input to representation layer of the multiagent system. This file contains all the composite factual semantic features describing successive steps of a given game of Risk. The following is an example of a CFSF from the game of Risk. Every territory is represented by a CFSF with the player who owns the territory and the number of armies. In this example, player red owns the territory called Quebec, at step 14, with 2 armies:

```
<cfsf>
    <type>     territory </type>
    <name>     Quebec    </name>
    <step>     14        </step>
    <player>   red       </player>
    <nbArmies> 2         </nbArmies>
</cfsf>
```

Using CFSFs as input, representation layer represents the global current situation and its dynamic tendency of ulterior evolutions. *The challenge is to represent both a whole situation and its dynamic tendency of evolution.*

The most difficult part is not to represent a static view of the current situation, i.e., a complete and accurate map of the board game at a given step of the game. Indeed, the main challenge consists in adding dynamic tendency of evolutions. In other words, the problem is how to make a static data structure such as a CFSF dynamic? From the successive static representations, the main problem is to compute information describing evolutions used by the characterization layer. One answer is to use factual agents in the representation layer. A factual agent is made of two parts: a knowledge part—its CFSF—and a computed behavior part due to semantic comparison measure between factual agents. Semantic comparisons lean on a dedicated domain ontology. Factual agents are more detailed in the next section.

Characterization layer partitions subsets of factual agents of representation layer according to levels of internal activity, and then, computes a synthetic measure characteristic for each subset.

Interpretation and decision are parts of a case-based reasoning system [10] where the interpretation layer associates subsets of the characterization layer with sections of a scenario and the decision layer finds scenarii closed to the current situation and chooses the most appropriate one to propose to users. The representation of a given case and definition of the chosen similarity measure are defined in the section concerning the interpretation layer.

3.2 *Intelligent Tutoring System*

Wenger [19] considers the main goal of ITS to be communicating knowledge efficiently. The model contains four components: domain model, student model, teaching model, and user interface. Woolf uses a similar model to the one presented by Wenger, with updated computer architectures, one of which being based on multiagent systems [23]. In the challenge for capturing intelligence, Luger also mentions this possible choice [12].

Our ITS (Fig. 4) follows the four components proposed by Wenger:

1. Knowledge of the domain contains domain ontology, semantic proximity measure, and composite factual semantic features. The users' inputs from the 3D virtual campus are transformed into composite factual semantic features;
2. Student model uses case-based reasoning [1]. During the first step, supervised learning with a group of students generates input from the interpretation layer to the base of scenarii;

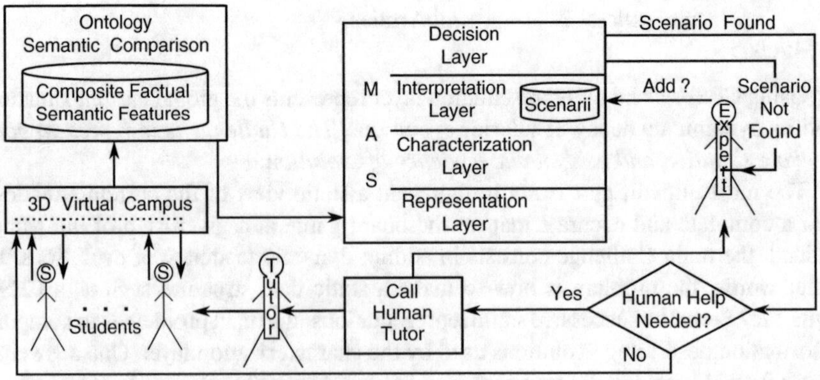

Fig. 4 Architecture of the intelligent tutoring system

3. Teaching models are associated to student models during the supervised learning steps. Decision layer must produce the most appropriate answer for the learner from the data retrieved in the base of scenarii. That layer may also call a human tutor when needed;
4. GE3D virtual campus provides interfaces for students [7].

Architecture of our ITS differs from other multiagent intelligent tutoring systems [16], because we do not ask agents to mimic traditional human roles as teachers, experts or peers. Instead, we choose to represent, characterize and interpret a sequence of learners' activities and compare them to previous recorded sequences. In that respect, we are close to the question asked by Craw [4]: "what happened when we saw this pattern before?"

We also foresee—as a special pattern—the situation when only a human tutor can answer, because the learner is so lost that an ITS could not analyze and help him or her. This is why we include in our ITS a possibility of calling a human tutor when needed. The idea is to make a team with—on the one hand, the intelligent tutor, and on the other hand the human tutor. The intelligent tutor automatically deals with students who learn easily. Thus the human tutor will have more time to focus on students with difficulties.

Figure 5 shows the enriching process of the base of scenarii with students' inputs. One or a few scenarii which are similar to the current situation and associated to the given profile of the student emerge from the system. The system will either adapt the chosen scenario or ask the expert for creating a brand-new one.

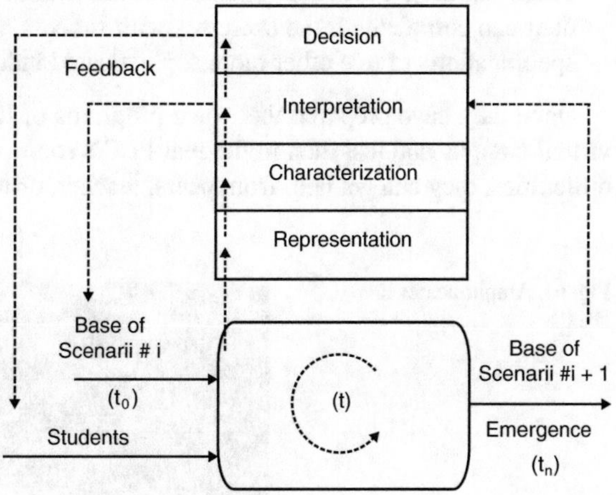

Fig. 5 Enriching process of the base of scenarii

4 3D Virtual Campus

We have built the GE3D virtual campus that will be connected to the ITS. This platform is a virtual site in 3D similar to university premises [7] with several rooms:

1. The patio, which is the entrance of the site. Inside, the students can consult their schedules, their assessments, and other information. The connected person is represented in 3D by a simple photo as an avatar.
2. The amphitheater (Fig. 6) which is generic. There, different kinds of lectures (mathematics, physics, ...) can be given. Then the material displayed on the screen will vary accordingly.
3. The examination room for e-assessments. The usual communication tools are not available in this room. Their examination taken, students can immediately get their marks.
4. The library for e-books and media.
5. The meeting room. There, any student can simultaneously collaborate using a whiteboard and interact with the other connected people. This room is also a generic room where any discipline of teaching can be provided (Fig. 7).
6. The room for practical workshops (Fig. 8) using Programmable Logic Controllers (PLC's). A simple actuator is presented here by a cylinder and its Sequential Function Chart (SFC) program. With the 3D environment users can undertake activities in a synchronous way. For example, when one student activates the SFC, the rod of the cylinder moves according to its evolution cycle. At the same time, all connected users can see the cylinder movements and the SFC evolutions synchronously. Other simple exercises are available in the same room. If a student can complete all the exercises with success, then he can access videos and specifications of five other more sophisticated industrial processes.

Once they have prepared their own programs of the process, they can leave the virtual campus and test then in the real PLC's room. As always, if they meet some difficulties, they can get help from peers, teacher, or intelligent tutoring system.

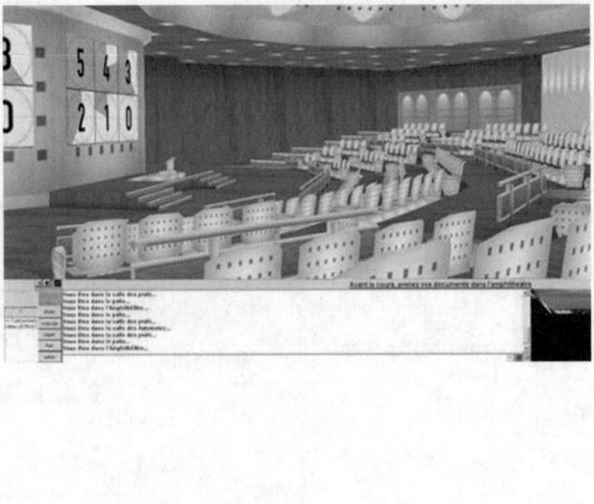

Fig. 6 Amphitheater in GE3D

Fig. 7 Meeting room in GE3D

Fig. 8 Programmable logic controllers room in GE3D

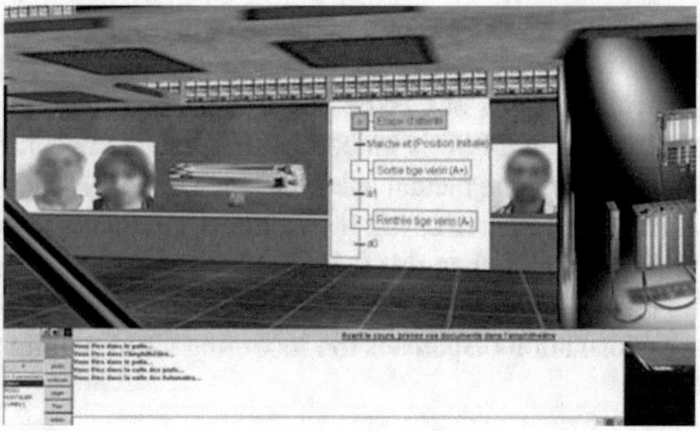

5 Multiagent Representation Layer

Figure 9 represents the successive steps of multiagent internal processing. Layer L0 indicates external input of data represented by two composite factual semantic features SF_1 and SF_2, later followed by SF_i. L1, L2, and L3 are short notations for representation, characterization, and interpretation layers. The arrival of SF_1 triggers processing P_1 in representation layer L1. The span of time used for processing changes according to internal activity needed to take the new data into account. The processing ends when no more message is sent nor received by agents and when no more internal evolution takes place within agents. During experimentation, no deadlock nor infinite loop happened, thus representation process always ended after finding an equilibrium.

Characterization layer L2 waits until representation layer processes completion to start process P'. In the same way interpretation layer L3 waits until characteriza-

Fig. 9 Successive steps of multiagent processing

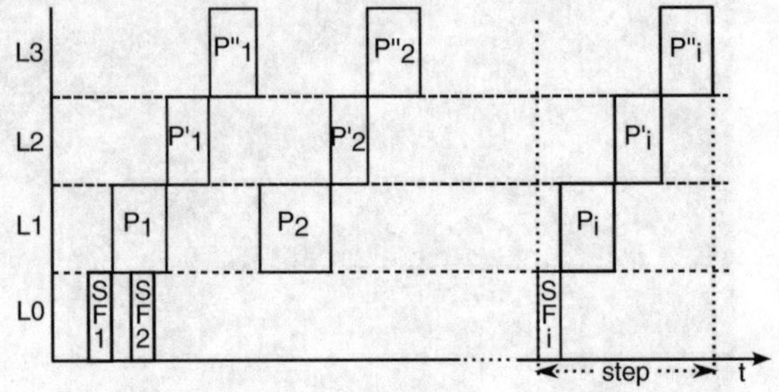

tion layer processes termination for beginning process P". When $P"_1$ is completed, SF_2—which has been put in the queue—is processed. As shown in Fig. 9 a step is defined as the whole process initiated by the arrival of a composite factual semantic feature followed by the successive processes of representation, characterization, and interpretation layers. Our prototype is coded with Java Agent DEvelopment framework [2].

The successive transformations and uses of a given incoming CFSF will be detailed below to explain factual agents [17]. The first transformation occurs when users' inputs are converted into CFSFs and sent to the representation layer from predetermined types of CFSFs stored in the ontology. The second transformation starts when a given CFSF is embedded in a factual agent. Each CFSF is associated with one and only one factual agent. When a CFSF reaches the representation layer of the MAS, there are two cases. If a factual agent contains a CFSF with the same key (couple *type-name*), then this factual agent is updated with the incoming CFSF. There is no conservation of the previous one. Else, a new factual agent is created. The third transformation dispatches the incoming CFSF within the whole multiagent system.

5.1 Internal Indicators

Figure 10 shows the structure of a factual agent where the knowledge part is the composite factual semantic feature, and where the computed behavior part is made of four internal indicators, an automaton describing the level of internal activity and an acquaintance network. The three components of the behavior part are strongly bound.

The four indicators are *position p*, *celerity c*, *acceleration a*, and *satisfaction s*. The computation of *position p* depends on the type of CFSF. This indicator represents the agent in the representation space. Definitions and computations of all the other parts of the behavior are independent on the type of CFSF. *Celerity c* and *acceleration a* describe the internal dynamics of the agent between two evolutions *t* and *t'*:

$$c_{t'} = p_{t'} - p_t \tag{1}$$

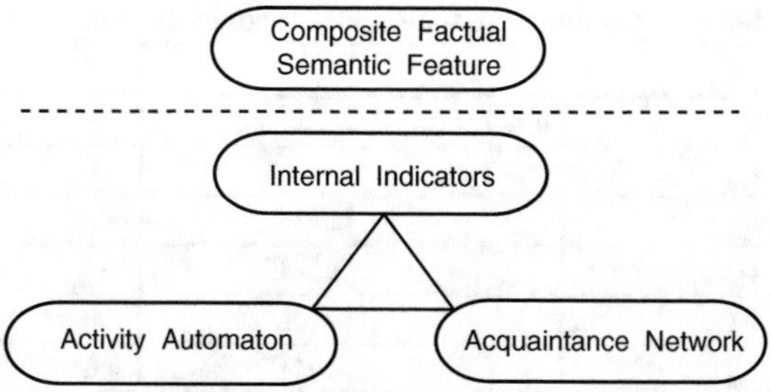

Fig. 10 Structure of a factual agent

$$a_{t'} = c_{t'} - c_t \quad (2)$$

Satisfaction s is linked to the automaton, and so, will be defined after the presentation of the automaton.

5.2 Activity Automaton

Figure 11 shows the activity automaton with its five states. This automaton is set to the *initial* state when a factual agent is created. Reaching the *end* state causes the death of its factual agent. The states are typical of the significance of its associated CFSF in the whole representation, from minimal to maximal activity. There is always one and only one transition from the current state to another one or to the same state. As a notation, t_{ij} stands for the transition from state *i* to state *j*. Transitions t_{23} and t_{34}—associated with acquaintance network—are explained below in the paragraph about semantic proximity. Conditions of transitions depend on values of internal indicators. Table 1 contains the transitions and associated conditions.

The satisfaction indicator measures whether a transition occurs from one state to the same one or from a state to a different state. This indicator keeps the last ten values in memory and adds them up to give a single value. Thus, satisfaction indicates the level of activity in [0 .. 20].

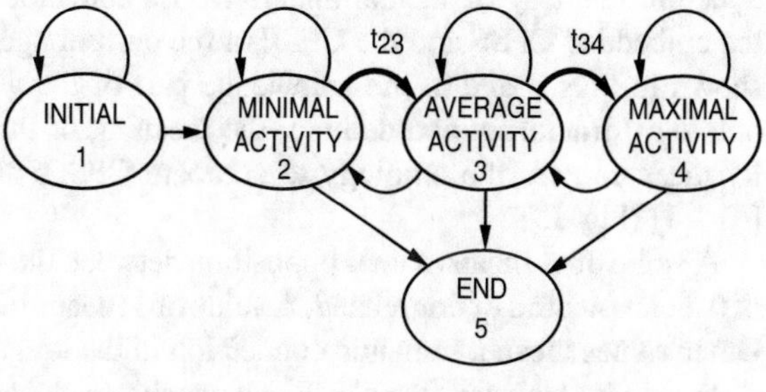

Fig. 11 Activity automaton of a factual agent

Table 1 Conditions of transitions in activity automaton

	p	c		a	s
t_{11}	$p < 1$				
t_{12}	$p \geqslant 1$				
t_{22}		$c \leqslant 0$			$s \geqslant 0$
t_{23}		$c > 0$			$s \geqslant 0$
t_{25}					$s < 0$
t_{32}		$c \leqslant 0$	and	$a \leqslant 0$	$s \geqslant 0$
t_{33}		$c \leqslant 0$	xor	$a \leqslant 0$	$s \geqslant 0$
t_{34}		$c > 0$	and	$a > 0$	$s \geqslant 0$
t_{35}					$s < 0$
t_{43}		$c \leqslant 0$	or	$a \leqslant 0$	$s \geqslant 0$
t_{44}		$c > 0$	and	$a > 0$	$s \geqslant 0$
t_{45}					$s < 0$

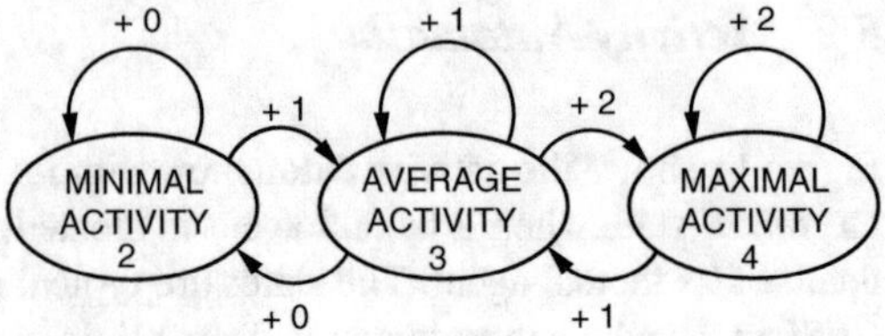

Fig. 12 Transitions and values used for computation of satisfaction

Figure 12 shows the transitions which are active for the computation of satisfaction; i.e., when the given transition is activated, the associated value of the arrow is the new value added to satisfaction.

5.3 Acquaintance Network

The third component of the behavior part is an acquaintance network. This is a dynamic memory of factual agents whose semantic proximity measure, between the embedded CFSF and the CFSF of the current agent, is semantically related. As shown in Figs. 3 and 4, the knowledge part of the whole system contains domain ontology structuring and defining the meaning of the observed facts. The proximity measure uses the ontology to compare CFSFs and returns a value in interval [–1 .. 1] (Fig. 13).

A value of -1 means a total opposition between the two compared CFSFs. A value of 0 means neutral or nonrelated. A value of 1 means identity between the two CFSFs. Other values mean a semantic connection in the range from opposite to close.

Proximity between two CFSFs—opposite or close—is either strong or weak. As previously described, transition occurs according to predefined values of internal

Fig. 13 Semantic proximity measure

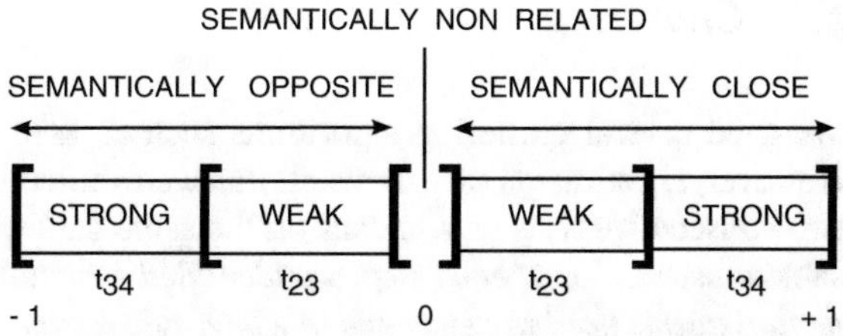

indicators. Two transitions—t_{23} and t_{34}—cause particular actions as displayed in Fig. 11: transition from state "minimal activity" to state "average activity," and transition from state "average activity" to state "maximal activity." The first one, t_{23}, triggers messages to agents of the acquaintance network with weak proximity. The second one triggers messages to agents of the acquaintance network with strong proximity.

A multiagent system is made of several agents interplaying. Systematically, when a factual agent is created or updated, it broadcasts messages containing its own CFSF to all the others but itself. Then, each receiving agent uses semantic proximity measure to compare the CFSF carried by the message with its private CFSF. If, and only if, the result returned by semantic proximity measure is semantically related, i.e., different from 0, then the receiving agent is activated. In this case, its acquaintance network is kept up-to-date and its position can be modified. As a consequence, that could cause changes in internal indicators and lead to a new transition in the activity automaton which might in its turn generate an evolution inside the set of semantically connected agents. This third transformation dispatches the incoming CFSF beyond its own factual agent consistent with the strength of this new CFSF in the global representation.

6 Multiagent Characterization Layer

The aim of the characterization layer is to determine the clusters of the agents of the representation layer. Each characterization agent of this layer:

1. is associated with one cluster;
2. receives data from factual agents of this cluster;
3. computes global parameters identifying the associated cluster.

When ready, characterization agents send their data to the interpretation layer. Below we explain first the clustering of the factual agents and second the characterization of each subset computed by characterization agents.

6.1 Clustering

We tried several clustering algorithms such as K-Means [14] and DBSCAN [6]. However, experiments gave us results that were difficult to interpret or to reproduce—two consecutive tests of K-means on the same data would not necessarily give the same clusters. So, *at each step, we identified structural and dynamic properties of factual agents used to determine to which subset every factual agent belonged.*

The following notations are used to define the partition:

$\mathbf{E}_{FA}$ is the set of all factual agents.

$\mathbf{E}_{C}$ is the singleton containing only the factual agent whose composite factual semantic feature has just been updated. Here, "C" is a short notation for CFSF.

$\mathbf{E}_{AN}$ is the subset of factual agents semantically related to E_C. This subset contains all the factual agents of the acquaintance network of factual agent whose CFSF has just been updated.

The acquaintance network subset E_{AN} could be partitioned into two smaller subsets, where $\mathbf{E}_{AN+}$ is the subset of agents of the acquaintance network which are semantically close and $\mathbf{E}_{AN-}$ the subset of agents of the acquaintance network which are semantically opposite:

$$E_{AN} = E_{AN+} \cup E_{AN-} \tag{3}$$

$\mathbf{E}_{\overline{CAN}} = \{\mathbf{E}_{FA} - \mathbf{E}_{C} - \mathbf{E}_{AN}\}$ is the subset of all the other factual agents, that is to say, all factual agents except the one whose composite factual semantic feature has just been updated together with those of its own acquaintance network as displayed in Fig. 14.

By definition, the intersection of the subsets of the partition is empty. From a structural point of view, the partition of the whole set of factual agents is defined in (4):

$$E_{FA} = E_C \cup E_{AN+} \cup E_{AN-} \cup E_{\overline{CAN}} \tag{4}$$

A characteristic of all factual agents is to contain internal indicators: position, celerity, acceleration, and satisfaction whose shortened notations are "p", "c", "a" and "s".

That leads to complementary notations for subset $E_{\overline{CAN}}$:

$\mathbf{E}_{\Delta PCA} = \{E_{\overline{CAN}} : \exists \Delta\, p \text{ or } \exists \Delta\, c \text{ or } \exists \Delta\, a\}$ where at least one position, celerity, or acceleration variation exists.

$\mathbf{E}_{\Delta S} = \{E_{\overline{CAN}} : \nexists \Delta\, p \text{ and } \nexists \Delta\, c \text{ and } \nexists \Delta\, a \text{ and } \exists \Delta\, s\}$ for which at least one satisfaction indicator variation exists but no variation for position, celerity, and acceleration.

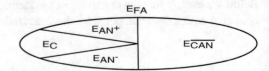

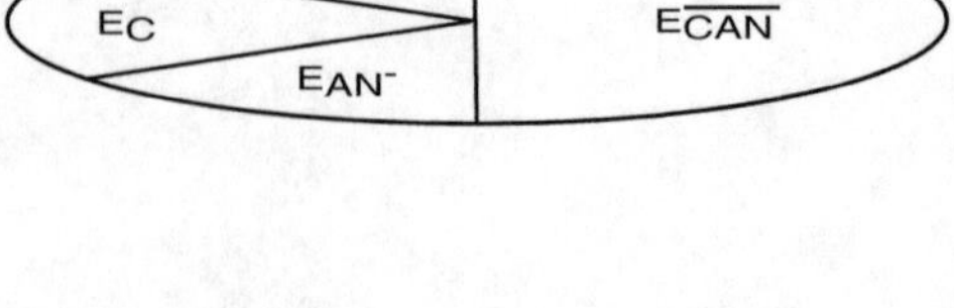

Fig. 14 Partition of the set of factual agents

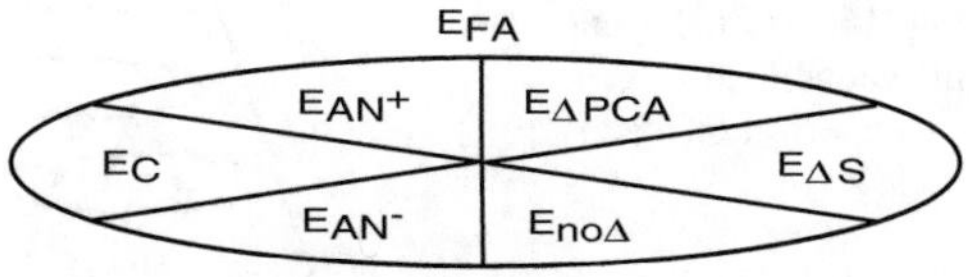

Fig. 15 Detailed partition of the set of factual agents

$\mathrm{E}_{no\Delta} = \{E_{\overline{CAN}} \ : \ \nexists\ \Delta\, p \text{ and } \nexists\ \Delta\, c \text{ and } \nexists\ \Delta\, a \text{ and } \nexists\ \Delta\, s\}$ for which there is no variation for either position, celerity, acceleration, or satisfaction indicators.

Subset $\mathrm{E}_{\overline{CAN}}$ could be partitioned into three subsets:

$$E_{\overline{CAN}} = E_{\Delta PCA} \ \cup \ E_{\Delta S} \ \cup \ E_{no\Delta} \tag{5}$$

Now, using (4) and (5), the partition of the factual agents of the representation layer could be defined by (6) and shown in Fig. 15:

$$\begin{aligned} E_{FA} = E_C \ \cup \ E_{AN^+} \ \cup \ E_{AN^-} \ \cup \\ E_{\Delta PCA} \ \cup \ E_{\Delta S} \ \cup \ E_{no\Delta} \end{aligned} \tag{6}$$

Expression (6) of E_{FA} is adequate for showing the impact of a given composite factual semantic feature on the decision support system by highlighting the strength of the new data for the factual agent directly concerned by composite factual semantic feature (E_C) and for its two semantically related subsets (E_{AN^+}, E_{AN^-}). The analysis of data from representation layer shows that the impact of a given composite factual semantic feature could be important. In some cases, a CFSF influences factual agents beyond agents belonging to its acquaintance network:

1. The case when $\mathrm{E}_{\Delta PCA}$ is not empty, is called a *propagation/diffusion* beyond acquaintance network.
2. The other particular situation, when $\mathrm{E}_{\Delta S}$ is not empty means a *trail effect* affecting some factual agents which are not directly linked to the new or updated factual semantic feature but have been recently quite active.

6.2 Subset Characterization

Each subset E_C, E_{AN^+}, E_{AN^-}, $\mathrm{E}_{\Delta PCA}$, $\mathrm{E}_{\Delta S}$ and $\mathrm{E}_{no\Delta}$ is represented by one characterization agent (Fig. 16). When representation layer processing ends after reaching an equilibrium, factual agents exchange messages between themselves and with characterization agents for:

1. asking each factual agent to find to which subset it belongs;
2. informing each characterization agent of the beginning of a new step by sending the names of the factual agents which are members of the subset;

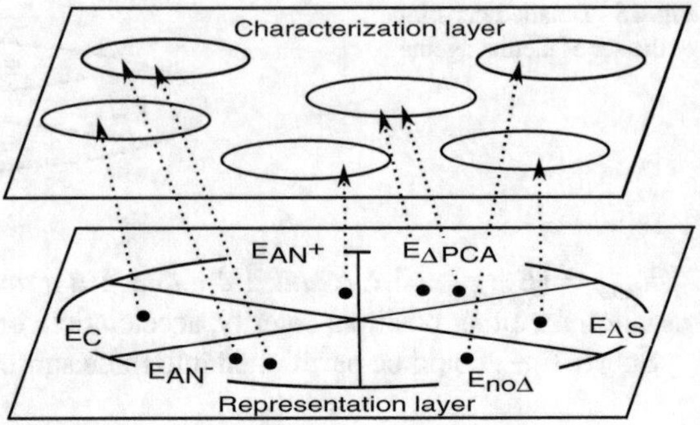

Fig. 16 Factual agents membership to characterization agent

3. requesting each factual agent to send to its associated characterization agent a list of all variations of indicators—position, celerity, acceleration, satisfaction—that occurred during the last representation processing. This information is the input of the characterization layer.

When all characterization agents have received messages from all its factual agents—with individual list of variations of internal indicators, each characterization agent needs to compute the global signature of each subset from the received data. Before explaining *how* to compute characterization parameters, the key point is to identify *why*. The results of the characterization layer are inputs of the interpretation layer. The interpretation layer relies on similarity measure to retrieve stored past situations with a similar pattern as the target situation being analyzed. The similarity measure of the interpretation layer uses characterization parameters defined as signature during characterization layer processing. So, the challenge is to design characterization parameters before being able to test them in the interpretation layer. As a consequence, we chose to define several characterization parameters to anticipate the needs of the interpretation layer: cardinal number of subsets determined by (6), subset variation ratio, dispersion and shape parameters.

Figure 17 shows that cardinal number of subsets vary depending on steps, but the cardinal number of E_C which is always equal to one and gives scale of this figure. As displayed, cardinal numbers of E_{AN^-}, $E_{\Delta PCA}$, $E_{\Delta S}$ could be zero, which could also be true for E_{AN^+}, even if it is not the case in this sequence of steps coming from the experiments of our prototype.

For nonempty subsets, variation ratio describes the density of evolution of a given subset by counting numbers of variations of position indicator for each agent belonging to this subset and dividing this sum by the cardinal number of the subset. This ratio is not computed for celerity and acceleration indicators, to avoid duplication, nor for satisfaction which is already a cumulative indicator.

Our first approach to compute dispersion and shape parameters was to choose descriptive statistics as standard deviation and skewness. But, definitions of standard deviation or even absolute deviation use an average value: the mean for standard deviation and the median or the mode for absolute deviation. Using an average value is relevant for a sample of data where there is no particular data to be given

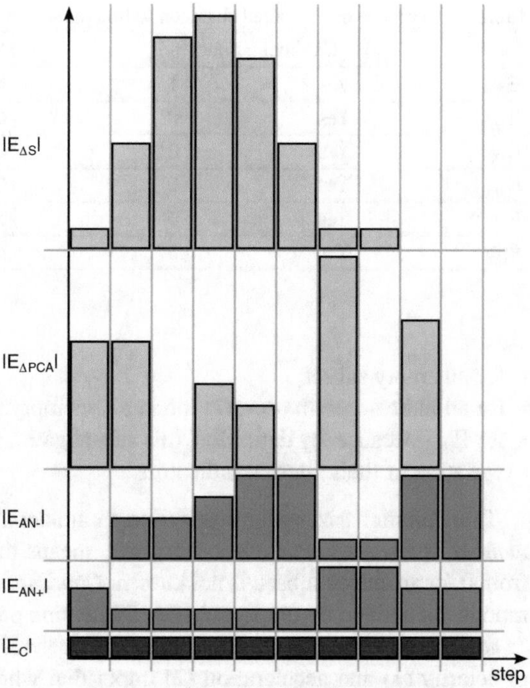

Fig. 17 Cardinal number of subsets on several successive steps

priority. This is not the case with our data because, at the end of representation layer processing, each factual agent determines one last value summarizing successive values for each of its internal indicators. So, to keep the benefit of representation layer treatment giving last values as references, we chose to define dispersion and shape parameters to compute global characterization of subsets with the last value obtained when variations occurred instead of an average value. Last value notation is x_n. The computed deviation, referring to last value n, is called ${}^n\sigma$ to avoid confusion with standard deviation σ. In the same way, ${}^n\gamma$ is the computed skewness:

$$ {}^n\sigma = \sqrt{\frac{1}{n}\sum_{i=1}^{n}(x_i - x_n)^2} \tag{7} $$

$$ {}^n\gamma = \frac{\frac{1}{n}\sum_{i=1}^{n}(x_i - x_n)^3}{{}^n\sigma^3} \tag{8} $$

Using (7) and (8), each characterization agent computes, for its associated subset, parameters ${}^n\sigma$ and ${}^n\gamma$ for internal indicators position, celerity, acceleration, and satisfaction. There is no parameter to compute:

Table 2 Existence of computed characterization parameters

Subset	Cardinal number		Variation ratio	Dispersion, shape
E_C	Yes	1	Yes	p, c, a, s
E_{AN^+}	Yes	0^+	Conditional	p, c, a, s
E_{AN^-}	Yes	0^+	Conditional	p, c, a, s
$E_{\Delta PCA}$	Yes	0^+	Conditional	p, c, a, s
$E_{\Delta S}$	Yes	0^+	No	s
$E_{no\Delta}$	Yes	0^+	No	No

- for an empty subset;
- for an indicator with no variation in a nonempty subset;
- for $E_{no\Delta}$ because, by definition, this subset gathers together factual agents with no variation of their internal indicators.

Table 2 summarizes characterization parameters. As E_C is a singleton, its *cardinal number* is always 1. For other subsets 0^+ means that the cardinal number, starting from 0 for an empty subset, is not known. *Conditional* refers to the fact that variation ratio is not defined for empty subsets. *Dispersion* parameter $^n\sigma$ and *shape* parameter $^n\gamma$ could have values for position, celerity, acceleration, and satisfaction. Definitions of celerity (1) and acceleration (2) imply that when variation occurs for position, values are computed for celerity and acceleration.

The following example illustrates the computation of parameters. A subset with two factual agents has successive values for position indicator:

agent#i	Values				x_n#i
agent#1	1	2	4		x_n#1 = 4
agent#2	3	12	7	15	x_n#2 = 15

There are three values for agent#1 and four values for agent#2, so variation ratio is: $\frac{3+4}{2} = 3.5$

A property of standard deviation σ which is also true for $^n\sigma$ is that adding a constant to every value does not change the deviation. For this example, x_n#2 is the chosen constant for agent#1 and x_n#1 is the chosen constant for agent#2. These choices give x_n#1 equal to x_n#2:

agent#i	New values				x_n#i
agent#1	16	17	19		x_n#1 = 19
agent#2	7	16	11	19	x_n#2 = 19

Using $|16|17|19|7|16|11|19$ with (7) and (8), rounded results are $^n\sigma = 5.73$ and $^n\gamma = -1.75$.

After the completion of characterization layer processing, characterization agents send their data to the interpretation layer, which is detailed in the next section.

7 Multiagent Interpretation Layer

Characterization layer data are sent to the interpretation layer for retrieving past cases which are similar to the target case being analyzed. To do so, the interpretation layer uses a similarity measure. Outputs of the interpretation layer—past cases retrieved—must be sent to the decision layer. Below we detail how to represent a given case and the definition of the similarity measure.

The characterization layer sends the characterization parameters of each subset to the interpretation layer. Thus, the input of the interpretation layer is a vector containing computed parameters described in Table 3. The size of this vector depends on the content of the subsets. The existence of a value for a parameter is marked by ✓ as in E_C for variation ratio. The satisfaction indicator evolves only when the activity of the agent implies a change of state in the internal automaton, however, it is not always the case and it is then represented by ? in Table 3.

When the cardinal number of a subset is equal to zero, there is no value for parameters which are indicated by x. In several cases we observe that E_{AN} contains agents (positive cardinal number) but the impact of the incoming CFSF is not important enough to impulse any change for position, celerity or acceleration indicators resulting in a variation ratio equal to zero. By definition $E_{\Delta PCA}$ only contains agents with evolutions of indicators. So variation ratio is strictly positive when this subset is nonempty. Variation ratio depends on position indicator and therefore is always equal to zero for subset $E_{\Delta S}$.

Results obtained with a sample of experimental cases give 100 % for E_C (A, Fig. 18). A third of cases is when E_{AN} has a strictly positive variation ratio. E_{AN} defined in (3) regroups cases where at least E_{AN+} or E_{AN-} is nonempty with a strictly

Table 3 Variable number of computed parameters

			n_σ				n_γ			
Subset	c	r	p	c	a	s	p	c	a	s
E_C	1	✓	✓	✓	✓	?	✓	✓	✓	?
E_{AN+} or E_{AN-}	0	x	x	x	x	x	x	x	x	x
	1^+	= 0	x	x	x	?	x	x	x	?
	1^+	>0	✓	✓	✓	?	✓	✓	✓	?
$E_{\Delta PCA}$	0	x	x	x	x	x	x	x	x	x
	1^+	>0	✓	✓	✓	?	✓	✓	✓	?
$E_{\Delta S}$	0	x	x	x	x	x	x	x	x	x
	1^+	= 0	x	x	x	✓	x	x	x	✓

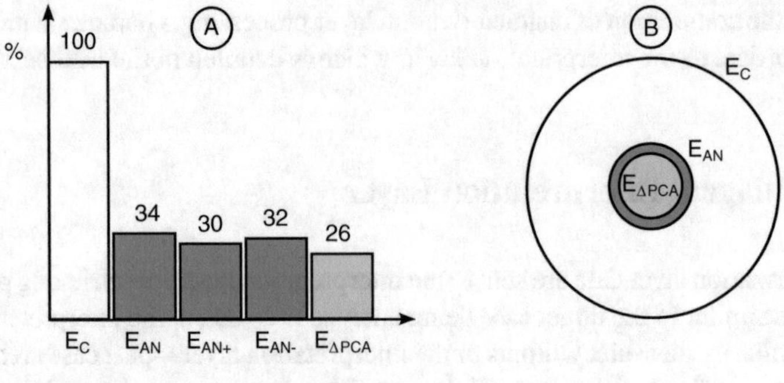

Fig. 18 Percentage of strictly positive variation ratio for tested cases

positive variation ratio. $E_{\Delta PCA}$ contains agents when activity in E_{AN} is important enough to spread beyond E_{CAN}. Consequently percentage of nonempty subset $E_{\Delta PCA}$ is always inferior to percentage of E_{AN} (A, Fig. 18) and only happens when E_{AN} is nonempty with a strictly positive variation ratio (B, Fig. 18).

Figure 19 summarizes the content of input computed parameter vector for a given case. Similarity measure relies on those data and must give priority to E_C parameters as they are the only data available in two-thirds of cases (A, Fig. 19).

Similarity measure must also take into account parameters describing evolution of E_{AN} and $E_{\Delta PCA}$. Data from $E_{\Delta S}$ are not used in the similarity measure. Instead they indicate a global level of activity from the whole multiagent system. Similarity is a comparison between two cases computed by:

$$\begin{aligned} sim(target, c\#i) = {} & simE_C(target, c\#i) \\ & and\ simE_{AN}(target, c\#i) \\ & and\ simE_{\Delta PCA}(target, c\#i) \end{aligned} \tag{9}$$

As described in Table 4 simE$_C$(target, c#i) compares $^n\sigma$ and $^n\gamma$ for position, celerity, and acceleration; simE$_{AN}$(target, c#i) compares values of cardinal number and

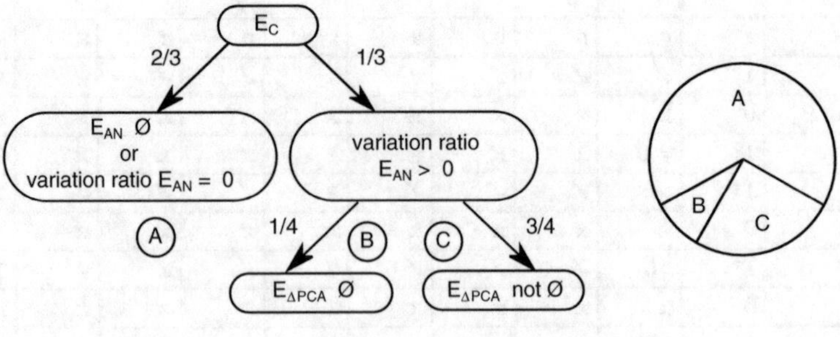

Fig. 19 Frequency of having value for subsets in computed parameter vector

Table 4 Compared values in similarity measure

simE_C(target, c#i)	c#i	target	c#i
	${}^n\sigma(p) - 1\,\% \leq$	${}^n\sigma(p) \leq$	${}^n\sigma(p) + 1\,\%$
and	${}^n\sigma(c) - 1\,\% \leq$	${}^n\sigma(c) \leq$	${}^n\sigma(c) + 1\,\%$
and	${}^n\sigma(a) - 1\,\% \leq$	${}^n\sigma(a) \leq$	${}^n\sigma(a) + 1\,\%$
and	${}^n\gamma(p) - 1\,\% \leq$	${}^n\gamma(p) \leq$	${}^n\gamma(p) + 1\,\%$
and	${}^n\gamma(c) - 1\,\% \leq$	${}^n\gamma(c) \leq$	${}^n\gamma(c) + 1\,\%$
and	${}^n\gamma(a) - 1\,\% \leq$	${}^n\gamma(a) \leq$	${}^n\gamma(a) + 1\,\%$
simE_{AN}(target, c#i)	c#i	target	
	$(c = 0)$ *or* $(r = 0)$	*and* $(c = 0)$ *or* $(r = 0)$	
or	$(r > 0)$	*and* $(r > 0)$	
$\text{simE}_{\Delta PCA}$	c#i	target	
	$(r = 0)$	*and* $(r = 0)$	
or	$(r > 0)$	*and* $(r > 0)$	

variation ratio; $\text{simE}_{\Delta PCA}$(target, c#i) compares variation ratio only when $\text{E}_{\Delta PCA}$ is nonempty (Fig. 19). Some compound parameters are not used in the similarity measure described above. However, we keep them all as we are still testing alternative similarity measures.

The interpretation layer processes data coming from the characterization layer:

1. to store incoming compound parameter vector;
2. to retreive stored cases and compute a similarity measure with the current target case.

8 Multiagent Decision Layer

The decision layer receives the result of the interpretation layer which is a list of past cases which are similar to the target case. Even if this layer is still under development, we assume to follow these hypotheses:

- the term *case* refers to a *step* previously defined (Fig. 9) *as the whole process initiated by the arrival of a composite factual semantic feature followed by the successive processes of representation, characterization, and interpretation.*
- the special pattern where only a human tutor can answer, is when the list of past cases is empty. In that situation a call on an human tutor is needed. During the development phase, the expert and the tutor are the same person.
- a *scenario* is a list of successive steps.

The expected result of the decision layer is to return a case chosen to help the student by offering her or him a personalized answer.

When the expert is called to decide to add or not a new scenario, the system provides a description of the successive actions of the student. The expert does not see at all the computed process described in this article except for tests. The expert must judge to add a scenario and he associates to it actions to be activated in the students' interface. By doing so, the expert will enrich the work of the decision support system—the result of decision layer which is the aim of the whole MAS—with concrete actions to help the student.

9 Conclusion

Here we have presented an intelligent tutoring system with its 3D students' interface and its internal decision support system designed to face the challenge of processing the evolution of a large amount of data. Our prototype was tested and used to improve the design of agents. The GE3D campus was well approved by students on programmable logic controllers course.

A Ph.D. thesis focuses on the interpretation layer to propose a dynamic case-based reasoning architecture able to merge incoming data while looking for suitable scenarii [24, 25].

Our architecture is designed to be as independent as possible on learning fields. The key point of our current work is the development of the knowledge part, i.e., ontology, semantic proximity, and predetermined composite factual semantic features, for programmable logic controllers course to connect the intelligent tutoring system to the 3D virtual campus. Courses on other topics such as English, computer science, or sociology are also planned. Combining Bloom's taxonomy and Felder-Silverman's index of learning styles as a guide for intelligent tutoring system course designers could be a way to create the knowledge part regardless of the topic to teach [11].

Finally, we project to elaborate a 3D visualization to display and follow evolutions of agents and messages exchanged inside the multiagent system.

References

1. Aamodt, A.: Case-based reasoning and intelligent tutoring. In: SAIS-SSLS Proceedings, pp. 8–22. Vasteras, Sweden (2005)
2. Bellifemine, F., Caire, G., Greenwood, D.: Developing Multi-agent Systems with JADE. Wiley (2007)
3. Clemens, M.: The art of complex problem solving. http://www.idiagram.com/CP/cpprocess.html
4. Craw, S.: Case based reasoning lecture. Robert Gordon University (2013). http://www.comp.rgu.ac.uk/staff/smc/teaching/cm3016/Lecture-1-cbr-intro.ppt
5. Denning, P.: Computer science: the discipline. In: Ralston, A., Hemmendinger, D. (eds.) Encyclopedia of Computer Science. Wiley (2000)

6. Ester, M., Kriegel, H.P., Sander, J., Xu, X.: A density-based algorithm for discovering clusters in large spatial databases with noise. In: Simoudis, E., Han, J., Fayyad, U.M. (eds.) Proceedings of the 2nd International Conference on Knowledge Discovery and Data Mining (KDD), pp. 226–231. AAAI Press (1996)
7. Grieu, J., Lecroq, F., Person, P., Galinho, T., Boukachour, H.: GE3D: a virtual campus for technology-enhanced distance learning. Int. J. Emerg. Technol. Learn. **5**(3), 12–17 (2010). doi:10.3991/ijet.v5i3.1388. http://online-journals.org/i-jet/article/view/1388
8. Hafner, K.: Software tutors offer help and customized hints (2004). http://www.nytimes.com/2004/09/16/technology/circuits/16tuto.html?_r=0
9. Knuth, D.E.: The Art of Computer Programming, Volume I: Fundamental Algorithms, 3rd edn. Addison-Wesley (1997)
10. Kolodner, J.: An introduction to case-based reasoning. Artif. Intell. Rev. **6**, 3–34 (1992)
11. Lecroq, F., Grieu, J., Person, P., Galinho, T., Boukachour, H.: Intelligent tutoring system in GE3D virtual campus. Int. J. Comput. Sci. Artif. Intell. 6 (2012). http://www.jcsai.org/paperInfo.aspx?ID=20
12. Luger, G.: Artificial Intelligence: Structures and Strategies for Complex Problem Solving. Addison-Wesley (2005)
13. Lyne, O.: Risk faq—version 5.61. http://www.kent.ac.uk/smsas/personal/odl/riskfaq.htm
14. MacQueen, J.: Some methods for classification and analysis of multivariate observations. In: Proceedings of 5th Berkeley Symposium on Mathematical Statistics and Probability pp. 281–297 (1967)
15. Murr, M.: How forensic tools recover digital evidence (data structures) (2007). http://www.forensicblog.org/how-forensic-tools-recover-digital-evidence-data-structures
16. Paviotti, G., Rossi, P., Zarka, D.: Intelligent tutoring systems: an overview (2012). http://www.intelligent-tutor.eu/files/2012/06/2012_Intelligent_Tutoring_Systems_overview.pdf
17. Person, P., Galinho, T., Lecroq, F., Boukachour, H., Grieu, J.: Intelligent tutor design for a 3D virtual campus. In: IEEE Conference of Intelligent Systems (IS12), pp. 74–79. Sofia, Bulgaria (2012). http://dx.doi.org/10.1109/IS.2012.6335194
18. Vigicrues: (2013). http://www.vigicrues.ecologie.gouv.fr/niveau3.php?idspc=7Źidstation=742
19. Wenger, E.: Artificial Intelligence and Tutoring Systems: Computational Approaches to the Communication of Knowledge. Morgan Kaufmann (1987)
20. Willging, S.: Factors that influence students' decision to dropout of online courses. J. Asynchronous Learn. Networks **8**(4), 105–118 (2004)
21. Wolf, M.: An intelligent artificial player for the game of Risk. Master's thesis, TU Darmstadt, Knowledge Engineering Group (2005). http://www.ke.informatik.tu-darmstadt.de/lehre/arbeiten/diplom/2005/Wolf_Michael.pdf. Diplom
22. Wooldridge, M.J.: An Introduction to MultiAgent Systems, 2nd edn. Wiley (2009)
23. Woolf, B.: Building Intelligent Interactive Tutors: Student-centered strategies for revolutionizing e-learning. Elsevier Science (2010)
24. Zouhair, A., En-Naimi, E.M., Amami, B., Boukachour, H., Person, P., Bertelle, C.: Multiagent case-based reasoning and individualized follow-up of learner in remote learning. In: The 2nd International Conference on Multimedia Computing and Systems, (ICMCS'11), pp. 1–6. Ouarzazate , Morocco (2011). 10.1109/ICMCS.2011.5945644
25. Zouhair, A.: Raisonnement à Partir de cas dynamique multi-agents : application à un système de tuteur intelligent. PhD thesis, Tanger, Morocco (2014)

Generalized Net Model of the Scapulohumeral Rhythm

Simeon Ribagin, Vihren Chakarov and Krassimir Atanassov

Abstract The dynamics of the upper extremity can be modeled as the motion of an open kinematic chain of rigid links, attached relatively loosely to the trunk. The upper extremity or upper limb is a complex mechanism which includes many bones, joints, and soft tissues allowing various movements in space. The shoulder is one of the most complex musculoskeletal units not only in the upper limb but also in the entire human body. Codman [7] understood the complex and dependent relationships of the structures of the shoulder when he coined the term "*scapulohumeral rhythm*" to describe the coordinated motion. The purpose of the paper is to present a simple mathematical model of the scapulohumeral rhythm using the apparatus of the Generalized Nets Theory [1, 2]. The presented model is a part of a series of generalized net models of the upper limb developed by the authors (see [22, 23]).

Keywords Generalized net · Modeling · Scapulohumeral rhythm · Upper limb

Abbreviations

GN Generalized nets
CVS CardioVascular System
ENS ENdocrine System
LMS LyMphoid System
CNS Central Nervous System
PNS Peripheral Nervous System
MSS Muscle–Skeletal System

S. Ribagin (✉) · V. Chakarov · K. Atanassov
Institute of Biophysics and Biomedical Engineering, Bulgarian Academy of Sciences,
Acad. G. Bonchev Str., Block 105, 1113 Sofia, Bulgaria
e-mail: sim_ribagin@mail.bg

V. Chakarov
e-mail: vihren@clbme.bas.bg

K. Atanassov
e-mail: krat@bas.bg

V. Sgurev et al. (eds.), *Recent Contributions in Intelligent Systems*,
Studies in Computational Intelligence 657,
DOI 10.1007/978-3-319-41438-6_13

S Sagittal plane
F Frontal plane
R Rotation
T Transversal plane

1 Introduction

GN [1, 2] are extensions of Petri nets and their modifications. During the last 30 years, they have a lot of applications in medicine and biology. In [3] GN models of human body and of the separate systems in the human body are described. One of the modeled by GN systems is the muscle–skeletal (see [4]). In the papers [22, 23], the authors discuss GN models of the upper limb together with the circulatory system involved and also a simple example of involuntary movement of the upper extremity. For building a detailed GN model of the human upper limb the authors suggest a theoretical model which includes the basic biomechanical relations among the upper limb structures observed in a simple voluntary movement of the shoulder. The proposed GN model describes the movement from 0° to 180° abduction of the shoulder in the frontal plane.

2 Short Anatomical Description of the Upper Limb

The upper limb or upper extremity is a complex mechanism which includes many bones, joints, and soft tissues allowing various movements in space. The musculoskeletal anatomy of the upper limb is particularly well suited to illustrate and illuminate the anatomical basis of function. In general the upper limb can be divided into shoulder girdle, arm, wrist, and hand. The scapula, clavicle, sternum, and the proximal part of the humerus comprise the shoulder girdle. The shoulder girdle is a complex of five joints: glenohumeral joint, acromioclavicular joint, sternoclavicular joint, scapulothoracic joint, and suprahumeral [5] or subdeltoid joint [13]. The last two are not anatomical but physiological ("false") joints [13, 26]. Arm or "brachium" is composed of three bones: distal part of the humerus, radius, and ulnae. These bones form the elbow complex which includes humeroradial, humeroulnar, and superior radioulnar joints and also distal radioulnar articulation and the so-called "antebrachium" (composed of the radius and ulna). The wrist is a terminal link of the upper limb. The wrist complex includes three joints: radiocarpal, distal radioulnar, and midcarpal joints. The human hand is a multicomponent system not only with motor but also with sensory function. Bones and joint structures of the hand formed a mobile and stable segment [20]. There are carpometacarpal, metacarpophalangeal, and interphalangeal joints. The skeleton of the upper limb is attached relatively loosely to the trunk. That relatively loose attachment maximizes

Table 1 Ligaments and muscles of the upper limb

Upper limb segments	Ligaments	Muscles
Shoulder girdle	lig. interclaviculare, lig. sternoclaviculare, lig. costoclaviculare, lig. coracohumerale, lig. coracoacromiale, lig. transversum scapulae super. et infer., lig. acromioclaviculare, lig. coracoclaviculare	m. trapezius, m. latissimus dorsi, m. levator scapulae, m. rhomboideus, m. pectoralis major, m. pectoralis minor, m. subclavius, m. serratus anterior, m. coracobrachialis, m. deltoideus, m. supraspinatus, m. infraspinatus, m. teres minor, m. teres major and m. subscapularis
Arm	lig. collaterale radiale, lig. collaterale ulnare, lig. annulare radii, Chorda obliqua, Membrana interossea antebrachii	m. brachialis, m. biceps brachii, m. brachioradialis, m. triceps brachii, m. anconeus, m. pronator teres, m. supinator, m. pronator quadratus, m. flexor carpi ulnaris, m. flexor carpi radialis, m. extensor carpi radialis longus, m. extensor carpi radialis brevis and m. extensor carpi ulnaris
Wrist	lig. radiocarpeum palmare, lig. collaterale carpi radiale, lig. collaterale carpi ulnare, lig. radiocarpeum dorsale	m. extensor carpi ulnaris, m. extensor carpi radialis longus, m. extensor carpi radialis brevis, m. flexor carpi radialis, m. flexor carpi ulnaris, m.abductor pillicis longus and m. extensor pollicis brevis
Hand	lig. carpi radiatum, lig. pisohamatum, lig. pisometacarpeum, lig. carpomatacarpeum palmare, ligg. metacarpea palmaria, ligg. metacarpea transversa profunda, ligg. collateralia, ligg. palmaria, ligg. metacarpea dorsalia, ligg. carpometacarpea dorsalia	m. extensor digitorum, m. extensor indicis, m. extensor digiti minimi, m. flexor digitorum profundus, m.flexor digitorum superfacialis, mm. lumbricalis, mm interossei, m.flexor digiti minimi, m. abductor digiti minimi, m. extensor pollicis longus, m. extensor pollicis brevis, m. abductor pollicis longus, m. flexor pollicis longus, m. flexor pollicis brevis, m.opponens pollicis, m.abductor pollicis brevis, m. adductor pollicis and m. opponens digiti minimi

upper limb mobility and flexibility (movement is possible in all 3 planes). The mobility and stability of the upper limb is provided by the large number of ligaments and muscles (see Table 1).

The proper functioning of the upper limb depends entirely on the intactness and coordination of the composed segments together with the major structures of the nervous system involved.

The nerve supply of the upper limb is provided by the brachial plexus and some branches of the cervical plexus (see Table 2). The brachial plexus is formed by the anterior rami of C5 to T1 (the posterior roots give innervation for the skin and

Table 2 Innervation of the upper limb joints and muscles

Peripheral nerve	Innervated joint	Innervated muscle
n. axillaris	Glenohumeral joint	m. Deltoideus, m. teres minor
n. suprascapularis	Glenohumeral joint acromioclavicular joint	m. infraspinatus, m. supraspinatus
n. subclavius	Sternoclavicular joint	m. subclavius, m
n. dorsalis scapulae		m. levator scapulae, mm. rhomboidei
n. thoracalis longus		m. serratus anterior
nn. thoracales anteriores		m. pectoralis major et minor
nn. subscapularis		m. teres major, m. subscapularis
n. thoracodorsalis		m. latissimus dorsi
n. accessorius		m. trapezius
n. pectoralis lateralis et medialis	Glenohumeral joint	m. pectoralis major et minor
n. musculocutaneus	Glenohumeral joint, Humeroradial joint	m. biceps brachii, m. brachialis, m. coracobrachialis
n. medianus	Elbow complex, wrist, and hand	m. pronator teres, m. pronator quadratus, m. flexor carpi radialis, m. flexor digitorum superficialis, mm lumbricalis, m. flexor pollicis brevis, m. opponens pollicis, m. abductor pollicis brevis
n. ulnaris	Elbow complex, wrist, and hand	m. flexor carpi ulnaris, mm. interossei, mm. lumbricalis, m. flexor digiti minimi, m. abductor digiti minimi, m. abductor pollicis, m. opponens digiti minimi, m. flexor digitorum profundus
n. radialis	Elbow complex, wrist and hand	m. brachioradialis, m. triceps brachii, m. anconeus, m. supinator, m. extensor carpi radialis longus, m. extensor carpi radialis brevis
n. interosseus	Elbow comlex and wrist	m. extensor digitorum, m. extensor indicis, m. extensor digiti minimi, m. extensor carpi ulnaris, m. extensor pollicis longus et brevis, m. abductor pollicis longus, m

muscle of the paravertebral area). The anterior rami supply the upper (C5-6), middle (C7), and lower (C8-T1) trunks. At the level of the superior border of the first rib, each trunk divides into an anterior and posterior division. The six divisions combine to form tree cords—lateral, posterior, and medial. At the lower part of the axilla the tree cords split into the terminal branches which enter the arm and innervate the different segments. The major branches of the brachial plexus are n. axillaris, n. musculocutaneous, n. ulnaris, n. radialis, and n. medianus.

The PNS connects the brain and the spinal cord to the periphery and it includes the cranial nerves, the spinal nerves, the peripheral nerves, and the peripheral extension of the autonomic nervous system [25]. Within peripheral nerves are motor and sensory fibers. The sensory fibers receive information from the receptors in the muscles, tendons, joints, and skin. Through these special structures providing information about muscle length, muscle tension, joint angles, and indication of the distribution of forces at points of contact becomes possible (somatosensory receptors). That information is transmitted via afferent roots to the CNS. The processing and analysis of this information is subjective expression in the emergence of different senses. Its two submodalities are sense of stationary position (position sense) and sense of movement (kinesthetic sense) [16]. Apart from sensory, muscles are innervated by motor nerve fibers. Motor nerve fibers are called motoneurons and innervate the different parts of the muscle tissue. The relationship between a sensory and a motor neuron in the gray matter of the spinal cord is performed by an intermediate/inhibit/interneuron. The most routes from the higher centers of the CNS ended on these interneurons.

3 Short Description of the "Relaxed" (Resting) Position of the Upper Limb

For the purpose of the present paper we will describe shortly our concept on the "relaxed position" of the upper limb and inner-relationship between musculoskeletal and nervous systems.

In terms of the upright posture, which is a natural one (body position in space) in the human, the upper limbs are freely granted to the body as "volare" surface of the hand facing the body. Normally shoulders have a round contour due to prominence of the grater tuberosity beneath the deltoid muscle [14] and they are both with symmetrical height. (However, many people have lower shoulder on the dominant side [15].) The shoulders are slightly protracted but relatively relaxed [11]. There is a minimal flexion in the elbows and the forearms (antebrachium) which are in semi-pronated position. All of the five fingers are slightly flexed at all their joints. For the maintenance of this position it is not necessary to have voluntary movement or effort. In these conditions stability of the upper limb depends on static restraints (ligaments), muscular stabilizers, and intra-articular forces. Ligaments and joint structures not only provide mechanical support but also provide sensory feedback information (from sensory receptors) that regulates involuntary muscular activation for joint positioning and stability. By virtue of gravity and the weight of the upper extremity there is slight tension in the soft tissues (ligaments, tendons, and muscles). The tension activates the different receptors and they send the information to the regulatory structures of the CNS. The spinal cord controls the positioning through the muscle activity and condition by means of a closed circuit. This type of regulation is through the formation of so-called "reflex arcs."

4 The Shoulder Complex and Scapulohumeral Rhythm

As already mentioned above the shoulder complex is composed of five joints: glenohumeral joint, acromioclavicular joint, sternoclavicular joint, scapulothoracic "joint," and suprahumeral "joint." The glenohumeral joint is an example of a multiaxial "ball-and-socket" synovial joint. This joint is formed by the articulations of the rounded humeral head with the glenoid fossa of the scapula. Because the head of the humerus is larger than the glenoid fossa, only part of the humeral head can be in articulation with the glenoid fossa in any position of the joint. At any given time, only 25–30 % of the humeral head is in contact with the glenoid fossa [27]. The glenohumeral joint largely depends on the ligamentous and muscular structures for stability, provided by both static and dynamic components. The static components for stability include the glenoid labrum, joint capsule, and ligaments. The static stability of the glenohumeral joint is provided by two major mechanisms, the glenoid suction cup mechanism and the limited joint volume mechanism [20]. The glenohumeral suction cup provides stability by virtue of the seal of the labrum and capsule to the humeral head. A suction cup adheres to a smooth surface by expressing the interposed air or fluid and then forming a seal to the surface [28]. The limited joint volume is a stabilizing mechanism in which the humeral head is held to the socket by the relative vacuum created when they are distracted. Dynamic support of the glenohumeral joint is provided by a large number of muscles acting in a coordinated pattern.

The acromioclavicular joint is a plane synovial joint and is formed by the articulation of the distal end of the clavicula with the medial border of the acromion process of the scapula. Both articular surfaces are covered with fibrocartilage, and the joint line formed by the two bones slopes inferiorly and medially, causing the clavicle to tend to override the acromion [21]. Due to the relative flatness of the articulating surfaces, they held in place by strong ligaments.

The only point of skeletal attachment of the upper extremity to the trunk occurs at the sternoclavicular joint. The sternoclavicular joint is a gliding synovial joint with fibrocartilaginous disc which compensates for the mismatch of surfaces between the two saddle—shaped articular faces of the clavicle and manubrium sterni.

Both of physiological (or functional) articulations: scapulothoracic "joint" and suprahumeral "joint" are playing an important role in proper functioning of the shoulder complex. The suprahumeral "joint" contains subacromial and subdeltoid bursae. These tissues lie in a space called the "subacromial space" and that allows gliding between the acromion and the rotator cuff muscles. The scapulothoracic articulation occurs between the concave costal surface of the scapula and the convex surface of the thorax. The scapula actually rests on the two muscles, the serratus anterior and the subscapularis, both connected to the scapula and moving across each other as the scapula moves. Underneath these two muscles lies the thorax. Movements of the scapula on the thorax result from combined motions of the sternoclavicular and acromioclavicular joints.

The design of the shoulder complex provides the upper limb with an extensive range of movement. The complex is more mobile than any other joint mechanism of the body because of the combined movement at all the four articulations comprising the shoulder. This wide range of motion permits positioning of the hand in space, allowing performance of numerous gross and skilled functions [28]. Shoulder range of motion is traditionally measured in terms of flexion and extension (elevation or movement of the humerus away from the side of the thorax in the sagittal plane), abduction (elevation in the coronal plane), and internal–external rotation (axial rotation of the humerus with the arm held in an adducted position) [17]. The arm can move through approximately 165° to 180° of flexion to approximately 30° to 60° of hyperextension in the sagittal plane [29, 19].

$$\mathbf{S: 60^\circ - 0^\circ - 180^\circ}$$

The arm can also abduct through 150° to 180°. The abduction movement can be limited by the amount of internal rotation occurring simultaneously with abduction. The adduction of the arm is approximately 75° from 25° of flexion in the sagittal plane [19].

$$\mathbf{F: 180^\circ - 0^\circ - 0^\circ}$$
$$\mathbf{F_{(s25^\circ)}: 180^\circ - 0^\circ - 75^\circ}$$

The arm can rotate both internally and externally 60° to 90° for a total of 120° to 180° of rotation.

$$\mathbf{R: 75^\circ - 0^\circ - 90^\circ}$$

Finally, the arm can move across the body in an elevated position for 135° of horizontal flexion or adduction and 45° of horizontal extension or abduction [29].

$$\mathbf{T: 45^\circ - 0^\circ - 135^\circ}$$

As stated earlier, the four joints of the shoulder complex must work together in a coordinated action to create arm movements. Any time the arm is raised in flexion or abduction, accompanying scapular and clavicular movements take place. As the humerus moves in the glenoid fossa the scapula rotates on the thorax and the clavicle moves on the sternum. This coordinated and synchronous movement of the shoulder's structures driven by the muscular and ligament systems is called "**Scapulohumeral rhythm**". The synchronized movement of the shoulder girdle and humerus, is known since the first studies on the upper limb, carried out during the nineteenth century, but only in 1934 was described by the first time by Codman [7] as the "scapulohumeral rhythm." According to Codman, during arm elevation, for every 2° of glenohumeral motion, there is 1° of scapulothoracic motion. In 1944, Inman et al. [12], described the scapulohumeral rhythm as the bidimensional relationship between the scapular spine (projected on the frontal plane) and the

Table 3 Scapulohumeral rhythm ratios

Study	Ratio	Range of motion
Innman et al. (1944) [12]	2:1	over entire range (180°)
Freedman and Munro (1966) [10]	3:2	From 0° to 135°
Doody et al. (1970) [8]	1.74:1	over entire range (180°)
Poppen and Walker (1976) [18]	4.3:1 5:4	during first 30° from 30° to 180°
Saha (1961) [24]	2.3:1	from 30° to 135°

humeral angle during arm flexion and abduction. They suggest that the scapular upward rotation and humeral abduction to be a 2:1 ratio between the humerus and scapula throughout flexion and abduction (entire range of motion from 0° to 180°). Further research has shown that scapulohumeral rhythm might not be as simple as it was first described. Many studies have evaluated scapular and arm motion and have reported different ratios of glenohumeral motion to scapulothoracic upward rotation depending on how the study was measured, different planes of elevation, and anatomic variations between the individuals (see Table 3).

The relationship between glenohumeral and scapulothoracic motion is critical and is generally considered to be 2:1, culminating in 120° and 60°, respectively [12, 19, 20, 28]. During the arm elevation in frontal plane (abduction) from 0° to 180° there are four phases of scapulohumeral rhythm.

- Phase 1: The upper limb is in relaxed position (see above).
- Phase 2 (30° abduction): The supraspinatus muscle "unlocks" the arm and together with deltoid muscle slowly elevates the upper limb. In the first phase of 30° of elevation through abduction, the scapula is said to be "setting". This setting phase means that the scapula moves either toward the vertebral column or away from it to seek a position of stability on the thorax. During this phase the main contributor to movement is the glenohumeral joint. The clavicle elevates minimally (0°–15°) during this stage [19].
- Phase 3 (during the next 60° of elevation): The contraction of the deltoid muscle is accompanied by the supraspinatus, which produces abduction while at the same time compressing the humeral head and resisting the superior motion of the humeral head by the deltoid. The rotator cuff muscles contract as a group to compress the humeral head and maintain its position in the glenoid fossa. The teres minor, infraspinatus, and subscapularis muscles stabilize the humerus in elevation by applying a downward force. The antagonistic action is produced by the mm. pectoralis major et minor, m. levator scapulae, and mm. rhomboidei. The latissimus dorsi also contracts eccentrically to assist with the stabilization of the humeral head and increases in activity as the angle increases. At this stage there is a contraction of the upper and lower trapezius and the serratus anterior muscles and they work as a force couple to create the lateral, superior, and rotational motions of the scapula to maintain the glenoid fossa in the optimal position. The scapula rotates about 20° and the humerus elevates 40° [15, 19] with minimal

protraction or elevation of the scapula. Because of the scapular rotation the clavicle elevates approximately 30°–35° through the sternoclavicular joint.

- Phase 4 (final 90° of elevation): During the final stage of abduction the humerus elevates another 60° and the scapula continues to rotate and now begins to elevate. The scapula rotates 30° through acromioclavicular joint. It is in this stage that the clavicle rotates posteriorly 30°–50° on a long axis and elevates up to a further 15°. As the arm abducts to 90°, the greater tuberosity on the humeral head approaches the coracoacromial arch, compression of the soft tissue begins to limit further abduction. External rotation of the humerus to 90° places the grater tuberosity posteriorly, allowing the humerus to move freely under the coracoacromial arch. The external rotation of the humerus is produced by the infraspinatus and the teres minor muscles. The final degrees of elevation are achieved trough contralateral trunk flexion and/or trunk extension [6] by contraction of the paravertebral muscles.

For the total range of motion through 180° of abduction, the glenohumeral to scapula ratio is 2:1; thus, the 180° range of motion is produced by 120° of glenohumeral motion and 60° of scapular motion. The contributing joint actions to the scapular motion are 20° produced at the acromioclavicular joint, 40° produced at the sternoclavicular joint, and 40° of posterior clavicular rotation [9]. From the discussion of scapulohumeral rhythm, it becomes apparent that restriction or disorder in movement at any of the joints of the shoulder complex will limit the ability of proper position the hand for functioning.

5 A Reduced Generalized Net Model of the Upper Limb

Here we represent a simplified GN model of the upper limb in relaxed position.

The GN model (Fig. 1) has 4 transitions and 11 places with the following meaning.

- Transition Z_1 represents the function of the CNS.
- Transition Z_2 represents the function of the PNS (sensory and motor fibers of brachial plexus branches).
- Transition Z_3 represents the function of the striated muscles and tendons of the upper limb.
- Transition Z_4 represents the function of the joints and ligaments of the upper limb.

Each of these transitions contains a special place to collect and keep information about the current status of the respective structures which it represents, as follows.

- In place l_2 token α stays permanently and it collects information about the current status of the CNS.
- In place l_5 token β collects information about the current status of the PNS.

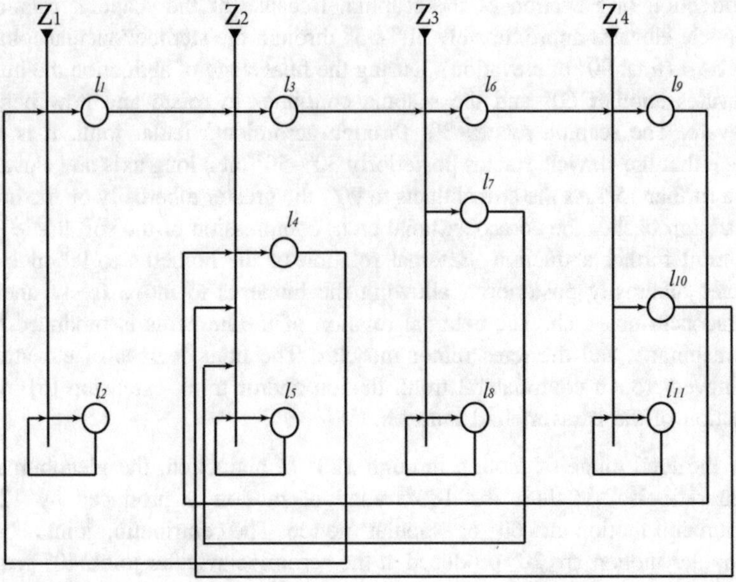

Fig. 1 GN model of the upper limb

- In place l_{18} token μ collects information about the current status of the striated muscles and tendons.
- In place l_{11} token ν collects information about the current status of the joints and ligaments.

Tokens α, β, μ, ν that permanently stay, respectively, in these places obtain as current characteristic the corresponding information. At the time of duration of the GN functioning, some of these tokens can split, generating new tokens, that will transfer in the net obtaining respective characteristics, and also in some moments they will unite with some of tokens α, β, μ, ν.

When some of these tokens splits, e.g., token $\omega \in \{\alpha, \ldots, te\}$, let us assume that it generates two or more tokens that we shall note by ω, ω_1, ω_2, … and the first of them (ω) will continue to stay in its place, while the other ones will go somewhere in the net.

The four GN transitions have the following forms:

$$Z_1 = \langle \{l_2, l_4\}, \ \{l_1, l_2\}, r_1 \rangle,$$

where

$r =$

	l1	l2
	W1,2	*true*
l4	*false*	*true*

and $W_{2,1}$ = *"efferent impulses from the CNS are necessary for the maintenance and regulation of the muscles."*

The tokens from all input places enter place l_2 and unite with token α that obtains the above mentioned characteristic. On the other hand, token α splits to two tokens, the same token α, and α_1 that enters place l_1, when predicate $W_{2,1}$ has truth value "*true*". In the model, place l_1 corresponds to the kind of the efferent impulse from CNS.

$$Z_2 = \langle \{l_1, l_5, l_7, l_9, l_{10}\}, \{l_3, l_4, l_5\}, r_2 \rangle,$$

where

$r_2 =$

	l3	l4	l5
l1	*false*	*false*	*true*
l5	W5,3	*true*	*true*
l7	*false*	*false*	*true*
l9	*false*	*false*	*true*
l10	*false*	*false*	*true*

and $W_{5,3}$ = *"efferent impulse from CNS was transmitted* via *motor fibers of the PNS branches to the muscles of the upper limb"*

The tokens from all input places enter place l_5 and unite with token β that obtains the above-mentioned characteristic. On the other hand token β splits to three tokens, the same token β and tokens β_1, β_2, that enter, respectively, in places l_3, l_4. When predicate $W_{5,3}$ has truth value *"true"*, a token enters place l_3.

Token in place l_3 obtains characteristics

"efferent impulse to the muscles of the upper limb."

Token in place l_4 enters with characteristics

"afferent (sensory) impulse from the joints, ligaments and the extrafusal muscle fibers of the upper limb."

$$Z_3 = \langle \{l_3, l_8\}, \{l_6, l_7, l_8\}, r_3 \rangle,$$

where

$r_3 =$

	l6	l7	l8
l3	*false*	*false*	*true*
l8	W8,6	*true*	*true*

and $W_{8,6}$ = *"there is involuntary muscular activation."*

The tokens from all input places enter place l_{10} and unite with token μ that obtains the above-mentioned characteristic. On the other hand, token μ splits to three tokens, the same token μ and tokens μ_1, μ_2 that enter, respectively, in places l_6, l_7.

When predicate $W_{8,6}$ has truth value *"true"*, a token enters place l_6. There it obtains characteristics

"influence of the muscular activation on the upper limbs joints and ligaments (joints positions)."

Place l_7 corresponds to the sensory receptors in muscle fibers of upper limb muscles.

$$Z_4 = \langle \{l_6, l_{11}\}, \{l_9, l_{10}, l_{11}\}, r_4 \rangle,$$

where

$r_4 =$

	l_9	l_{10}	l_{11}
l_6	*false*	*false*	*true*
l_{11}	*true*	*true*	*true*

The tokens from the two input places enter place l_{11} and unite with token ν that obtains the above-mentioned characteristic. On the other hand, token ν splits to three tokens: the same token ν and tokens ν_1, ν_2 that enter, respectively, in places l_9, l_{10}.

Token in place l_9 obtains characteristics

"the position of individual joints of upper limb segments in space."

Place l_{10} corresponds to the sensory receptors in ligaments and joints of the upper limb.

6 Generalized Net Model of the Scapulohumeral Rhythm

Here we represent a GN model of the scapulohumeral rhythm observed in arm abduction in the frontal plane from 0° to 180°. The initial phase of the abduction begins with the upper limb in relaxed position.

The GN model (Fig. 2) has 7 transitions and 24 places with the following meaning:

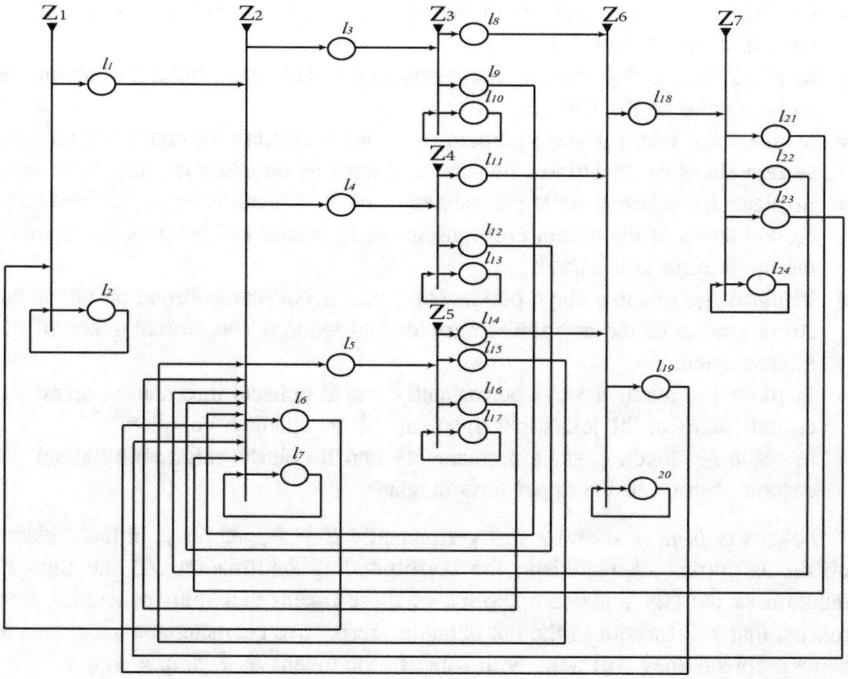

Fig. 2 GN model of the scapulohumeral rhythm

- Transition Z_1 represents the function of the CNS.
- Transition Z_2 represents the function of the PNS (sensory and motor fibers of brachial plexus branches).
- Transition Z_3 represents the function of the striated muscles and tendons initiating the arm elevation.
- Transition Z_4 represents the function of the rotator cuff muscles and function of the striated muscles and tendons that connect the upper limb to the trunk of the body from which it buds
- Transition Z_5 represents the function of the antagonist muscles and the lumbar part of the spine.
- Transition Z_6 represents the joints and ligaments of the shoulder complex.
- Transition Z_7 represents the position of the upper limb in space

The net contains seven types of tokens: α, β, μ, η, κ, π, and ϕ tokens. Each of these transitions contains a special place to collect and keep information about the current status of the respective structures which it represents, as follows.

- In place l_2, token α stays permanently and it collects information about the current status of the CNS.
- In place l_7, token β stays permanently and it collects information about the current status of the PNS.
- In place l_{10}, token μ stays permanently and it collects information about the current status of the striated muscles and tendons initiating the arm elevation.
- In place l_{13}, token η stays permanently and it collects information about the current status of the rotator cuff muscles and muscles and tendons that connect the upper limb to the trunk.
- In place l_{17}, token κ stays permanently and it collects information about the current status of the antagonist muscles and tendons and current status of the lumbar spine.
- In place l_{20}, token π stays permanently and it collects information about the current status of the joints and ligaments of the shoulder complex.
- In place l_{24}, token ϕ stays permanently and it collects information about the current position of the upper limb in space.

Tokens α, β, μ, η, κ, and π that permanently stay, respectively, in these places obtain as current characteristic the corresponding information. At the time of duration of the GN functioning, some of these tokens can split, generating new tokens, that will transfer in the net obtaining respective characteristics, and also in some moments they will unite with some of the tokens α, β, μ, η, κ, and π.

The six GN transitions have the following forms:

$$Z_1 = \langle \{l_2, l_6\}, \{l_1, l_2\}, r_1 \rangle,$$

where

$$r_1 = \begin{array}{c|cc} & l_1 & l_2 \\ \hline l_2 & W_{2,1} & true \\ l_6 & false & true \end{array}$$

and $W_{2,1}$ = *"in the higher centers of the brain is generated a conscious command for elevation of the upper limb."*

The tokens from the two input places enter place l_2 and unite with token α that obtains the above-mentioned characteristic. On the other hand, token α splits to two tokens, the same token α and α_1 that enters place l_1, when predicate $W_{2,1}$ has truth value "*true*". The token α_1 enters place l_1 with characteristics *"efferent motor impulse to the PNS"*

$$Z_2 = \langle \{l_1, l_7, l_9, l_{12}, l_{15}, l_{16}, l_{19}, l_{21}, l_{22}, l_{23}\}, \{l_3, l_4, l_5, l_6, l_7, \}, r_2 \rangle,$$

where

$r_2 =$

	l_3	l_4	l_5	l_6	l_7
l_1	*false*	*false*	*false*	*false*	$W_{7,1}$
l_7	*true*	*true*	*true*	*true*	*true*
l_9	*false*	*false*	*false*	*false*	*true*
l_{12}	*false*	*false*	*false*	*false*	*true*
l_{15}	*false*	*false*	*false*	*false*	*true*
l_{16}	*false*	*false*	*false*	*false*	*true*
l_{19}	*false*	*false*	*false*	*false*	*true*
l_{21}	*false*	*false*	*false*	*false*	*true*
l_{22}	*false*	*false*	*false*	*false*	$W_{22,7}$
l_{23}	*false*	*false*	*false*	*false*	$W_{23,7}$

and $W_{1,7}$ = *"efferent impulse from CNS was transmitted* via *motor fibers of the PNS branches to the muscles of the upper limb."*

$W_{22,7}$ = *"there is a 30° of humeral elevation and the humerus begins to elevate in the range of 30° to 90° in the frontal plane"*

$W_{23,7}$ = *"there is a 90° of humeral elevation and the humerus begins to elevate in the range of 90° to 180° in the frontal plane"*

When predicate $W_{1,7}$ has a true value "true", token α_1 from place l_1 enters in place l_7 and the transition Z_2 becomes active.

The tokens from all input places enter place l_7 and unite with token β that obtains the above mentioned characteristic. On the other hand, token β splits to five tokens, the same token β and β_1, β_2, β_3, β_4.

Token β_1 enters in place l_3 with characteristics:

"motor impulses travelling trough n. axilaris and n.suprascapularis"

When predicate $W_{22,7}$ has a true value "*true*", token β_2 enters in place l_4 with characteristics:

"motor impulses travelling trough n. axillaris, n. suprascapularis, n. subclavius, n. thoracalis longus, n. subscapulares, n. thoracodorsalis and n.accesorius"

When predicate $W_{23,7}$ has a true value "*true*", token β_3 enters in place l_5 with characteristics:

"motor impulses travelling trough n. axillaris, n. suprascapularis, n. subclavius, n. thoracalis longus, n. subscapulares, n. thoracodorsalis, n. accesorius, n. dorsalis scapulae, nn. thoracales anteriores, n. pectoralis lateralis et medialis"

Token β_4 enters in place l_6 with characteristics

"afferent (sensory) impulse from the joints, ligaments and the extrafusal muscle fibers of the shoulder complex"

$$Z_3 = \langle\{l_3, l_{10}\}, \{l_8, l_9, l_{10}, \}, r_3\rangle,$$

where

$$r_3 = \begin{array}{c|ccc} & l_8 & l_9 & l_{10} \\ \hline l_3 & false & false & W_{3,10} \\ l_{10} & true & true & true \end{array}$$

and $W_{3,10}$ = *"the afferent motor impulses are reached the motoneurons of the muscles initiating the upper limb elevation"*

When predicate $W_{3,11}$ has a true value "*true*", token β_1 enters in place l_{10} and unites with token μ that obtains the above-mentioned characteristics. On the other hand, token μ splits to three tokens, the same token μ and tokens μ_1, μ_2 that enter, respectively, in places l_8 and l_9.

Token μ_1 enters in place l_8 and there it obtains characteristics

"contraction of the m. supraspinatus and m. deltoideus"

Place l_9 corresponds to the sensory receptors in muscle fibers of the muscles initiating the upper limb elevation.

$$Z_4 = \langle\{l_4, l_{13}\}, \{l_{11}, l_{12}, l_{13,} \}, r_4\rangle,$$

where

$$r_4 = \begin{array}{c|ccc} & l_{11} & l_{12} & l_{13} \\ \hline l_4 & false & false & W_{4,13} \\ l_{13} & true & true & true \end{array}$$

and $W_{4,13}$ = *"the afferent motor impulses are reached the motoneurons of the rotator cuff muscles, m.trapezius and m.serratus anterior."*

When predicate $W_{4,15}$ has a true value "*true*", token β_2 enters in place l_{15} and unites with token η that obtains the above mentioned characteristics. On the other hand, token η splits to three tokens, the same token η and tokens η_1, η_2.

Token η_1 enters in place l_{11} and there it obtains characteristics

"contraction of the rotator cuff muscles, m.trapezius and m.serratus anterior"

Place l_{12} corresponds to the sensory receptors in muscle fibers of the rotator cuff muscles, m. trapezius and m. serratus anterior.

$$Z_5 = \langle \{l_5, l_{17}\}, \{l_{14}, l_{15}, l_{16}, l_{17}\}, r_5 \rangle,$$

where:

$$r_4 = \begin{array}{c|cccc} & l_{14} & l_{15} & l_{16} & l_{17} \\ \hline l_5 & false & false & false & W_{5,17} \\ l_{17} & true & true & true & true \end{array}$$

and $W_{5,17}$ = *"the afferent motor impulses are reached the motoneurons of mm. pectoralis major et minor, m. levator scapulae, mm. rhomboidei, m. latissimus dorsi and paravertebral muscles of the lumbar spine."*

When predicate $W_{5,17}$ has a true value "*true*", token β_3 enters in place l_{17} and unites with token κ that obtains the above-mentioned characteristics. On the other hand, token κ splits to four tokens, the same token κ and tokens κ_1, κ_2, κ_3.

Token κ_1 enters in place l_{14} and there it obtains characteristics

"contraction of the mm.pectoralis major et minor, m.levator scapulae, mm.romboidei, m. latissimus dorsi"

Token κ_2 enters in place l_{15} and there it obtains characteristics

"contraction of the paravertebral muscles of the lumbar spine"

Place l_{16} corresponds to the sensory receptors in muscle fibers of the antagonist muscles and the lumbar part of the spine.

$$Z_6 = \langle \{l_8, l_{11}, l_{14}, l_{20}\}, \{l_{18}, l_{19}, l_{20}, \}, r_6 \rangle,$$

where

$$r_6 = \begin{array}{c|ccc} & l_{18} & l_{19} & l_{20} \\ \hline l_8 & false & false & true \\ l_{11} & false & false & true \\ l_{14} & false & false & true \\ l_{20} & true & true & true \end{array}$$

The tokens from all input places enter place l_{20} and unite with token π that obtains the above-mentioned characteristic. On the other hand, token π splits to five tokens, the same token π and tokens π_1, π_2, π_3 and π_4.

Tokens π_1, π_2, and π_3 enter place l_{18} with characteristics

"30° of abduction in the glenohumeral joint"

"60° of abduction in the glenohumeral joint, 30° rotation of the scapula and 30° elevation of the clavicle"

"30° of abduction and 90° of internal rotation in the glenohumeral joint, 30° rotation of the scapula and 30° to 50° posterior rotation of the clavicle"

Place l_{19} corresponds to the sensory receptors in the joints and ligaments of the shoulder complex.

$$Z_7 = \langle \{l_{18}, l_{24}\}, \{l_{21}, l_{22}, l_{23}, l_{24}\}, r_7 \rangle,$$

where:

$$r_7 = \begin{array}{c|cccc} & l_{21} & l_{22} & l_{23} & l_{24} \\ \hline l_{18} & \mathit{false} & \mathit{false} & \mathit{false} & \mathit{true} \\ l_{24} & \mathit{true} & \mathit{true} & \mathit{true} & \mathit{true} \end{array}$$

The tokens from the two input places enter place l_{24} and unite with token ϕ that obtains the above-mentioned characteristic. On the other hand, token ϕ splits to four tokens, the same token ϕ and ϕ_1, ϕ_2, ϕ_3.

Token ϕ_1 enters place l_{21} and there it obtains characteristics

"the upper limb is abducted in the range from 0° to 30° in the frontal plane"

Token ϕ_2 enters place l_{22} and there it obtains characteristics

"the upper limb is abducted in the range from 30° to 90° in the frontal plane"

Token ϕ_3 enters place l_{23} and there it obtains characteristics

"the upper limb is abducted in the range from 90° to 180° in the frontal plane"

7 Conclusion

The so-described GN model can be used for simulation of different situations, related to activities of the upper limb. It can help for the studying of different real processes, related to the mechanism of its movement from biological point of view. In future, the model can be used for a basis of new models, describing functioning of an artificial hand and appropriate rehabilitation treatments.

References

1. Atanassov, K.: Generalized Nets. World Scientific, Singapore (1991)
2. Atanassov, K.: On Generalized Nets Theory. "Prof. M. Drinov" Academic Publishing House, Sofia (2007)
3. Atanassov, K., Chakarov, V., Shannon, A., Sorsich, J.: Generalized Net Models of the Human Body. "Prof. M. Drinov" Academic Publishing House, Sofia (2008)
4. Chakarov, V., Atanassov, K., Shannon, A.: Generalized net model of the human muscular-sceletal system. In: Choy, E.Y.H., Krawczak, M., Shannon, A., Szmidt, E. (eds.) A Survey of Generalized Nets, Raffles KvB Monograph No. 10, Sydney, pp. 127–140 (2007)

5. Cailliet, R.: Shoulder Pain, edition 2, F. A. Davis Company (1981)
6. Clarkson, H.: Joint Motion and Function Assessment. Lippincott Williams and Wilkins; 2nd Revised edition, pp. 345–346 (2005)
7. Codman, E.A.: The Shoulder. Thomas Todd Company, Boston (1934)
8. Doody, S.G., et al.: Shoulder movements during Abduction in the Scapular plane. Arch Phys. Med. Rehabil. **51**, 595 (1970)
9. Einhorn, A.R.: Shoulder rehabilitation: equipment modifications. J. Orthop. Sports Phys. **6**, 247–253 (1985)
10. Freedman, L., Munro, R.R.: Abduction of the arm in the scapular plane: Scapural and glenohumeral movements. J. Bone Joint Surg. Am. 48-A, 1503–1510 (1966)
11. Gjelsvik, N.: The Bobath Concept in Adult Neurology. Thieme, Stuttgart (2008)
12. Inman, V. et. al.: Observations on the function on the shoulder joint. J. Bone Joint Surg. **26** (1944) (San Francisco)
13. Kapandji, I.A.: The Physiology of the Joints, vol. 1, 5th edn. Churchill Livingstone, New York (1982)
14. Kotwal, P.P., Natarajan, M.: Textbook of Orthopedics. Elsevier, New Delhi (2005)
15. Magee, D.: Orthopedic Physical Assessment. Sounders, Philadelphia (1999)
16. Noback, C.R., et al.: The Human Nervous System. Humana Press, New Jersey (2005)
17. Nordin, M., Frankel, V.H.: Basic Biomechanics of the Musculoskeletal System, 3rd Revised edition, vol. 12, pp. 319–320. Lippincott Williams and Wilkins (2001)
18. Poppen, N.K., Walker, P.S.: Normal and abnormal motion of the shoulder. J. Bone Joint Surg. (PDF copy) **58**, 195–201 (1976)
19. Popov, N.: Clinical Pathophysiology. NSA Press (in Bulgarian), Sofia (2002)
20. Popov, N., Dimitrova, E.: Kinesitherapy in Orthopedic and Traumatic Conditions of the Upper Limb. NSA Press (in Bulgarian), Sofia (2007)
21. Rees, N., Bandy, W.: Joint Range of Motion and Muscle Length Testing, vol. 3, pp. 49–50, 2 edn. Saunders, Canada (2009)
22. Ribagin, S., Chakarov, V., Atanassov, K.: New developments in fuzzy sets, intuitionistic fuzzy sets, generalized nets and related topics. Applications **2**, 201–210 (2012)
23. Ribagin, S., Chakarov, V., Atamassov, K.: Generalized net model of the upper limb vascular system. In: Proceedings of IEEE 6th conference on intelligent systems, vol. 2, pp. 221–224. Sofa, 6–8 Sept 2012
24. Saha, A.K.: Theory of Shoulder Mechanism: Descriptive and Applied, Charles C. Thomas, Springfield, Illinois (1961)
25. Shepard, G.M.: Neurobiology. Oxford University Press, New York (1994)
26. Terri, M., et al.: Rehabilitation of the Hand and Upper extremity. Elsevier, Philadelphia (2002)
27. Terry, G.C., Chopp, T.M.: Functional anatomy of the shoulder. J. Athletic Training **35**, 248–255 (2000)
28. Wilk, K., Rainold, M., Andrews, J.: The Athlete's Shoulder, 2 edn. Churchill Livingstone, 13:14 (2008)
29. Zuckerman, J.D., Matsea III, F.A.: Biomechanics of the shoulder. In: Nordin, M., Frankel, V. H. (eds.) Biomechanics of the Musculoskeletal System, pp. 225–248. Lea & Febiger, Philadelphia (1989)

Method for Interpretation of Functions of Propositional Logic by Specific Binary Markov Processes

Vassil Sgurev and Vladimir Jotsov

Abstract The current paper proposes a method for interpretation of propositional binary logic functions using multi-binary Markov process. This allows logical concepts 'true' and 'false' to be treated as stochastic variables, and this in two ways —qualitative and quantitative. In the first case, if the probability of finding a Markov process in a definitely true state of this process is greater than 0.5, it is assumed that the Markov process is in state 'truth.' Otherwise the Markov process is in state 'false.' In quantitative terms, depending on the chosen appropriate binary matrix of transition probabilities it is possible to calculate the probability of finding the process in one of the states 'true' or 'false' for each of the steps $n = 0, 1, 2, \ldots$ of the Markov process. A single-step Markov realization is elaborated for standard logic functions of propositional logic; a series of analytical relations are formulated between the stochastic parameters of the Markov process before and after the implementation of the single-step transition. It has been proven that any logical operation can directly, uniquely, and consistently be described by a corresponding Markov process. Examples are presented and a numerical realization is realized of some functions of propositional logic by binary Markov processes.

Methods and means of propositional logic perform a fundamental role in series of modern scientific fields for decision-making such as artificial intelligence and intelligent systems, perform a fundamental role methods and means of propositional logic [1]. Therefore, different formal mathematical structures are recently proposed to describe this logic and to expand its applications, namely probabilistic Bayesian logic [2], matrix logic [3], network-stream logic [4], fuzzy logic [5]. Of these, fuzzy logic was most widely spread in different classes of technical systems and control systems.

V. Sgurev (✉)
Institute of Information and Communication Technologies, Bulgarian Academy of Sciences, Acad. G. Bonchev str., bl. 2, 1113 Sofia, Bulgaria
e-mail: vsgurev@gmail.com

V. Jotsov
University of Library Studies and IT (ULSIT), P.O.Box 161, Sofia 1113, Bulgaria
e-mail: bgimcssmc@gmail.com

V. Sgurev et al. (eds.), *Recent Contributions in Intelligent Systems*,
Studies in Computational Intelligence 657,
DOI 10.1007/978-3-319-41438-6_14

A general integration of propositional logic using a discrete Markov process with two states was proposed in [6]. In the same work it was shown that via Markov processes it is possible to achieve a new stochastic realization of propositional logic with special features missing in other classes of logic—from [1–4].

In this work are introduced some new results concerning the stochastic interpretation of propositional logic based on a binary Markov process.

Let us define a set of two states: $N=\{a_1, a_2\}$, the first of which is related to truth, and the second to false meanings. Markov process may be in one of these two states; for a discrete time interval it may be either in state a_1 or in state a_2 with respective probabilities p_{11} or p_{22} or it may change its state to the opposite one a_2 or a_1 with respective probabilities p_{12} or p_{21}. At that

$$0 \le p_{ij} \le 1; i,j \in I = \{1,2\}; \tag{1}$$

$$p_{11} + p_{12} = 1 \text{ and } p_{21} + p_{22} = 1. \tag{2}$$

We introduce the following denotations:

$$0 \le P_1 \le 1; 0 \le P_2 \le 1; 0 \le P_3 \le 1 \tag{3}$$

$$\pi_1 = (P_1, 1-P_1);\ \pi_2 = (P_2, 1-P_2);\ \pi_3 = (P_3, 1-P_3), \tag{4}$$

where π_1, π_2 and π_3 are line vectors with two elements.

Then, if at the moment of time n the Markov process is in state a_1 with probability P_1 or in state a_2 with probability P_2 and if we apply the Markov transition matrix

$$\left\| p_{ij} \right\| = \left\| \begin{matrix} p_{11} & p_{12} \\ p_{21} & p_{22} \end{matrix} \right\|, \tag{5}$$

the general transition probability P_2 is embedded in the elements of it, then the one-step Markov process will be implemented in the following matrix-vector equation:

$$\pi_1 \times \left\| p_{ij} \right\| = \pi_3 . \tag{6}$$

Below we consider two cases where probabilities $\{P_{ij}\}$ receive specific meanings s and p—respectively $1-s$ and $1-p$. These two cases are related to two types of binary matrices $P(1)$ and $P(2)$.

It is assumed that the following dependencies hold for the matrix $P(1)$:

$$p_{21} = s;\ p_{22} = 1-s;\ p_{11} = 1-p;\ p_{12} = p \tag{7}$$

and there are analogous dependencies for matrix $P(2)$:

$$p_{21} = 1 - s;\ p_{22} = s;\ p_{11} = p;\ p_{12} = 1 - p. \tag{8}$$

Then both matrices $P(1)$ and $P(2)$ satisfy (5) and they take the following form:

$$P(1) = \left\| \begin{matrix} 1-p & p \\ s & 1-s \end{matrix} \right\|;\ P(2) = \left\| \begin{matrix} p & 1-p \\ 1-s & s \end{matrix} \right\|, \tag{9}$$

and the vector matrix Eq. (6) can be written, respectively, for $P(1)$ and $P(2)$ as

$$\pi_1 \times P(1) = \pi_3 \text{ and } \pi_1 \times P(2) = \pi_3\,. \tag{10}$$

At every step of the Markov process it will be determined with what probability this process is at state a_1 (true) or at state a_2 (false).

We shall denote the probability for the process to be in state a_1 by P_1 and the probability to be in state a_2 by $1 - P_1$. Then, if

$$0 \le 1 - P_1 < P_1 \le 1, \tag{11}$$

then the Markov process is with prevailing probability P_1 in the true state a_1. Otherwise, if

$$0 \le P_1 < 1 - P_1 \le 1, \tag{12}$$

then the process at the same step is with prevailing probability $1 - P_1$ in the false state a_2. The case

$$P_1 = 1 - P_1 = 0.5 \tag{13}$$

Fig. 1 .

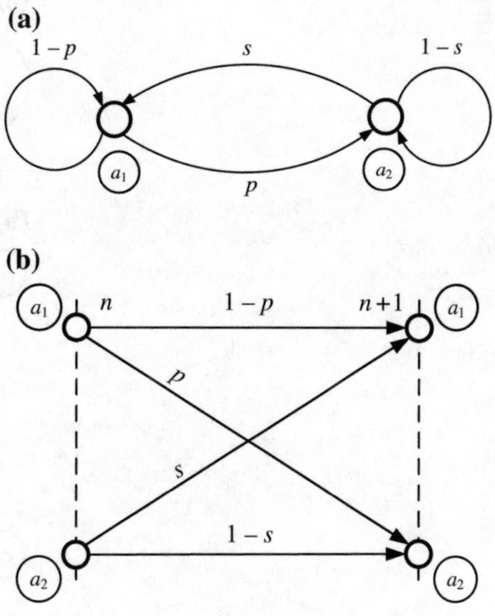

is boundary and it is possible to accept that the process is at any of the two states a_1 and a_2 with equal probability.

Since now, for brevity the word 'prevailing' will be meant by default and it will be omitted.

Analogously to P_1, from (11) and (12) it is possible to interpret parameters P_2 and P_3 defined by (3) and (4).

Graphically the possibilities and the respective probabilities for a single-step transition in the matrix $P(1)$ from one state to the next state can be represented by arcs and nodes of these two graphs in Fig. 1—static (a) and dynamic (b):

For the transition matrix $P(2)$ by (9), respectively, the static and dynamic graphs will have the following form:

Further, the element s in $P(1)$ and $P(2)$ is considered as the general probability P_2 of the transition in matrices (9), i.e.,

$$P_2 = s. \tag{14}$$

Each of the values P_1, P_2, P_3, s and p can be represented by its positive or negative difference ε_i to the value 0.5, which distinguishes the probabilities of being in true—a_1 or in false state a_2. For this purpose, taking

$$P_1 = 0.5 + \varepsilon_1;\ P_1 = 0.5 + \varepsilon_2;\ P_3 = 0.5 + \varepsilon_3;\ S = 0.5 + \varepsilon_s;\ P = 0.5 + \varepsilon_P, \tag{15}$$

where $\varepsilon_i \ = \gtrless 0$; $|\varepsilon_i| \leq 0.5$ and $|\varepsilon_i|$ is the absolute value of ε_i.

Interesting is the case when the transition probabilities s and p coincide with each other, i.e.,

$$s = p. \tag{16}$$

Fig. 2 .

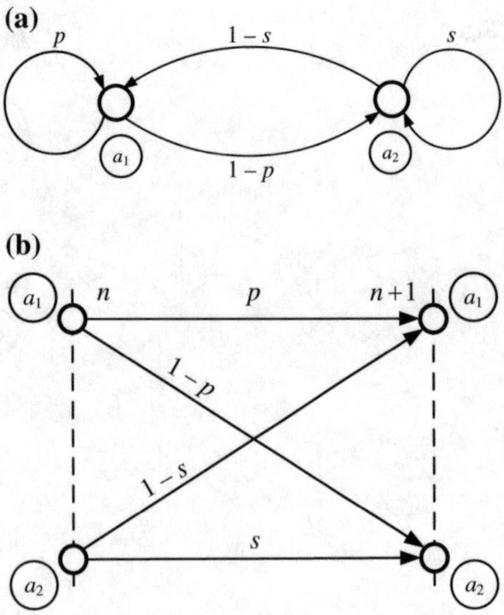

This equation provides the necessary symmetry using P_2—otherwise the value P_2 a priori and genetically will 'favor' one of the two states—a_1 or a_2 and the Markov process will be directed with a high probability to it (Fig. 2).

Lemma 1 *For the matrix $P(1)$ by (9) and for the vector matrix Eq. (10) the following dependencies exist:*

$$\varepsilon_3 = -2\varepsilon_1\varepsilon_2; \tag{17}$$

$$P_3 = 0.5 - 2\varepsilon_1\varepsilon_2;\; 1 - P_3 = 0.5 + 2\varepsilon_1\varepsilon_2. \tag{18}$$

Proof The product of the vector line π_1 from (4) and the vector columns of the matrix $P(1)$ of (9) leads to the same subject:

$$P_1(1 - (s + p)) + s = P_3. \tag{19}$$

Assumptions (14)–(16) lead to

$$P_1(1 - 2s) + s = P_3;\; \varepsilon_s = \varepsilon_p = \varepsilon_2. \tag{20}$$

If (20) is carried by the substitutions (15), it is true that

$$(0.5 + \varepsilon_1)(1 - 2(0.5 + \varepsilon_2)) + 0.5 + \varepsilon_2 = 0.5 + \varepsilon_3. \tag{21}$$

After the appropriate transformations we obtain (17). Addictions (18) are a direct consequence of (4) and (17).

Next eight corollaries can be derived directly from the results (17) and (18) of Lemma 1:

Corollary 1.1 *If $\varepsilon_1 = 0$ or $\varepsilon_2 = 0$ or both ε_1 and ε_2 are zero, then*

$$P_3 = 1 - P_3 = 0.5.$$

Corollary 1.2 *If $\varepsilon_1 = 0.5$, then*

$$P_3 = 0.5 - \varepsilon_2 \text{ and } 1 - P_3 = 0.5 + \varepsilon_2.$$

Corollary 1.3 *If $\varepsilon_1 = -0.5$, then*

$$P_3 = 0.5 + \varepsilon_2 \text{ and } 1 - P_3 = 0.5 - \varepsilon_2.$$

Corollary 1.4 *If $\varepsilon_2 = 0.5$, then*

$$P_3 = 0.5 - \varepsilon_1 \text{ and } 1 - P_3 = 0.5 + \varepsilon_1.$$

Corollary 1.5 *If* $\varepsilon_2 = -0.5,$ *then*

$$P_3 = 0.5 + \varepsilon_1 \text{ and } 1 - P_3 = 0.5 - \varepsilon_1.$$

Corollary 1.6 *If* $\varepsilon_1 = \varepsilon_2 = 0.5,$ *then*

$$P_3 = 0 \text{ and } 1 - P_3 = 1.$$

Corollary 1.7 *If* $\varepsilon_1 = \varepsilon_2 = -0.5,$ *then*

$$P_3 = 0 \text{ and } 1 - P_3 = 1.$$

Corollary 1.8 *If* $\varepsilon_1 = -0.5$ *and* $\varepsilon_2 = 0.5$ *or* $\varepsilon_1 = 0.5$ *and* $\varepsilon_2 = -0.5,$ *then*

$$P_3 = 1 \text{ and } 1 - P_3 = 0.$$

The results from the next Lemma 2 can be considered as opposite in some aspects to those of Lemma 1.

Lemma 2 *For the matrix* $P(2)$ *by (9) and for the vector matrix Eq. (10) the following relations exist:*

$$\varepsilon_3 = 2\varepsilon_1\varepsilon_2; \tag{22}$$

$$P_3 = 0.5 + 2\varepsilon_1\varepsilon_2;\ 1 - P_3 = 0.5 - 2\varepsilon_1\varepsilon_2. \tag{23}$$

These two results can be obtained in a manner analogous to the proof of Lemma 1. *The product* $\pi_1 \times P(2)$ *leads to:*

$$P_1((s + p) - 1) + 1 - s = P_3. \tag{24}$$

From assumptions (14)–(16) it follows that

$$P_1(2s - 1) + 1 - s = P_3, \varepsilon_s = \varepsilon_p = \varepsilon_2. \tag{25}$$

Then by substituting (15) in (25)

$$(0.5 + \varepsilon_1)(2(0.5 + \varepsilon_2) - 1) + 1 - 0.5 - \varepsilon_2 = 0.5 + \varepsilon_3. \tag{26}$$

The above results directly to (22), and this, in turn, together with (18) proves (22) and (23).

Analogously to Lemma 1 eight consequences of Lemma 2 can be derived from 2.1 to 2.7. Corollary 2.1 coincides entirely with Corollary 1.1.

Corollary 2.1 *If* $\varepsilon_1 = 0.5$, *then*

$$P_3 = 0.5 + \varepsilon_2 \text{ and } 1 - P_3 = 0.5 - \varepsilon_2.$$

Corollary 2.2 *If* $\varepsilon_1 = -0.5$, *then*

$$P_3 = 0.5 - \varepsilon_2 \text{ and } 1 - P_3 = 0.5 + \varepsilon_2.$$

Corollary 2.3 *If* $\varepsilon_2 = 0.5$, *then*

$$P_3 = 0.5 + \varepsilon_1 \text{ and } 1 - P_3 = 0.5 - \varepsilon_1.$$

Corollary 2.4 *If* $\varepsilon_2 = -0.5$, *then*

$$P_3 = 0.5 - \varepsilon_1 \text{ and } 1 - P_3 = 0.5 + \varepsilon_1.$$

Corollary 2.5 *If* $\varepsilon_1 = \varepsilon_2 = 0.5$, *then*

$$P_3 = 1 \text{ and } 1 - P_3 = 0.$$

Corollary 2.6 *If* $\varepsilon_1 = \varepsilon_2 = -0.5$, *then*

$$P_3 = 1 \text{ and } 1 - P_3 = 0.$$

Corollary 2.7 *If* $\varepsilon_1 = -0.5$ *and* $\varepsilon_2 = 0.5$ *or* $\varepsilon_1 = 0.5$ *and* $\varepsilon_2 = -0.5$, *then*

$$P_3 = 0 \text{ and } 1 - P_3 = 0.$$

In the last three corollaries of both lemmas the stochastic values P_1, P_2, and P_3 can actually be considered as deterministic parameters.

Lemma 2 can be inferred directly by Lemma 1. To do this, instead $s = 0.5 + \varepsilon_2$ it is necessary to make

$$s = p = 1 - P_2 = 1 - (0.5 + \varepsilon_2).$$

Then $s = 0.5 - \varepsilon_2$. If this relationship is made in (20), it will lead to results (22) and (23) of Lemma 2.

Values $\varepsilon_1, \varepsilon_2$, and ε_3 are convenient for a logical interpretation, since by (15) they not only provide quantitative indicators for the transition probabilities P_1, P_2 and P_3 but their sign (plus or minus) can be placed directly under the prevailing true state in a_1 or under the prevailing false state—in a_2. The positive sign for $\varepsilon_i > 0$ always shows a predominantly positive value of the variable $P_i, i \in \{1, 2, 3\}$, i.e., $0.5 \le P_i \le 1$; the negative sign for $\varepsilon_i < 0$ corresponds to the case $0 \le P_i \le \; \le 0.5$.

The sign of the value ε_i—plus or minus will be denoted by $\bar{\varepsilon}_i, i \in \{1, 2, 3\}$.

The results of Lemma 1 by sign $\{\bar{\varepsilon}_i\}$ are shown in Table 1, and their logical interpretation is reflected in Table 2.

Table 1 .

$\bar{\varepsilon}_1$	$\bar{\varepsilon}_2$	$\bar{\varepsilon}_3 = -\bar{\varepsilon}_1\bar{\varepsilon}_2$
+	+	−
+	−	+
−	+	+
−	−	−

Table 2 .

X	Y	$Z \equiv X + Y$
T	T	F
T	F	T
F	T	T
F	F	F

Logical variables *X*, *Y*, and *Z* in Table 2 are a relevant interpretation of the sign parameters $\bar{\varepsilon}_1, \bar{\varepsilon}_2$ and $\bar{\varepsilon}_3$ from Table 1. It should be noted that instead the sign parameters $\bar{\varepsilon}_1, \bar{\varepsilon}_2$ and $\bar{\varepsilon}_3$ in Table 1 there can be used also the values $\varepsilon_1, \varepsilon_2$ and ε_3 or the corresponding stochastic quantities $P_1, P_2$and P_3 by (15). The results in Table 2 will be the same.

Similarly, Table 3 shows the sign parameters $\bar{\varepsilon}_1, \bar{\varepsilon}_2$ and $\bar{\varepsilon}_3$ based on the results (22) and (23) of Lemma 2. The following Table 4 shows the logical variables *X*, *Y*, and *Z*, corresponding to $\bar{\varepsilon}_1, \bar{\varepsilon}_2$ and $\bar{\varepsilon}_3$ by Lemma 2.

Both Tables 2 and 4 show two possible binary logic functions out of 16 possible ones. The next six logic functions can be realized by the matrix structures of the transitional probabilities $P(1)$ and $P(2)$ as it follows:

1. Table 2 corresponds to the binary logic function 'Exclusive OR,' which can be indicated by logical variables *X*, *Y*, and *Z* as it follows:

$$Z \equiv X + Y, Z \equiv ((X \wedge \neg Y) \vee (\neg X \wedge Y)), \tag{27}$$

Table 3 .

$\bar{\varepsilon}_1$	$\bar{\varepsilon}_2$	$\bar{\varepsilon}_3 = \bar{\varepsilon}_1\bar{\varepsilon}_2$
+	+	+
+	−	−
−	+	−
−	−	+

Table 4 .

X	Y	$Z \equiv X \sim Y$
T	T	T
T	F	F
F	T	F
F	F	T

where $\equiv$ means equivalency and $\wedge$, $\vee$, and $\neg$ are, respectively, the symbols of conjunction, disjunction, and negation.
This logical function is realized uniquely and directly via the vector matrix Eq. (10) and also by the stochastic matrix $P(1)$ from (9).

2. Logical function 'equivalence' corresponds to the truth Table 4 and it can be denoted as it follows:

$$Z \equiv X \sim Y, Z \equiv ((X \wedge Y) \vee (\neg X \wedge \neg Y)), \tag{28}$$

Logical function 'equivalence' is uniquely and directly realized by the vector matrix Eq. (10) and also by the transition matrix $P(2)$ from (9).

3. The following Table 5 for truth corresponds to the logical function 'disjunction'. If we compare it to the truth Table 2 then the only difference will be found in the row with number 1. This allows us to move from the logical function 'Exclusive OR' to the logical function 'disjunction' via the following Rule 1, namely:

$$\varepsilon_3 := \begin{cases} 2\varepsilon_1\varepsilon_2 \text{ iff } \varepsilon_1 > 0 \text{ and } \varepsilon_2 > 0; \\ -2\varepsilon_1\varepsilon_2 \text{ otherwise.} \end{cases} \tag{29}$$

In this case we use matrix $P(1)$ and relations (9) and (10).

4. Table 6 for truth corresponds to the logical function 'conjunction' ($\wedge$). When compared to the truth Table 4 the only difference will be found in the line with number 4 on both matrices; the other three lines coincide. This allows us to move from the logical function 'equivalence' to the similar function 'conjunction' with the following Rule 2, namely:

$$\varepsilon_3 := \begin{cases} -2\varepsilon_1\varepsilon_2 \text{ iff } \varepsilon_1 < 0 \text{ and } \varepsilon_2 < 0; \\ 2\varepsilon_1\varepsilon_2 \text{ otherwise.} \end{cases} \tag{30}$$

Table 5 .

X	Y	$X \vee Y \equiv Z$
T	T	T
T	F	T
F	T	T
F	F	F

Table 6 .

X	Y	$X \wedge Y \equiv Z$
T	T	T
T	F	F
F	T	F
F	F	F

Table 7 .

X	Y	$X \rightarrow Y \equiv Z$
T	T	T
T	F	F
F	T	T
F	F	T

Matrix $P(2)$ is used in (30) and also the relations (9) and (10).

5. The logical function 'implication' corresponds to the truth Table 7:
The comparison of Tables 4 and 7 leads to the conclusion that the only difference is in the row with number 3. This allows us to move from the logical function 'equivalence' to the logical function 'implication' by following Rule 3:

$$\varepsilon_3 := \begin{cases} -2\varepsilon_1\varepsilon_2 \text{ iff } \varepsilon_1 < 0 \text{ and } \varepsilon_2 > 0; \\ 2\varepsilon_1\varepsilon_2 \text{ otherwise }. \end{cases} \tag{31}$$

The above rule uses matrix $P(2)$ and also the relations (9) and (10).

6. Unary logical operation 'negation' is realized in a natural way, changing the locations of the elements of the vectors π_1, π_2 and π_3 as it is shown in Table 8. The negation for the three vectors can be defined as follows: for each $i \in \{1, 2, 3\}$

$$\varepsilon_i := -\varepsilon_i \tag{32}$$

or in another way

$$P_i := 1 - P_i. \tag{33}$$

Table 9 shows a logical variable X and the corresponding function 'negation,' realized through $\{\varepsilon_i\}$ from Table 8.

It is possible to define also the remaining 10 binary logic functions by the described above binary Markov process in an analogous way.

Table 8 .

ε_i	$-\varepsilon_i$
$-\varepsilon_i$	ε_i

Table 9 .

X	$\neg X$
T	F
F	T

Using the above-mentioned six logical functions (only three of them—disjunction, conjunction, and negation, are necessary for completeness) all formulas of classical propositional logic can be derived as binary Markov processes. The so-constructed Markov processes give the way for direct and unambiguous interpretation of the formulas of propositional logic, without any ambiguity and logical contradictions.

This should take into account the specifics of the stochastic interpretation of truth and false—the process is in the true state iff at the moment n its probability to be in state a_1 is greater than 0.5. Otherwise the process is in the false state.

In those results attention was paid basically to the qualitative nature in precise and unambiguous distinction between true or false states of the binary Markov process. But the results of Lemmas 1, 2 and their corollaries show that also qualitative changes occur at any single-step transition in the stochastic parameters that can be best represented by (17) and (22), namely:

$$\varepsilon_3 = \pm 2\varepsilon_1\varepsilon_2.$$

The research of stochastic behavior of the so-defined binary Markov process is the subject of a separate investigation.

Numerical examples for the implementation of Markov processes described by six logical functions: +, ~, ∨, ∧, →, ¬ will be presented below.

The raw data for these examples are:

a) $\varepsilon_1 = 0.2$ iff $\bar{\varepsilon}_1 = (+)$ and $\varepsilon_1 = -0.3$ iff $\bar{\varepsilon}_1 = (-)$; (34)

b) $\varepsilon_2 = 0.3$ iff $\bar{\varepsilon}_2 = (+)$ and $\varepsilon_2 = -0.1$ iff $\bar{\varepsilon}_2 = (-)$. (35)

The relevant logic functions will have the following form:

1. Based on Table 1, equality (17) and the data from (34) and (35), 'Exclusive OR' function can be described by the elements of Table 10.
2. Table 11 shows items corresponding to the 'equivalence' function. They are calculated by formula (22), Table 3 and the data from (34) and (35).
3. The 'disjunction' function may be determined by the data from Table 1, rule (29) and relations (34) and (35). The results are shown in Table 12.
4. The elements of Table 13 are calculated via Table 3, Rule (30) and data (34) and (35). They describe the logical function 'conjunction.'

Table 10 .

ε_1	ε_2	ε_3
0.2	0.3	–0.12
0.2	–0.1	0.04
–0.3	0.3	0.18
–0.3	–0.1	–0.06

Table 11 .

ε_1	ε_2	ε_3
0.2	0.3	0.12
0.2	–0.1	–0.04
–0.3	0.18	–0.18
–0.3	–0.1	0.06

Table 12 .

ε_1	ε_2	ε_3
0.2	0.3	0.12
0.2	–0.1	0.04
–0.3	0.3	0.18
–0.3	–0.1	–0.06

Table 13 .

ε_1	ε_2	ε_3
0.2	0.3	0.12
0.2	–0.1	–0.04
–0.3	0.3	–0.18
–0.3	–0.1	–0.06

5. The logical function 'implication' is shown in Table 14. The elements of this table are determined by the data from Table 3, Rule (36) and (34) and (35).
6. Quantitative data of the logical 'negation' are given in Table 15. They are based on the elements of Table 8 and also on the dependencies (32).

Analogous results can be obtained for different values of ε_1 and ε_2,, respectively, also for P_1 and P_2.

Table 14 .

ε_1	ε_2	ε_3
0.2	0.3	0.12
0.2	–0.1	–0.04
–0.3	0.3	0.18
–0.3	–0.1	0.06

Table 15 .

ε_3	$-\varepsilon_3$
0.12	–0.12
–0.12	0.12

1 Conclusion

1. A method for interpreting binary logic functions with a multistep binary Markov process is introduced. It allows logical concepts 'true' and 'false' be treated not as deterministic but as stochastic variables that are 'mostly true', with a probability between 0.5 and 1, to enter state a_1 or 'mostly false' with a probability between 0.5 and 1 to enter state a_2. This enables a Markov process to deal with different degrees of truth or false.
2. A single-step Markov realization is elaborated for standard logic functions of propositional logic; a series of analytical relations are formulated between the stochastic parameters of the Markov process before and after the implementation of the single-step transition. It has been proven that any logical operation can directly, uniquely, and consistently be described by a corresponding Markov process. It is shown that it is possible via the proposed method to implement a Markov interpretation of different formulas of propositional logic. The results allow precise studies of the behavior of the used class of Markov processes.
3. Examples are presented and a numerical realization is realized of some functions of propositional logic by binary Markov processes.

References

1. Kleene, S.C.: Mathematical Logic. Wiley, N.Y. (1977)
2. Zadeh, L.A.: The Concept of a Linguistic Variable and Its Applications to Approximate Reasoning. Elsevier Publ. C, N.Y. (1973)
3. Stern, A.: Matrix Logic and Mind. North Holland, N.Y. (1992)
4. Sgurev, V.S.: Network flow approach in logic for problem solving. Int. J. Inf. Theor. Appl. **7** (1995)
5. Gorodetsky, V.I.: Bayes inference and decision making in artificial intelligence systems. In: Industrial Applications of AI, Elsevier S. Publ. B.V., Amsterdam, pp. 276–281 (1991)
6. Sgurev, V., Jotsov, V.: Discrete Markov process interpretation of propositional logic. In: Proceedings of IEEE Conference, IS'10, London (2010)
7. Neapolitan R.E.: Learning Bayesian Networks. Prentice Hall (2004)
8. Kemeny, J.G., Snell, J.K., Knapp, A.W.: Denumerable Markov Chain. Springer, Heidelberg, Berlin (1976)
9. Mine, H., Osaki, S.: Markovian Decision Processes. American Elsevier Publ. Co, N.Y. (1970)
10. Domingos, P., Richardson, M.: Markov Logic: A Unifying Framework for Statistical Relational Learning. In: Getoor, L., Taskar, B. (eds.) Introduction to Statistical Relational Learning, pp. 339–371. MIT Press, Cambridge, MA (2007)
11. Singla, P., Domingos, P.: Discriminative training of Markov logic networks. In: AAAI-2005, pp. 868–873 (2005)
12. Huang, S., Zhang, S., Zhou, J. Chen, J.: Coreference Resolution using Markov Logic Network. Advances in Computational Linguistics, Research in Computing Science vol. 41, pp. 157–168 (2009)
13. Mihalkova, L., Huynh, T., Mooney, R.J.: Mapping and revising Markov logic networks for transfer learning. In: Proceedings of the 22nd Conference on Artificial Intelligence (AAAI-07). pp. 608–614, Vancouver, Canada, July 2007

Generalized Net Models of Academic Promotion and Doctoral Candidature

Anthony G. Shannon, Beloslav Riecan, Evdokia Sotirova, Krassimir Atanassov, Maciej Krawczak, Pedro Melo-Pinto, Rangasamy Parvathi and Taekyun Kim

Abstract In a series of research papers, the authors have studied some of the most important features of the principal operations within universities and have constructed Generalized Net (GN) models to describe them. The main focus in this paper is to analyse the process of academic promotion through the hierarchy in higher education and the preparation of PhD candidates.

Keywords Generalized nets · University · Modelling · Intuitionistic fuzzy estimations

A.G. Shannon (✉)
Faculty of Engineering and IT, University of Technology, Sydney, NSW 2007, Australia
e-mail: Anthony.Shannon@uts.edu.au; tshannon38@gmail.com

B. Riecan
Faculty of Natural Sciences, Department of Mathematics,
Matej Bel University Mathematical Institute of Slovak Academy of Sciences, Tajovskeho 40, Banska Bystrica, Slovakia
e-mail: riecan@mat.savba.sk; riecan@fpv.umb.sk

E. Sotirova · K. Atanassov
Asen Zlatarov University, 8000 Burgas, Bulgaria
e-mail: esotirova@btu.bg

K. Atanassov
e-mail: krat@bas.bg

K. Atanassov
Bioinformatics and Mathematical Modelling Department,
Institute of Biophysics and Biomedical Engineering Bulgarian
Academy of Sciences, 1113 Sofia, Bulgaria

M. Krawczak
Systems Research Institute—Polish Academy of Sciences,
Wyzsza Szkola Informatyki Stosowanej i Zarzadzania, Warsaw, Poland
e-mail: krawczak@ibs.pan.waw.pl

V. Sgurev et al. (eds.), *Recent Contributions in Intelligent Systems*,
Studies in Computational Intelligence 657,
DOI 10.1007/978-3-319-41438-6_15

1 Introduction

The processes of administrative functions within an idealized university have been described within the framework of Generalized Nets (GNs, see [1, 2]) in a series of papers (subsequently collated in a book [6]). The rapid growth of university education in general, and the onset of the Bologna Process in particular, have made the consideration of these functions more urgent. Moreover, the preparation of doctoral candidates in this era of growth necessitates more detailed analyses of the preparation of these research students [8].

In Chap. 5 of [6], the process of promotion through the higher education hierarchy (universities and scientific institutes) was described. The information we have used about the processes involved is derived from our own countries. While there are certain small differences between university staff and scientific institute staff in different countries, for the sake of brevity and simplicity, we shall ignore these differences.

The more important fact is that the scientific degrees and titles in the separate countries are different. In the monograph [6], we have provided schematic summaries to illustrate the order of these titles and degrees, as stipulated in Argentina, Australia, Belgium, Bulgaria, Greece, Korea, Lebanon, Poland, Portugal, Romania, Slovakia, the UK and the USA.

In this paper in Sect. 2 we present a GN-model of academic promotion G (see Fig. 1), which is based on the model from Chapter Five in [6]. In Sect. 3, we extend the GN G with the hierarchical operator H_1 over GNs [1, 2], so that we can replace the place l_{13} of GN G with a new GN from Fig. 2 that represents PhD preparation.

Throughout the discussion, by "Academic Institution" (AI) we shall mean either a university or a scientific institute.

P. Melo-Pinto
CITAB—UTAD Quinta de Prados, Apartado 1013, 5001-801 Vila Real, Portugal
e-mail: pmelo@utad.pt

R. Parvathi
Department of Mathematics, Vellalar College for Women, Erode 638 009, Tamilnadu, India
e-mail: paarvathis@rediffmail.com

T. Kim
I Department of Mathematics, College of Natural Science Kwangwoon University, Seoul 139-704, South Korea
e-mail: tkkim@kw.ac.kr

2 GN-Model of Academic Promotion

The GN-model in Fig. 1 describes the procedure and time scheduling to obtain scientific titles and degrees in one of the aforementioned countries (Bulgaria). In Sect. 4, we discuss the changes needed to extend the model in order to describe similar processes in each of the remaining countries. The basic time step of the GN is assumed to be one year. If we need to describe the processes involved in more detail, we can decrease the duration of this time step.

The GN-model consists of six transitions that represent, respectively:

- the process of studying in a university (transition Z_1),
- the process of competition for starting a procedure of PhD dissertation or of starting a job as an Assistant Professor (transition Z_2),
- the process of preparing of PhD dissertations (transition Z_3),
- the set of Assistant Professors and their activities (transition Z_4),
- the set of Associated Professors and their activities (transition Z_5),
- the set of the (Full) Professors and their activities (transition Z_6).

For the different stages of the process of obtaining scientific titles and degrees, we include some possibilities for potential evaluation procedures. In order to do so, we utilize estimations of intuitionistic fuzzy sets [3–5].

During the whole period of the GN-functioning, tokens α_{Year} will enter the net with initial characteristic

x^{α}_{Year} = "number of students who enrolled a university after finishing secondary school in year *Year.*"

Let the number of the students who have started study in year *Year* be *SN*(*Year*).

Therefore, the formal form of an α-token's initial characteristic for time step that corresponds to year *Year* is

$$x^{\alpha}_{\text{Year}} = {}''SN(\text{Year})''.$$

Initially, the tokens β, γ, δ, ε, ζ, η stay in places l_4, l_{13}, l_{17}, l_{20}, l_{21} and l_{24}. They will be in their own places during the whole time during which the GN functions. While they may split into two or more tokens, the original token will remain in its own place the whole time. The original tokens have the following initial and current characteristics:

- token β in place l_4:
  x^{β}_{cu} = "number of students who currently study in the universities,"
- token γ in place l_{13}:
  x^{γ}_{cu} = "number of PhD students who currently work on their dissertations in AIs",
- token δ in place l_{17}:
  x^{δ}_{cu} = "number of Assistant Professors, who currently work in AIs",
- token ε in place l_{20}:

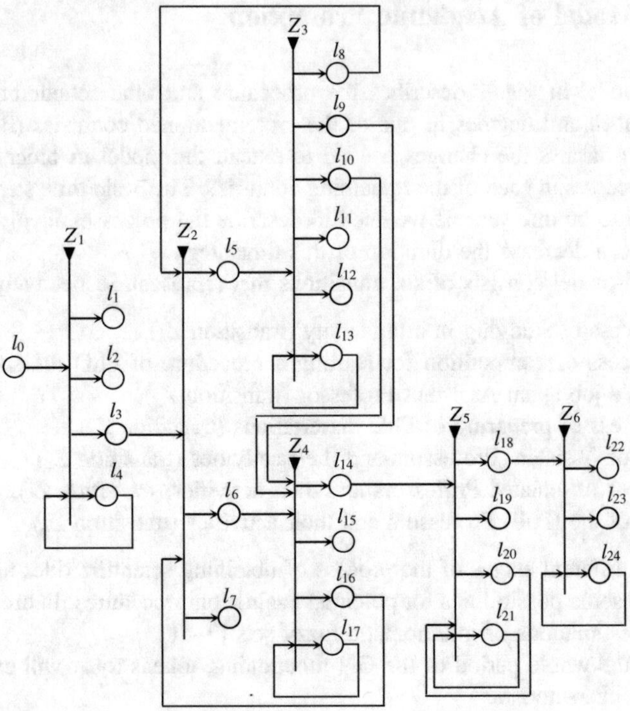

Fig. 1 GN-model of the process of academic promotion (Bulgaria)

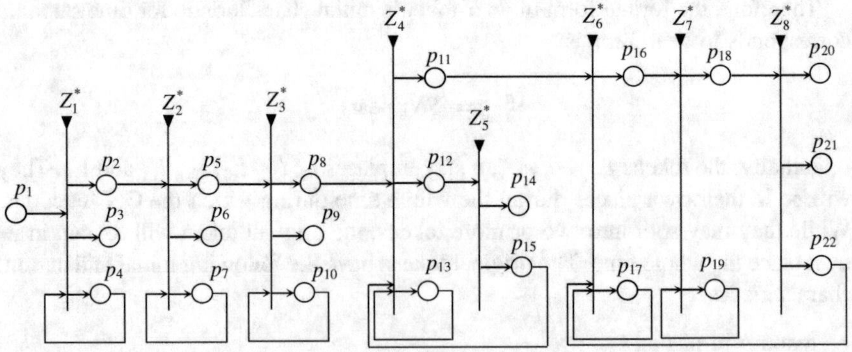

Fig. 2 GN-model of the PhD development

x_{cu}^{ε} = "number of Associate Professors, who currently work in AIs and who simultaneously prepare their DSc-dissertations,"

- token ζ in place l_{21}:

  x_{cu}^{ζ} = "number of Associate Professors, who have not started yet or who had finished their DSc-dissertations,"

- token η in place l_{24}:
  x^{η}_{cu} = "number of (Full) Professors, who currently work in AIs."

Obviously, if the duration of the full university course is D years, then the number $AN(Year)$ of all ordinary students in year $Year$ is

$$AN(\text{Year}) = \sum_{i=\text{Year}-D+1}^{\text{Year}} SN(i).$$

Transition Z_1 has the form:

$$Z_1 = \langle \{l_0, l_4\}, \{l_1, l_2, l_3, l_4\}, r_1, \vee (l_0, l_4) \rangle,$$

where

$$r_1 = \begin{array}{c|cccc} & l_1 & l_2 & l_3 & l_4 \\ \hline l_0 & false & false & false & true \\ l_4 & W_{4,1} & W_{4,2} & W_{4,3} & true \end{array},$$

where

$W_{4,1}$ = "there are students who stop studying,"
$W_{4,2}$ = "there are students who finish studying and start work outside scientific institutions,"
$W_{4,3}$ = "there are students who finish studying and start work in scientific institutions."

On each time step, token α from l_0 enters place l_4 and unites with token β. The β token can split to four tokens. As we mentioned above, the original β token continues to stay in place l_4.

- Token β_1 enters place l_1 with characteristic
  x^{β}_{1} = "number of students who stop their studies in a university,"
- token β_2 that enters place l_2 with characteristic
  x^{β}_{2} = "number of students who finish their studies in a university and start working outside some AIs,"
- token β_3 that enters place l_4 with characteristic
  x^{β}_{3} = "number of students who finish their studies in a university and start working in some AIs."

From the point of view of the process of obtaining scientific titles and degrees, the intuitionistic fuzzy estimation of the results of this transition is

$$\langle \frac{x^{\beta}_{2} + x^{\beta}_{3}}{x^{\alpha}_{Year-D+1}}, \frac{x^{\beta}_{1}}{x^{\alpha}_{Year-D+1}} \rangle. \qquad (1)$$

The degree of uncertainty in estimation (1) corresponds to the number of students who have been learning in year *Year*.

Transition Z_2 has the form:

$$Z_2 = \langle\{l_3, l_8, l_{16}\}, \{l_5, l_6, l_7\}, r_2, \vee(l_3, l_8, l_{16})\rangle,$$

where

$$r_2 = \begin{array}{c|ccc} & l_5 & l_6 & l_7 \\ \hline l_3 & W_{3,5} & W_{3,6} & W_{3,7} \\ l_8 & false & W_{8,6} & W_{8,7} \\ l_{16} & W_{16,5} & W_{16,6} & false \end{array},$$

where

$W_{3,5}$ = "there are participants in a competition for PhD doctorates, who win the competition,"
$W_{3,6}$ = "there are participants (without PhD) in a competition for Assistant Professors, who win the competition,"
$W_{3,7} = \neg W_{3,5}$ OR $\neg W_{3,6}$,
$W_{8,6}$ = "there are participants with PhD in a competition for Assistant Professors, who win the competition,"
$W_{8,7} = \neg W_{8,6}$,
$W_{16,5}$ = "there are Assistant Professors who win the competition for PhD doctorates,"
$W_{15,6} = \neg W_{16,5}$.

Token β_3 splits to tree tokens:

- $\beta_{3,1}$ that enters place l_5 with a characteristic $x^{\beta}_{3,1}$ = "number of candidates for PhD doctorates who win the competition,"
- $\beta_{3,2}$ that enters place l_6 with a characteristic $x^{\beta}_{3,2}$ = "number of candidates for title Assistant Professor, who win the competition"
- $\beta_{3,3}$ that enters place l_7 with a characteristic $x^{\beta}_{3,3}$ = "number of candidates who lose the competition."

On the other hand, token δ_3 enters place l_5 from place l_{16} with a characteristic x^{δ}_3 = "number of Assistant Professors that want to participate for doctorates."

Transition Z_3 has the form:

$$Z_3 = \langle\{l_5, l_{13}\}, \{l_8, l_9, l_{10}, l_{11}, l_{12}, l_{13}\}, r_3, \vee(l_5, l_{13})\rangle,$$

where

$$r_3 = \begin{array}{c|cccccc} & l_8 & l_9 & l_{10} & l_{11} & l_{12} & l_{13} \\ \hline l_5 & false & false & false & false & false & true \\ l_{13} & W_{13,8} & W_{13,9} & W_{13,10} & W_{13,11} & W_{13,12} & true \end{array},$$

where

$W_{13,8}$ = "the dissertation has been successfully defended and the person want to start work as an Assistant Professor in an AI,"
$W_{13,9}$ = "the dissertation of an Assistant Professor has been successfully defended,"
$W_{13,10} = \neg W_{13,9}$,
$W_{13,11}$ = "the dissertation has been successfully defended and the person starts working outside an AI,"
$W_{13,12}$ = "the dissertation has not been successfully defended and the person starts working outside an AI."

Token $\beta_{3,1}$ unites with token γ in place l_{13} with the above mentioned characteristic. Token γ splits to six tokens. The original token γ remains in place l_{13}.

- token γ_1 enters place l_8 with a characteristic
  x_1^{γ} = "number of persons with defended PhD, who want to work as Assistant Professors in some AI,"
- token γ_2 enters place l_9 with a characteristic
  x_2^{γ} = "number of Assistant Professor, who works in AIs and have successfully defended dissertation,"
- token γ_3 enters place l_{10} with a characteristic
  x_3^{γ} = "number of Assistant Professor, who works in AIs and has not been successfully defended dissertation,"
- token γ_4 that place l_{11} with a characteristic
  x_4^{γ} = "number of persons with defended PhD, who start working outside AIs,"
- token γ_5 that enters place l_{12} with a characteristic
  x_5^{γ} = "number of persons who had not defended their PhD."

From the point of view of the process of obtaining scientific titles and degrees, the intuitionistic fuzzy estimation of the results of this transition is

$$\langle \frac{x_1^{\gamma} + x_2^{\gamma} + x_4^{\gamma}}{x_{\text{Year}-E+1}^{\alpha}}, \frac{x_3^{\gamma} + x_5^{\gamma}}{x_{\text{Year}-E+1}^{\alpha}} \rangle, \quad (2)$$

where E is the duration of PhD thesis preparation.

The degree of uncertainty in estimation (2) corresponds to the number of the PhD students who have been preparing their dissertations.

Transition Z_4 has the form:

$$Z_4 = \langle \{l_6, l_9, l_{10}, l_{17}\},\ \{l_{14}, l_{15}, l_{16}, l_{17}\}, r_4,\ \vee (l_6, l_9, l_{10}, l_{17})\rangle,$$

where

$$r_4 = \begin{array}{c|cccc} & l_{14} & l_{15} & l_{16} & l_{17} \\ \hline l_6 & false & false & false & true \\ l_9 & false & false & false & true \\ l_{10} & false & false & false & true \\ l_{17} & W_{17,14} & W_{17,15} & W_{17,16} & true \end{array},$$

where

$W_{17,14}$ = "there are Assistant Professors, who participate in competition for Associate Professor,"
$W_{17,15}$ = "there are Assistant Professors, who leave AIs,"
$W_{17,16}$ = "there are Assistant Professors (without PhD), who would like to prepare it."

Tokens $\beta_{3,1}$, γ_2, and γ_3 unite with token δ in place l_{17} with the above-mentioned characteristic. Token δ splits to four tokens, with the original staying in place l_{17}, while the other tokens enter places l_{14}, l_{15}, and l_{16}, respectively:

- token δ_1 that enters place l_{14} with a characteristic
  x_1^δ = "number of Assistant Professors (with PhD), who want to take part in competition for Associate Professor."
- token δ_2 that enters place l_{15} with a characteristic
  x_2^δ = "number of Assistant Professors (without PhD), who stopped work in an AI,"
- token δ_3 that enters place l_{16} with a characteristic
  x_3^δ = "number of Assistant Professors (without PhD), who want to start to prepare PhD dissertations."

From the point of view of the process of obtaining of scientific titles and degrees, the intuitionistic fuzzy estimation of the results of this transition is

$$\langle \frac{AP(\text{Year})}{x^\delta_{\text{Year}}}, \frac{AWP(\text{Year})}{x^\delta_{\text{Year}}} \rangle, \tag{3}$$

where $AP(Year)$ is the number of the Assistant Professors with PhD, $AWP(Year)$ is the number of Assistant Professors who had not then started preparing a PhD.

The degree of uncertainty in estimation (3) corresponds to the number of Assistant Professors who currently prepare their (free or external) dissertations.

Transition Z_5 has the form:

$$Z_5 = \langle \{l_{14}, l_{20}, l_{21}\}, \{l_{18}, l_{19}, l_{20}, l_{21}\}, r_5, \vee(l_{14}, l_{20}, l_{21})\rangle,$$

where

$$r_5 = \begin{array}{c|cccc} & l_{18} & l_{19} & l_{20} & l_{21} \\ \hline l_{14} & false & false & false & true \\ l_{20} & false & W_{20,19} & W_{20,20} & W_{20,21} \\ l_{21} & W_{21,18} & W_{21,19} & W_{21,20} & true \end{array},$$

where

- $W_{20,20} = W_{21,20}$ = "there are Associated Professors without DSc, who would like to prepare or currently prepare it,"
- $W_{20,19} = W_{21,19}$ = "there are Associated Professors, who leave AIs,"
- $W_{21,18}$ = "there are Associated Professors, who want to participate in competition for Full Professor."

Token δ_1 unite with token ζ in place l_{21} with the above-mentioned characteristic. Token ζ splits to four tokens:

- token ζ_1 that enters place l_{18} with a characteristic
  x_1^{ζ} = "number of Associated Professors, who want to take part in competition for Full Professor,"
- token ζ_2 that enters place l_{19} with a characteristic
  x_2^{ζ} = "number of Associated Professors, who stopped working in a AI,"
- token ζ_3 that enters place l_{20} with a characteristic
  x_3^{ζ} = "number of Associated Professors without DSc, who want to start a DSc procedure."

Tokens ζ_3 unites with token ε in place l_{20} with the above-mentioned characteristic.

From the point of view of the process of obtaining scientific titles and degrees, the intuitionistic fuzzy estimation of the results of this transition is

$$\langle \frac{APD(\text{Year})}{x^{\zeta}}, \frac{AWPS(\text{Year})}{x^{\zeta}} \rangle, \quad (4)$$

where *APD*(Year) is the number of Associated Professors with a DSc, *APWS*(*Year*) is the number of Associated Professors who had not started preparing a DSc dissertation.

The degree of uncertainty in estimation (4) corresponds to the number of the Associated Professors who currently prepare their DSc dissertations.

Transition Z_6 has the form:

$$Z_6 = \langle \{l_{18}, l_{24}\}, \{l_{22}, l_{23}, l_{24}\}, r_6, \vee(l_{18}, l_{24})\rangle,$$

where

$$r_6 = \begin{array}{c|ccc} & l_{22} & l_{23} & l_{24} \\ \hline l_{18} & false & false & true \\ l_{24} & W_{24,22} & W_{24,23} & true \end{array},$$

where

$W_{24,22}$ = "there are Professors who have left AIs,"
$W_{24,23}$ = "there are Professors, who have retired."

Tokens ζ_1 and η unite with token η in place l_{24} with the above mentioned characteristic. Token η splits to three tokens:

- token η_1 that enters place l_{22} with a characteristic x_1^η = "number of Professors, who have left AIs,"
- token η_2 that enters place l_{23} with a characteristic x_2^η = "number of Assistant Professors, who have retired," and token η that stays in place l_{24}.

From the point of view of the process of obtaining scientific titles and degrees, the intuitionistic fuzzy estimation of the results of this transition is

$$\langle \frac{PWIS(\text{Year})}{x^\eta}, \frac{PP(\text{Year})}{x^\eta} \rangle, \tag{5}$$

where *PWIS*(Year) is the number of the Professors working in AIs and *PP*(*Year*) is the number of the Professors who have retired.

The degree of uncertainty in estimation (5) corresponds to the number of Professors who are currently not working in any AI, but can potentially return to work there, because they have not yet reached the age of retirement.

3 A GN-Model of the PhD Candidature

The GN-model (see Fig. 2) contains 8 transitions and 22 places, and it is a set of transitions:

$$A = Z_1^*, Z_2^*, Z_3^*, Z_4^*, Z_5^*, Z_6^*, Z_7^*, Z_8^*$$

in which the transitions represent:

- Z_1^*—Submission of documents for PhD examination;
- Z_2^*—Examination for enrollment in the PhD;

- Z_3^*—The process of evaluation, ranking the candidates' evaluations and determining the candidates;
- Z_4^*—Preparation for the PhD exams;
- Z_5^*—Check the deadline for taking the PhD exams;
- Z_6^*—Submission of the PhD thesis;
- Z_7^*—Selection of thesis reviewers;
- Z_8^*—A thesis defense.

The forms of the transitions are the following:

The tokens ω_1, ω_2, …, ω_n enter the GN through place p_1 with characteristics Candidate: name, date, competition documents.

$$Z_1^* = \langle \{p_1, p_4\}, \{p_2, p_3, p_4\}, R_1^* \rangle,$$

where

$$R_1^* = \begin{array}{c|ccc} & p_2 & p_3 & p_4 \\ \hline p_1 & False & False & True \\ p_4 & w_{4,2} & w_{4,3} & w_{4,4} \end{array},$$

$w_{4,2}$ = "The time for submission documents has expired and commission has assessed that submitted documents are accurate,"
$w_{4,3}$ = "The time for submission documents has expired and commission has assessed that submitted documents are not accurate,"
$w_{4,4}$ = "The time for submission documents has not expired."

The tokens do not take on any characteristic in place p_4 and they obtain the characteristic, respectively:

"The candidate is approved" in place p_2,
"The candidate fails" in place p_3.

$$Z_2^* = \langle \{p_2, p_7\}, \{p_5, p_6, p_7\}, \vee R_2^* \rangle,$$

where

$$R_2^* = \begin{array}{c|ccc} & p_5 & p_6 & p_7 \\ \hline p_2 & False & False & True \\ p_7 & w_{7,5} & w_{7,6} & w_{7,7} \end{array},$$

$w_{7,5}$ = "Examination has taken place and the estimation is positive,"
$w_{7,6}$ = "Examination has taken place and the estimation is negative,"
$w_{7,7}$ = "The moment of examination has not yet taken place."

The tokens do not have any characteristic in place p_7 and they obtain in place p_5 and p_6 the characteristic, respectively:
Name of the candidate, positive estimation,
Name of the candidate, negative estimation.

$$Z_3^* = \langle \{p_5, p_{10}\}, \{p_8, p_9, p_{10}\}, R_3^* \rangle,$$

where

$$R_3^* = \begin{array}{c|ccc} & p_8 & p_9 & p_{10} \\ \hline p_5 & False & False & True \\ p_{10} & w_{10,8} & w_{10,9} & w_{10,10} \end{array},$$

$w_{10,8}$ = "A final decision has been made by the Commission, and it is positive,"
$w_{10,9}$ = "A final decision has been made by the Commission, and it is negative,"
$w_{10,10}$ = "The Commission has not yet made its final decision."

The tokens do not obtain any characteristic in place p_{10} and they obtain in places p_8 and p_9 the characteristic, respectively:
The candidate is finally approved,
The candidate is rejected.

$$Z_4^* = \langle \{p_8, p_{13}, p_{15}\}, \{p_{11}, p_{12}, p_{13}\}, R_4^* \rangle,$$

where

$$R_4^* = \begin{array}{c|ccc} & p_{11} & p_{12} & p_{13} \\ \hline p_8 & False & False & True \\ p_{13} & w_{13,11} & w_{13,12} & w_{13,13} \\ p_{15} & False & False & True \end{array},$$

- $w_{13,11}$ = "The PhD student has passed all of his/her examinations,"
- $w_{13,12}$ = "The PhD student has some of his/her examinations untaken,"
- $w_{13,13}$ = "Exam session has not yet finished."

The tokens do not obtain any characteristic in place p_{13} and they obtain in places p_{11} and p_{12} the characteristic, respectively:
The PhD student takes all exams and obtains all its educational credit units,
The candidate has untaken exam (exams).

$$Z_5^* = \langle \{p_{12}\}, \{p_{14}, p_{15}\}, R_5^* \rangle,$$

where

$$R_5^* = \begin{array}{c|cc} & p_{14} & p_{15} \\ \hline p_{12} & w_{12,14} & w_{12,15} \end{array},$$

$w_{12,14}$ = "The time of*** education of PhD student is finished,"
$w_{12,15} = \neg\, w_{12,14}$,

where $\neg P$ is the negation of predicate P.

The tokens that enter places p_{14} and p_{15} obtain the characteristic, respectively:

The time of education of PhD student has finished but he/she has not passed all exams,

The time of education of PhD student has not finished but he/she has not passed all exams.

$$Z_6^* = \langle \{p_{11}, p_{17}, p_{24}\}, \{p_{16}, p_{17}\}, R_6^* \rangle,$$

where

$$R_6^* = \begin{array}{c|cc} & p_{16} & p_{17} \\ \hline p_{11} & False & True \\ p_{17} & w_{17,16} & w_{17,17} \\ p_{24} & False & True \end{array},$$

$w_{17,16}$ = "The thesis is ready,"
$w_{17,17} = \neg\, w_{17,16}$.

The tokens do not have any characteristic in place p_{17} and they obtain in place p_{16} the characteristic:

The PhD student has prepared his/her thesis.

$$Z_7^* = \langle \{p_{16}, p_{19}\}, \{p_{18}, p_{19}\}, R_7^* \rangle,$$

where

$$R_7^* = \begin{array}{c|cc} & p_{18} & p_{19} \\ \hline p_{16} & False & True \\ p_{19} & w_{19,18} & w_{19,19} \end{array},$$

$w_{19,18}$ = "The reviewers are determined,"
$w_{19,19} = \neg\, w_{19,18}$.

The tokens take on in places p_{18} and p_{19} the characteristic

The examination reports of the thesis,

The chosen reviewers for current thesis.

$$Z_8^* = \langle \{p_{18}\}, \{p_{20}, p_{21}, p_{22}\}, R_8^* \rangle$$

where

$$R_8^* = \begin{array}{c|ccc} & p_{20} & p_{21} & p_{22} \\ \hline p_{18} & w_{18,20} & w_{18,21} & w_{18,22} \end{array},$$

$w_{18,20}$ = "The defense is successful,"
$w_{18,21}$ = "The defense is not successful and the student does not have possibility for another one,"
$w_{18,22}$ = "The defense is not successful, and the student has possibility for another one."

The tokens that enter places p_{20}, p_{21} and p_{22} obtain, respectively, the characteristics:

The defense is successful,

The defense is not successful and the PhD student not has the possibility for next defense,

The defense is not successful, but the PhD student has the possibility for next defense.

4 Conclusion

In the present paper, we construct two GN-models to represent the procedure for obtaining scientific titles and degrees and the time order for obtaining scientific titles in different countries. The second GN-model is a subnet of the first, because it can replace the place l_{13} from the first GN-model by the hierarchical operator H_1. The GNs constructed in this way can be used to study the dynamics of AI staff development.

The first GN-model is for academic promotion. With this model we can compare the status of the scientific potential of different countries. It can also be detailed in order to trace the status of the separate AIs in a particular country. On the other hand, the model can be transformed with only small changes in order to be used in different countries. For example, in the GN-models for Australia, Greece, Korea, Portugal there will be no feedback relations between transitions Z_4 and Z_2, because the PhD thesis is generally to be prepared after obtaining the master's degree.

The second GN-model is the detailed description of the place $l13$ which represent the salient features of the preparation of doctoral candidates in AIs. This GN can be used to study the dynamics of PhD candidate development in AI. It can also be used for monitoring and evaluating the PhD development process in that it provides the possibility of tracing all stages of the students' education, including the timing of the selection of candidates. Moreover, this model can be utilized for simulation purposes, particularly in workshops to prepare and develop doctoral advisers and supervisors [7, 8].

Acknowledgments This work has been supported by the Bulgarian National Science Fund under Grant DID-02-29.

References

1. Atanassov, K.: On Generalized Nets Theory. Prof. M. Drinov Academic Publishing House, Sofia (2007)
2. Atanassov, K.: Generalized Nets. World Scientific, Singapore/New Jersey (1991)
3. Atanassov, K.: On Intuitionistic Fuzzy Sets Theory. Studies in Fuzziness and Soft Computing, vol. 283. Springer (2012)
4. Atanassov, K.: Intuitionistic Fuzzy Sets. Springer Physica-Verlag, Berlin (1999)
5. Atanassov, K.: Intuitionistic fuzzy sets. Fuzzy Sets Syst. **20**, 87–96 (1986)
6. Shannon, A., Atanassov, K., Sotirova, E., Langova-Orozova, D., Krawczak, M., Melo-Pinto, P., Petrounias, I., Kim, T.: Generalized Nets and Information Flow Within a University. Warszawa (2007)
7. Shannon, A.G.: Research degree supervision: 'More mentor than master'. In: Lee, Alison, Green, Bill (eds.) Postgraduate Studies, Postgraduate Pedagogy, pp. 31–41. UTS, Sydney (1998)
8. Trigwell, K., Shannon, A.G., Maurizi, R.: Research-Coursework Doctoral Programs in Australian Universities. Australian Government Publishing Service, Canberra (1997)

Modeling Telehealth Services with Generalized Nets

Maria Stefanova-Pavlova, Velin Andonov, Todor Stoyanov, Maia Angelova, Glenda Cook, Barbara Klein, Peter Vassilev and Elissaveta Stefanova

Abstract Generalized Net model of processes, related to tracking the changes in health status (diabetes) of adult patients has been presented. The contemporary state of the art of the telecommunications and navigation technologies allows this model to be extended to the case of active and mobile patient. This requires the inclusion of patient's current location as a new and significant variable of the model. Various opportunities are considered for the retrieval of this information, with a focus on the optimal ones, and a refined Generalized Net model is proposed.

M. Stefanova-Pavlova (✉) · V. Andonov · T. Stoyanov · P. Vassilev
Institute of Biophysics and Biomedical Engineering Bulgarian Academy of Sciences,
1113 Sofia, Acad. G. Bonchev Str., bl. 105, Sofia, Bulgaria
e-mail: stefanova-pavlova@citt-global.net

V. Andonov
e-mail: velin_andonov@yahoo.com

P. Vassilev
e-mail: peter.vassilev@gmail.com

M. Angelova · G. Cook
Northumbria University, Newcastle upon Tyne NE2-1XE, UK
e-mail: maia.angelova@northumbria.ac.uk

G. Cook
e-mail: glenda.cook@northumbria.ac.uk

B. Klein
Fachhochschule Frankfurt am Main University of Applied Sciences,
Nibelungenplatz 1, 60318 Frankfurt am Main, Germany
e-mail: bklein@fb4.fh-frankfurt.de

E. Stefanova
Medical University Sofia, 11 Ivan Geshov str., 1606 Sofia, Bulgaria
e-mail: elissaveta.stephanova@abv.bg

V. Sgurev et al. (eds.), *Recent Contributions in Intelligent Systems*,
Studies in Computational Intelligence 657,
DOI 10.1007/978-3-319-41438-6_16

1 Introduction

Ambient-Assisted Living, telecare, and telehealth belong to the framework of assistive technologies, the aim of which is to secure an independent life at home for elderly and/ or chronically ill people or persons with disabilities as long as possible. In addition, there was in recent years a number of national and international policy initiatives and projects to develop the necessary technologies in pilot projects to test or to support the implementation. In Germany, it is the BMBF/VDE Initiative Ambient-Assisted Living, which significantly contributed to the connection of all relevant social groups. In Great Britain, the Department of Health has developed the whole systems demonstrator program to promote large-scale telecare/telehealth and to carry out the world largest randomized study. In Australia were created an Independent Living Centers as the LifeTech center in Brisbane, in addition to the large areas of assistive technologies, specifically telecare and telehealth. Relevant services were developed and tested [1].

The most effective assistive technology mentioned in research in Australia and United Kingdom is when older people are provided with early intervention, careful assessment, the correct prescription, and home-based follow-up training in how to use assistive technologies. The most effective assistive technologies, identified in research [2, 3] are aids, devices, and equipment to improve quality of life, environmental adaptations to the home, telecare/telehealth, and smart technologies. Although only brief information is given of assistive technology policies and developments in other countries, there is work under way to expand the provision of assistive technologies to older people in a number of countries, including the United States, Japan, China, Spain, and many Scandinavian countries.

One of the main goals of the EU Project MATSIQEL (Models for Aging and Technological Solutions for Improving and Enhancing the Quality of Life (2011–2013), IRSES People Marie Curie Action) is the research on new technologies, used for concepts as Ambient Assisting Living, Telecare or Telehealth, and their contribution for improving the quality of life of older people worldwide. The research field is interdisciplinary. The partners in the project are from different countries and different research areas—Northumbria University in UK (the project coordinator), University of Applied Sciences in Frankfurt, Germany, the Griffith University in Brisbane Australia, die Universidad National Autonoma de Mexico, University Kapstadt in South Africa

The Bulgarian partner is the Institute of Biophysics and Biomedical Engineering at the Bulgarian Academy of Sciences. New knowledge for development of new devices should be developed on the base of Generalized Net approach. Here, we shall show the application of the apparatus of Generalized Nets (GNs, see [4–6]) to assistive technology, namely to telehealth (including the action of a medical doctor) services for diabetes, and the advantages of using such model.

Diabetes mellitus (DM) is a major cause of mortality and morbidity in every country. In 2011, more than 366 million people had DM worldwide. Due to the world's increasingly aging populations, increasingly unhealthy diets, sedentary lifestyles, and obesity, it is estimated that the prevalence of DM will increase to 552 million people by 2030. DM is an intractable condition in which blood glucose levels cannot be regulated normally by the body alone; it has many complications, including cardiovascular diseases, nephropathy, neuropathy, retinopathy, and amputations. The treatment methods include dietary regulation to control blood glucose levels, oral medication, and insulin injection; however, all of these have adverse effects on the patients' quality of life.

Type 1 diabetes, Type 2 diabetes, and gestational diabetes are three main types of diabetes, although there are some other forms of DM, including congenital diabetes, cystic fibrosis-related diabetes, and steroid diabetes, induced by high doses of glucocorticoids. Type 1 diabetes is an autoimmune disease with pancreatic islet beta cell destruction. It is an autoimmune disorder in which the body cannot produce sufficient insulin. Type 2 diabetes, the most prevalent form, results from insulin resistance with an insulin secretary defect. Both Type 1 and Type 2 diabetes are chronic conditions that usually cannot be cured easily. Gestational diabetes is the term used when a woman develops diabetes during pregnancy. Generally, it resolves after delivery, but it may precede development of Type 2 diabetes later in life [7].

Criteria for the diagnosis of diabetes: Fasting glucose: ≥7.0 mmol/l (126 mg/dl) Fasting is defined as no caloric intake for at least 8 h. Symptoms of hyperglycemia and a casual plasma glucose ≥11.1 mmol/l (200 mg/dl). Casual is defined as any time of day without regard to time since last meal. The classic symptoms of hyperglycemia include polyuria, polydipsia, and unexplained weight loss. In conclusion, when the fasting blood glucose is above 7 mmol/l or blood glucose after 2 h after eating is above 11,1 mmol/l the patient has diabetes [8]. In order to have a view on the state of the patient and to have a reaction by a doctor we should monitor the blood glucose. The control of blood glucose levels relies on blood glucose measurement. Diabetic patients, whether Type 1 or Type 2, are encouraged to check their blood glucose levels several times per day; currently, the most common means of checking is using a finger-prick glucose meter. In this way, diabetic patients can obtain a clear picture of their blood glucose levels for therapy optimization and for insulin dosage adjustment for those who need daily injections. Finger-pricking, however, has several disadvantages. Many people dislike using sharp objects and seeing blood, there is a risk of infection, and, over the long term, this practice can result in damage to the finger tissue. Given these realities, the advantages of a noninvasive technology are easily understood. Further, the finger-prick glucose meter is a discrete glucose measurement device that is not practical for continuous monitoring of blood glucose. Some incidences of hyperglycemia or hypoglycemia between measurements may not be recorded. Thus, the resultant monitoring cannot fully represent the blood glucose pattern. Noninvasive glucose measurement eliminates the painful pricking experience, risk of infection, and damage to finger tissue. The noninvasive concept was launched more than 30 years ago. Nevertheless, it can be said that most of the noninvasive technologies are still in their early stages of development. Many noninvasive technologies have been described in the literature, and there is an increasing volume of recent research results.

Table 1 Information regarding noninvasive glucose-monitoring devices

Company (or Device)	Technology	Status
BioSensors Inc.	Bioimpedance spectroscopy	Under development
Freedom Meditech	Fluorescent technology	Awaiting FDA approval
Cnoga Medical	Near-infrared spectroscopy	Awaiting FDA approval
C8 MediSensors	Raman spectroscopy	Investigational device
Positive ID	Chemical sensing in exhaled breath	Under development
EyeSense	Fluorescent technology	R&D phase
Calisto Medical Inc.	Bio-electromagnetic resonance	Under production
Integrity Applications Ltd.	Ultrasonic, conductivity and heat capacity	Clinical trials phase
Grove Instruments	NIR spectroscopy(optical bridge technology)	Clinical trials phase
SCOUT DS, VeraLight Inc.	Fluorescent spectroscopy	Approved

Noninvasive glucose-monitoring technologies

- Bioimpedance spectroscopy
- Electromagnetic sensing
- Fluorescence technology
- Mid-infrared spectroscopy
- Near-infrared spectroscopy
- Optical coherence tomography
- Optical polarimetry
- Raman spectroscopy
- Reverse iontophoresis
- Ultrasound technology

Table 1 shows the most recent developments concerning noninvasive glucose measurement (c.f. [7])

It is important to note that noninvasive monitoring will never be achieved without vigorous scientific and clinical evidence. Many technical issues should be still resolved in order to have a reliable, technically proven glucose measurement.

Further we consider a noninvasive glucose meter as a sensor capable of collecting, storing (to some extent), analyzing the obtained data, and consequently taking the most expected decision. In practice, two types of sensors are considered. The first type are the sensors which are attached to the patient's body. These sensors are looking for biomedical parameters, e.g., ECG signal, SPO2 (Saturation of Peripheral Oxygen). The second-type sensors which are stationary are placed in the rooms to monitor for CO(carbon monoxide) concentration. There are also life sensors which are similar to the first type but work in standby mode and are activated by patient—when event has occurred, e.g., extra beats, the patient pushes event button and the sensors collect the signal. The first and second life sensor types are patient-independent and can work autonomously [9].

For the considered sensor, alarm message is sent to the server and, if necessary, parameter value (or a series of them). The server can send requests to the sensor to confirm the alarm event or the parameter. With these sensors we can have the False positive event. For this reason, the server has to have very smart filter for False positive removal or translate the alarm event to human operator if the case is complicated. This type of sensors can work with a cheap smart module for connecting to the GSM network. Since this network allows more flexibility, the patient is free to go wherever he wants. These sensors can make communication to smart phone by Bluetooth or direct cable communication. Nowadays, the existing GSM network has enough speed and data translation capability via, e.g., network type 3G and 4G too. Also these GSM modules can have a GPS module. This GPS module is necessary in case that the medical center has to localize the person in urgent cases such as earthquake, fires, etc. The smart module can send the GPS coordinates to the rescue center for easy localization of the person or persons. In order to carry out the connection between GSM networks, the sensor should have a GSM module or a smart module. Another requirement to prevent connection break is that the GSM module has to be connected to at least two networks available or a WiFi network connection should be accessible [10].

Further for the purpose of discussion, we will assume that the sensor carrier is equipped with a GPS tracking unit (a device using the Global Positioning System to determine the precise location of a vehicle, person, to which it is attached and to record the position of the asset at regular intervals). The recorded location data can be stored within the tracking unit, or it may be transmitted to a central location data base, or internet-connected computer, using a cellular, radio, or satellite modem embedded in the unit. This allows the asset's location to be displayed against a map backdrop either in real time or when analyzing the track later, using GPS tracking software http://www.liveviewgps.com/. GPS personal tracking devices assist in the care of the elderly and vulnerable. Devices allow users to call for assistance and optionally allow designated carers to locate the user's position, typically within 5 to 10 m. Their use helps promote independent living and social inclusion for the elderly. Devices often incorporate either one-way or two-way voice communication which is activated by pressing a button. Some devices also allow the user to call several phone numbers using preprogrammed speed dial buttons. GPS personal tracking devices are used in several countries to help in monitoring people with early stage of dementia and Alzheimer http://www.eurogps.eu/bg/world-news/tracking/99-gps-tracking-alzheimer.

2 Generalized Net Model

The *GN* model developed on the base of the models from [9] and [10] (see Fig. 1) consists of

- eleven transitions: $Z_1, \dots, Z_{11}$;
- thirty-one places $l_1, \dots, l_{31}$;

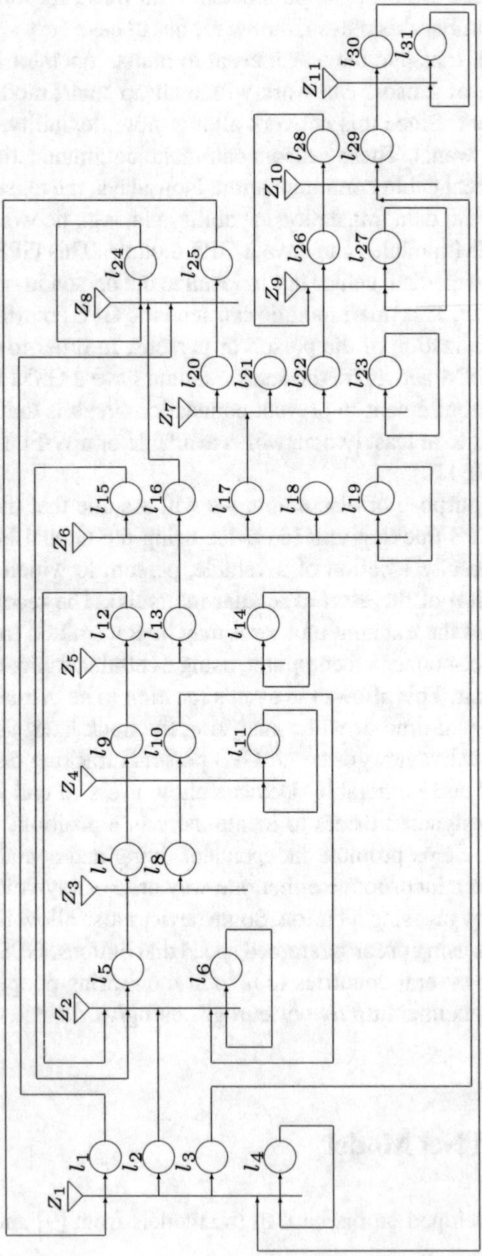

Fig. 1 The generalized net model

- tokens representing the patients, the sensors, criteria for correctness of the signals, history, and previous actions taken, the dispatchers that monitor the signals from the sensors, the medical doctors who examine the patients and the medical specialists who transport the patient to the hospital;

Tokens $\pi_1, \pi_2, \dots, \pi_k$ which represent the patients enter the net in place l_4 with initial characteristic

"patient; name of the patient; current health status"

Tokens $\sigma_1, \dots, \sigma_m$ which represent the sensors enter the net in place l_6 with initial characteristic

"name of the patient; type of sensor"

As an example, we can include in the model the glucose meter which was discussed in the previous section by adding an additional σ-token.

Tokens α and β enter the net in places l_{11} and l_{14} respectively with initial characteristics

"criteria for the correctness of the signals"

Tokens $d_1, \dots, d_n$ enter the net in place l_{19} with initial characteristics

"name of the patient; previously recorded sensor data and respective action taken"

Tokens $\delta_1, \dots, \delta_k$ enter the net in place l_{23} with initial characteristics

"dispatcher; name of the dispatcher; information about all received signals"

Tokens $s_1, \dots, s_l$ enter the net in place l_{27} with initial characteristics

"medical specialist responsible for the transportation of the patient; name of the specialist"

Tokens $\mu_1, \mu_2, \dots, \mu_p$ which represent the medical doctors who examine the patients enter the net in place l_{24} with initial characteristic:

"medical doctor; name of the medical doctor; specialty"

Below is a formal description of the transitions of the net.

$$Z_1 = \langle \{l_4, l_{30}, l_{25}\}, \{l_1, l_2, l_3, l_4\}, \begin{array}{c|cccc} & l_1 & l_2 & l_3 & l_4 \\ \hline l_4 & \mathit{false} & W_{4,2} & W_{4,3} & \mathit{true} \\ l_{25} & \mathit{true} & \mathit{false} & \mathit{false} & \mathit{false} \\ l_{30} & \mathit{false} & \mathit{false} & \mathit{false} & \mathit{true} \end{array} \rangle,$$

where
$W_{4,2}$ = "there is a change in the current patient's status.";
$W_{4,3}$ = "the current patient must be transported to hospital";

When the truth value of the predicate $W_{4,2} = true$ the token π_i representing the i-th patient (here and below $1 \leq i \leq k$) splits into two tokens the original token π_i that continues to stay in place l_4 with the above-mentioned characteristic, and token π_i' that enters place l_2 where it does not obtain new characteristics. When the truth value of the predicate $W_{4,3} = true$ the current π_i token enters place l_3. In place l_1 the tokens obtain the characteristic

"duration of the examination of the patient"

$$Z_2 = \langle \{l_2, l_6, l_{10}, l_{12}, l_{15}\}, \{l_5, l_6\}, \begin{array}{c|cc} & l_5 & l_6 \\ \hline l_2 & false & true \\ l_6 & W_{6,5} & W_{6,6} \\ l_{10} & false & true \\ l_{12} & false & true \\ l_{15} & false & true \end{array} \rangle,$$

where
$W_{6,5}$ = "the sensor detected the patient's body signals";
$W_{6,6} = \neg W_{6,5}$,
where $\neg P$ is the negation of the predicate P.

When the truth value of predicate $W_{6,5} = true$ the corresponding σ token splits into two tokens—the original and a new one that enters place l_5 with characteristic

"signal of the sensor about the current patient"

$$Z_3 = \langle \{l_5\}, \{l_7, l_8\}, \begin{array}{c|cc} & l_7 & l_8 \\ \hline l_5 & W_{5,7} & W_{5,8} \end{array} \rangle,$$

where
$W_{5,7}$ = "the signal comes from a stationary sensor",
$W_{5,8}$ = "the signal comes from a non-stationary sensor",

$$Z_4 = \langle \{l_7, l_8, l_{11}\}, \{l_9, l_{10}, l_{11}\}, \begin{array}{c|ccc} & l_9 & l_{10} & l_{11} \\ \hline l_7 & true & false & false \\ l_8 & W_{8,9} & W_{8,10} & false \\ l_{11} & false & false & true \end{array} \rangle,$$

where
$W_{8,9}$ = "the criterion shows that the signal of the sensor is correct and it must be further evaluated whether a medical doctor's reaction is necessary.",

$W_{8,10}$ = "the criterion shows that the current signal must be confirmed."
When the current σ token enters places l_9 or l_{10} it does not obtain any new characteristics.

$$Z_5 = \langle \{l_9, l_{14}\}, \{l_{12}, l_{13}, l_{14}\}, \begin{array}{c|ccc} & l_{12} & l_{13} & l_{14} \\ \hline l_9 & W_{9,12} & W_{9,13} & false \\ l_{14} & false & false & true \end{array} \rangle,$$

where
$W_{9,12}$ = "the criterion shows that the signal is incorrect."
$W_{9,13}$ = "the criterion shows that the signal is correct."
In place l_{12} the current σ token obtains the characteristic "there is a problem with the sensor." In place l_{13} the current σ token does not obtain any new characteristics.

$$Z_6 = \langle \{l_{13}, l_{19}\}, \{l_{15}, l_{16}, l_{17}, l_{18}, l_{19}\}, \begin{array}{c|ccccc} & l_{15} & l_{16} & l_{17} & l_{18} & l_{19} \\ \hline l_{13} & W_{13,15} & W_{13,16} & W_{13,17} & W_{13,18} & false \\ l_{19} & false & false & false & false & true \end{array} \rangle,$$

where
$W_{13,15}$ = "the history suggests that the signal must be confirmed";
$W_{13,16}$ = "the history suggests that a doctor should visit the patient";
$W_{13,17}$ = "the signal should be examined by dispatcher";
$W_{13,18}$ = "the patient should be sent to hospital"

$$Z_7 = \langle \{l_{17}, l_{23}\}, \{l_{20}, l_{21}, l_{22}, l_{23}\}, \begin{array}{c|cccc} & l_{20} & l_{21} & l_{22} & l_{23} \\ \hline l_{17} & W_{17,20} & W_{17,21} & W_{17,22} & false \\ l_{23} & false & false & false & true \end{array} \rangle,$$

where
$W_{17,20}$ = "a medical doctor should be sent to examine the patient";
$W_{17,21}$ = "no action is necessary";
$W_{17,22}$ = "the patient should be transported to a medical center";
When the truth-value of the predicate $W_{17,20} = true$ the current σ token enters place l_{20} with characteristic

"a decision to visit the patient has been taken"

When the truth-value of the predicate $W_{17,21} = true$ the current σ token enters place l_{21} with characteristic

"a decision to ignore the signal has been taken"

When the truth-value of the predicate $W_{17,22} = true$ the current σ token enters place l_{22} with characteristic

"a decision to transport the patient to a medical center has been taken"

$$Z_8 = \langle \{l_1, l_{16}, l_{20}, l_{24}\}, \{l_{24}, l_{25}\}, \begin{array}{c|cc} & l_{24} & l_{25} \\ \hline l_1 & true & false \\ l_{16} & true & false \\ l_{20} & true & false \\ l_{24} & W_{24,24} & W_{24,25} \end{array} \rangle,$$

where
$W_{24,25}$ = "a medical doctor should be sent to examine the patient",
$W_{24,24} = \neg W_{24,25}$.
In place l_{24} the σ tokens do not obtain new characteristics. When the truth value of the predicate $W_{24,25} = true$ the μ_i token representing the medical doctor who will visit the patient enters place l_{25} with characteristic

"name of the medical doctor who will visit the patient; name of the patient"

$$Z_9 = \langle \{l_{22}, l_{27}, l_{28}\}, \{l_{26}, l_{27}\}, \begin{array}{c|cc} & l_{26} & l_{27} \\ \hline l_{22} & false & true \\ l_{27} & W_{27,26} & W_{27,27} \\ l_{28} & false & true \end{array} \rangle,$$

where
$W_{27,26}$ = "specialists should be sent to transport the patient to the hospital";
$W_{27,27} = \neg W_{27,26}$. In place l_{26} the current token s_i receives the characteristic

"name of the patient that should be transported to the hospital"

In place l_{27} the tokens receive the characteristic

"names of the staff on duty"

$$Z_{10} = \langle \{l_3, l_{26}\}, \{l_{28}, l_{29}\}, \begin{array}{c|cc} & l_{28} & l_{29} \\ \hline l_3 & false & true \\ l_{26} & true & false \end{array} \rangle,$$

In place l_{28} the tokens receive the characteristics

"time for completing the transportation of the patient"

In place l_{29} the tokens receive the characteristics

"condition of the patient upon arrival at the hospital"

$$Z_{11} = \langle \{l_{29}, l_{31}\}, \{l_{30}, l_{31}\}, \begin{array}{c|cc} & l_{30} & l_{31} \\ \hline l_{29} & \mathit{false} & \mathit{true} \\ l_{31} & W_{31,30} & W_{31,31} \end{array} \rangle,$$

where
$W_{31,30}$ = "all medical procedures are completed";
$W_{31,31} = \neg W_{31,30}$
In place l_{30} the tokens receive the characteristics

"condition of the patient upon discharge from hospital"

In place l_{31} the tokens receive the characteristics

"condition of the patient during the procedures"

3 Conclusion

Telecare/telehealth is the remote or enhanced delivery of services to people in their own home by means of telecommunications and computerized systems. Telecare/telehealth ranges from basic community alarm services to more complex interventions involving fall detectors and sensors which monitor a range of physical behavior. The present GN-model describes the indirect (i.e., by life-sensors, glucosemetres) communication between patients in helpless condition and medical doctors from a telecare/telehealth center. It can be used, e.g., for simulation of different situations, related to increasing the number of emergent cases by diabetes mellitus to which the medical doctors/nurses or the person in the response center must react. The GN-model could show the necessary combinations of sensors used for the different patients on the basis of the simulations, we can determine the minimal number of the necessary professionals in the telecare/telehealth center.

Acknowledgments This work was partly supported by project 247541 MATSIQEL, European FP7 Marie Curie Actions-IRSES.

References

1. Klein, B., Horbach, A., Cook, G., Bailey, C., Moyle, W., Clarke, C.: Ambient Assisted Living, Telecare, Telehealth—Neue Technologieund Organisa-tionskonzepte. Projekte und Trends in Australien, Gro///britannien und Deutschland. Technik fuiir ein selbstbestimmtes Leben, Deutscher AAL-Kongress, pp. 24–25. Berlin (2012)
2. http://www.lifetec.org.au
3. Connell, J., Grealy, C., Olver, K., Power, J.: Comprehensive scoping study on the use of assistive technology by frail older people living in the community, Urbis for the Department of Health and Ageing (2008)
4. Alexieva, J., Choy, E., Koycheva, E.: Review and bibloigraphy on generalized nets theory and applications. In: Choy, E., Krawczak, M., Shannon, A., Szmidt, E. (eds.) A Survey of Generalized Nets, Raffles KvB Monograph No. 10, pp. 207–301 (2007)
5. Atanassov, K.: Generalized Nets. World Scientific, Singapore, London (1991)
6. Atanassov, K.: On Generalized Nets Theory. Prof. M. Drinov Academic Publ. House, Sofia (2007)
7. So, C.F., Choi, K.S., Wong, T.K.S.: Med. Devices: Evid. Res. Recent advances in noninvasive glucose monitoring. **5**, 45–52 (2012)
8. Genuth, S., Alberti, K., Bennett, P., Buse, J.: The Expert Committee on the Diagnosis and Classification of Diabetes Mellitus: Follow-up report on the Diagnosis of Diabetes Mellitus: Diabetes Care 26, pp. 3160–3167 (2003)
9. Andonov V., Stefanova-Pavlova, M., Stoyanov, T., Angelova, M., Cook, G., Klein, B., Atanassov, K., Vassilev, P.: Generalized Net Model for Telehealth Services. In: Proceedings of IEEE 6th Conference on Intelligent Systems, vol. 2, pp. 221–224. Sofia 6–8 Sept. (2012)
10. Atanassov, K., Andonov, V., Stojanov, T., Kovachev, P.: Generalized net model for telecommunication processes in telecare services. In: Proceedings of First International Conference on Telecommunications and Remote Sensing, pp. 158–162. Sofia 29–30 Aug. (2012)

State-Space Fuzzy-Neural Predictive Control

Yancho Todorov, Margarita Terziyska and Michail Petrov

Abstract The purpose of this work is to give an idea about the available potentials of state-space predictive control methodology based on fuzzy-neural modeling technique and different optimization procedures for process control. The proposed controller methodologies are based on Fuzzy-Neural State-Space Hammerstein model and variants of Quadratic Programming optimization algorithms. The effects of the proposed approaches are studied by simulation experiments to control a primary drying cycle in small-scale freeze-drying plant. The obtained results show a well-driven drying process without violation of the system constraints and accurate minimum error model prediction of the considered system states and output.

1 Introduction

Model Predictive Control (MPC) is an advanced control methodology that originates in the late 1970s. MPC represents an optimal control strategy that relies on dynamic model used to predict the future response of a plant. Afterwards, the MPC algorithm computes an optimal control policy by minimizing a prescribed cost function. One of the key advantages of MPC is its ability to deal with input and output constraints while it can be applied to multivariable process control. For this

Y. Todorov (✉)
Department of “Intelligent Systems”, Institute of Information and Communication Technologies, Bulgarian Academy of Sciences, Acad. G. Bontchev str., bl. 2, fl. 5, 1113 Sofia, Bulgaria
e-mail: yancho.todorov@iit.bas.bg

M. Terziyska
Department of “Informatics and Statistics”, University of Food Technologies, 27, Maritsa blvd., 4000 Plovdiv, Bulgaria
e-mail: m.terziyska@uft-plovdiv.bg

M. Petrov
Department of “Control Systems”, Technical Univeristy- Sofia, branch Plovdiv, 25, Ts. Dustabanov, st., 4000 Plovdiv, Bulgaria
e-mail: mpetrov@tu-plovdiv.bg

V. Sgurev et al. (eds.), *Recent Contributions in Intelligent Systems*,
Studies in Computational Intelligence 657,
DOI 10.1007/978-3-319-41438-6_17

purpose, MPC very quickly became popular and nowadays it is a well-known, classical control method [1, 2].

Since, most of the industrial processes are inherently nonlinear this implies the use of nonlinear models and, respectively, *Nonlinear Model Predictive Control* (NMPC) algorithms. NMPC is a variant of MPC that is characterized by the use of nonlinear system models in the prediction step. As in linear MPC, NMPC requires an iterative solution of optimal control problems on a finite prediction horizon. While these problems are convex in linear MPC, in nonlinear MPC they are not convex anymore. Because the nonlinear optimization task must be solved online, the success of a NMPC algorithm depends critically on the applied model. It is very important to find a predictive model that effectively describes the nonlinear behavior of the system and can easily be incorporated into NMPC algorithm. One possibility is to use first principle models such as nonlinear ordinary differential equations, partial differential equations, integro-differential equations and delay equations models. Such models can be accurate over a wide range of operating conditions, but they are difficult to develop for many industrial cases and may lead to numerical problems (e.g., stiffness, ill-conditioning). The other possibility is to use empirical or black-box models (e.g., neural networks, fuzzy models, polynomial models, Wiener, Hammerstein, and Volterra series models). How to select a nonlinear model for NMPC is described in detail in [3].

Fuzzy-neural (FN) systems have been proved to be a promising approach to solve complex nonlinear control problems. They have been proposed as an advantageous alternative to pure feed forward neural networks schemes for learning the nonlinear dynamics of a system from input–output data [4]. Also, any methods have been proposed in the literatures that combine fuzzy-neural network and model predictive control algorithm [5–7]. In the recent years, a general approach based on multiple LTI models around various function points has been proposed. The so-called multiple models, Takagi-Sugeno approach is a convex polytopic representation, which can be obtained either through mathematical transformation or through achieved linearization around various operating points [8].

In many situations, Hammerstein systems are seen to provide a good tradeoff between the complexity of general nonlinear systems and interpretability of linear dynamical systems [9]. They have been used, e.g., for modeling of biological processes [10, 11] chemical processes and signal processing applications. A lot of research has been carried out on identification of Hammerstein models. Hammerstein systems can be modeled by employing either nonparametric or parametric models. Parametric representations such as state-space models are more compact having fewer parameters and the nonlinearity is expressed as a linear combination of finite and known functions [12].

Bioprocesses are naturally involved in producing different pharmaceutical and food productions. Complicated dynamics, nonlinearity and non-stationarity make controlling them a very delicate task. The main control goal in this process is to get a pure product with a high concentration, which commonly is achieved by regulating a number of parameters. The MPC can fulfill these requirements [13] especially in industrial applications, where dynamics are relatively slow and hence it can accommodate online optimization easily [14].

The removal of the water at subfreezing temperatures, from the solution of a product that is potentially chemically and mechanically liable is a complex and costly operation [15]. The process is known as freeze-drying and it ensures conditions to induce sublimation process which help the product to dry and to keep its qualities for a long period of time without the degradative affect of either microbial or autolytic processes. Freeze-drying removes the water level below the level at which the microbial or enzymatic activity is possible. On the other hand, the process significantly reduces the total weight of the material by 8–9 folds. Removing water by about 90 % reduces the total volume and weight, facilitating easier transport [16, 17].

Nowadays, pharmaceutical industries are generating many pharmaceutical and bio products each year, thus creating pressure for reliable determination and control of the drying cycles during freeze-drying, as well as reduction of the operating costs. Several recent economical analyses conclude that a shorter drying regime gives lowest production cost and highest capacity. This issue is crucial for large scale drying apparatuses and depends on the nature of the material being dried and its robustness, aiming to provide high production capacity with high activity per unit product [15].

During the past years, many researchers propose the application of different innovative and advanced methodologies and procedures, in order to be achieved more reliable modeling/monitoring, control and parameter optimization of the drying processes. Effective solutions are already considered for different purposes: in thermodynamic modeling and control of grain dryers, in predicting drying kinetics [18], in development of optimal selection of freeze-drying operating conditions [19], in the framework of model based pharmaceuticals freeze-drying optimization [20] and innovation in monitoring of freeze-drying [21]. As well, in [22, 23] authors propose methods for monitoring and control of the freeze-drying of pharmaceuticals by applying an advanced algorithm for dynamic parameters estimation coupled with a controller which minimizes the drying time and monitoring observer approach which ensures that the maximum temperature of the product is maintained.

This latest developments in the area of drying technology have stimulated the active research in application of MPC strategies as effective tools for control and optimization of the drying cycles as well. For instance, in [24, 25] authors propose optimization and control strategy for the primary drying step, beside preserving product quality, to minimize the drying time and solution and to achieve robustness of the controller with respect to the main model parameter uncertainty. Recently, researchers have proposed different suitable applications of fuzzy logic and neural networks for drying processes, e.g. for rotary dryer control [26], to forecast freeze-drying parameters [27], to model intermittent drying of grains in a spouted bed [28], etc., but their proper use with appropriate optimization procedures into FN-MPC control schemes for effective process control of the freeze-drying process is not well studied area. For this purpose, the presented research is focused not only on the MPC algorithm development and to assess the potentials of such control methodologies and expected effects on a typical freeze-drying plant by carrying out a simulation study.

This chapter investigates the performances of a state-space MPC based on fuzzy-neural Hammerstein model and implementations of different Quadratic Programming optimization algorithms in order to be studied the effectiveness of the proposed methodologies for process control of a freeze-drying batch for pharmaceutical product. The used methodology assumes that the process states are fully estimable and they are used to predict steps ahead the product temperature. Afterwards, using the fuzzy-neural model predictions, the applied constrained optimization procedure compute an optimal control trajectory for the temperature of the heating shelves. The transient responses of the occurring processes, as well as the algorithm performances and their variations are studied by simulation experiments in MATLAB/SIMULINK environments.

2 Design of Hammerstein Fuzzy-Neural State-Space Model

Generally, the Hammerstein model represents a cascade connection of static nonlinearity and linear time invariant dynamics and during the past years it is widely used in practice for nonlinear system representation. Creating a hybrid structure combining the advantages of the Hammerstein model with the flexibility and robustness of a Takagi-Sugeno inference, gives the possibility to develop a dynamic predictive model which can be easily implemented in a MPC scheme.

In this contribution, the proposed idea in [29, 30] is adopted, so that the nonlinearity of the model is easily approximated as a set of local linear simple systems, while the linear part is flexibly introduced by mathematical transformations. Thus, the nonlinear part of the FN Hammerstein model is expressed in state-space as

$$\left| \begin{array}{l} x_1(k+1) = f_x(x_1(k), u(k)) \\ z(k) = f_z(x_1(k), u(k)) \end{array} \right. \tag{1}$$

where $x_1(k)$, $u(k)$ and $z(k)$ are vectors for the state, the input and the output of the nonlinear part. The nonlinear functions f_x and f_z are approximated by Takagi-Sugeno type fuzzy rules:

$$R^{(i)}\text{: if } r_1(k) \text{ is } M_1^{(i)} \ldots \text{and} \ldots r_p(k) \text{ is } M_p^{(i)} \text{ then} \\ \left| \begin{array}{l} x_1(k+1) = A_1 x_1(k) + B_1 u(k) \\ z(k) = C_1 x_1(k) + D_1 u(k) \end{array} \right|^{(i)} \tag{2}$$

where R is the ith rule of the rule base, r_p are the state regressors (outputs and inputs of the system), M_i is a membership function of a fuzzy set, $A^{(i)}$, $B^{(i)}$, $C^{(i)}$ and $D^{(i)}$ are the state-space matrices with dimensions in notion to ith fuzzy rule: $A^{(i)}(n \times n)$, $B^{(i)}$ $(n \times m)$, $C^{(i)}(q \times n)$ and $D^{(i)}(q \times m)$ where n is the number of the system states, q is the number of the system outputs and system inputs for m, respectively.

For each input vector, the output of the fuzzy model is computed by using the following equation:

$$\left| \begin{array}{l} x_1^{(i)}(k+1) = f_x^{(i)} g_u^{(i)} \\ z^{(i)}(k) = f_z^{(i)} g_u^{(i)} \end{array} \right. \quad \text{where } g_u^{(i)} = \prod_{i=1}^{N} \mu_{ui}^{(i)} \tag{3}$$

where μ_{ui} are the degrees of fulfillment in notion to the activated fuzzy membership function. On the other hand, the state-space matrices for the approximated nonlinear part of the model are calculated as a weighted sum of the local matrices using the normalized value of the membership function degree, $\bar{g}_{ui} = \mu_{ui} / \sum_{i=1}^{L} \mu_{ui}$ upon the ith activated fuzzy rule and L is the number of the activated rules. The fuzzification is performed by using Gaussian membership functions for nonlinear input approximation

$$g_{uj} = \exp - \frac{(r_i - c_{ij})^2}{2\sigma_{ij}^2} \tag{4}$$

Thereafter, the linear part of the model is introduced as

$$\left| \begin{array}{l} x_2(k+1) = A_2 x_2(k) + B_2 z(k) \\ y(k) = C_2 x_2(k) + D_2 z(k) + \vartheta \end{array} \right. \tag{5}$$

where ϑ is a free offset used to compensate possible disturbances in the process. In notion to each activated fuzzy rule, the general local fuzzy-neural model is expressed as combination of both approximated nonlinear and linear parts. Thus, the generalized model representation has the following form:

$$\begin{array}{l} \left| \begin{array}{c} \dot{x}_1 \\ \dot{x}_2 \end{array} \right| = \left| \begin{array}{cc} A_1^{(i)} & 0 \\ B_2 C_1^{(i)} & A_2 \end{array} \right| \times \left| \begin{array}{c} x_1 \\ x_2 \end{array} \right| + \left| \begin{array}{c} B_1^{(i)} \\ B_2 D_1^{(i)} \end{array} \right| u \\ y = \left| \begin{array}{cc} D_2 C_1^{(i)} & C_2 \end{array} \right| \times \left| \begin{array}{c} x_1 \\ x_2 \end{array} \right| + D_1^{(i)} D_2 u + \vartheta \end{array} \tag{6}$$

Finally, the designed model is described as

$$\left| \begin{array}{l} x_1(k+1) = \sum_{i=1}^{N} \bar{g}_{ui}(A_1 x_1(k) + B_1 u(k)) \\ x_2(k+1) = \sum_{i=1}^{N} (A_2 x_2(k) + B_2 \bar{g}_{ui}(C_1 x_1(k) + D_1 u(k)) \\ y(k) = \sum_{i=1}^{N} (C_2 x_2(k) + D_2 \bar{g}_{ui}(C_1 x_1(k) + D_1 u(k)) + \vartheta \end{array} \right. \tag{7}$$

as well as, with the following generalized scheme Fig. 1.

2.1 Learning Algorithm for the Designed Model

The learning procedure provides structure identification of the process and estimation of the values of the unknown parameters. The fuzzy-neural model structure depends on the type and the number of the chosen membership functions, their shape and the linear coefficients into the *f* functions in the consequent part of the fuzzy rules. Thus, the identification task implies the determination of two groups of parameters, the center and the deviation of the used Gaussian membership functions in the rules premise part and the linear coefficients in the rules consequent part.

A simple identification procedure is applied in this work in order to be facilitated the real-time implementation of the tuning procedure [31–33]. The learning algorithm for the fuzzy-neural model lies on the minimization of an error measurement function: $E(k) = \varepsilon^2/2$ where $\varepsilon(k) = y(k) - \hat{y}(k)$, is the instant error between the real plant output $y(k)$ and the model output $\hat{y}(k)$, calculated by the FN model. The algorithm performs two steps gradient learning procedure. Assuming that η is the considered learning rate and β_{si} is an adjustable *s*th coefficient for the functions *f* into the *i*th activated rule as a connection in the output neuron, the general parameter learning rule for the consequent parameters is: $\beta_{si}(k+1) = \beta_{si}(k) + \eta(\partial E/\partial \beta_{si})$. After calculating the partial derivatives, the final recurrent predictions for each adjustable coefficient β_{si} (a_s, b_s, c_s or d_s) are obtained by the following equations:

$$\begin{aligned} a_{si}(k+1) &= a_{si}(k) + \eta\varepsilon(k)\bar{g}_{ui}(k)r_p(k),\ s = 1 \div \tilde{n}\times\tilde{n} \\ b_{si}(k+1) &= b_{si}(k) + \eta\varepsilon(k)\bar{g}_{ui}(k)r_p(k),\ s = 1 \div \tilde{n}\times\tilde{m} \\ c_{si}(k+1) &= c_{si}(k) + \eta\varepsilon(k)\bar{g}_{ui}(k)r_p(k),\ s = 1 \div \tilde{q}\times\tilde{n} \\ d_{si}(k+1) &= d_{si}(k) + \eta\varepsilon(k)\bar{g}_{ui}(k)r_p(k),\ s = 1 \div \tilde{q}\times\tilde{m} \end{aligned} \tag{8}$$

where the dimensions of the general matrices are $\tilde{A}(\tilde{n}\times\tilde{n})$, $\tilde{B}(\tilde{n}\times\tilde{m})$, $\tilde{C}(\tilde{q}\times\tilde{n})$, $D(\tilde{q}\times\tilde{m})$. The output error E is used back directly to the input layer, where there are the premise (center-c_{pi} and the deviation-σ_{pi} of a Gaussian fuzzy set) adjustable parameters. The error E is propagated through the links composed by the corresponded membership degrees where the link weights are unit. Hence, the learning rule for the second group adjustable parameters in the input layer can be done by the same learning rule

$$\begin{aligned} c_{pi}(k+1) &= c_{pi}(k) + \eta\varepsilon(k)\bar{g}_{ui}(k)[f_x^{(i)}(k) - \hat{x}(k)]\frac{[r_p(k) - c_{pi}]}{\sigma_{pi}^2(k)} \\ \sigma_{pi}(k+1) &= \sigma_{pi}(k) + \eta\varepsilon(k)\bar{g}_{ui}(k)[f_x^{(i)}(k) - \hat{x}(k)]\frac{[r_p(k) - c_{pi}]^2}{\sigma_{pi}^3(k)} \end{aligned} \tag{9}$$

To improve the efficiency in the learning procedure of the nonlinear fuzzy-neural part of the model and adaptive learning rate scheduling algorithm has been introduced. For this purpose, at each sampling period the *Root Squared Error* of the predicted state is assumed as

$$\varepsilon = \sqrt{\sum_{j=1}^{M} \left(x_1(j) - \hat{x}_1(j)\right)^2} \tag{10}$$

Afterwards, following the rule at each sample step η is calculated as:

$$\begin{aligned} &\textit{if } \varepsilon_i > \varepsilon_{i-1} k_w \\ &\qquad \eta_{i+1} = \eta_i \tau_d \\ &\textit{if } \varepsilon_i \le \varepsilon_{i-1} k_w \\ &\qquad \eta_{i+1} = \eta_i \tau_i \end{aligned} \tag{11}$$

where $\tau_d = 0.7$ and $\tau_i = 1.05$ are scaling factors and $k_w = 1.41$ is the coefficient of admissible error accumulation [34].

3 Model Predictive Control Strategies

Using the designed Hammerstein model, the ***optimization algorithm*** computes the future control actions at each sampling period, by minimizing the following cost:

$$\begin{aligned} &J(k) = \sum_{i=N_1}^{N_2} \|\hat{y}(k+i) - r(k+i)\|^2 Q + \sum_{i=N_1}^{N_u} \|\Delta u(k+i)\|^2 R \\ &\text{subject to } \Omega \Delta \mathrm{U} \le \gamma \end{aligned} \tag{12}$$

which can be expressed in vector form as

$$J(k) = \|Y(k) - \mathrm{T}(k)\|^2 \mathbf{Q} + \|\Delta U(k)\|^2 \mathbf{R} \tag{13}$$

$$Y(k) = \begin{bmatrix} y(k+N_1) \\ \vdots \\ y(k+N_2) \end{bmatrix} \quad \mathrm{T}(k) = \begin{bmatrix} r(k+N_1) \\ \vdots \\ r(k+N_2) \end{bmatrix} \quad \Delta U(k) = \begin{bmatrix} \Delta u(k+N_1) \\ \vdots \\ \Delta u(k+N_u) \end{bmatrix} \tag{14}$$

where, Y is the matrix of the predicted plant output, T is the reference matrix, ΔU is the matrix of the predicted controls and Q and R are the matrices, penalizing the changes in error and control term of the cost function

$$Q = \begin{bmatrix} Q(N_1) & 0 & \cdots & 0 \\ 0 & Q(N_1+1) & \cdots & 0 \\ \vdots & \vdots & \ddots & \vdots \\ 0 & 0 & \cdots & Q(N_2) \end{bmatrix} \quad R = \begin{bmatrix} R(N_1) & 0 & \cdots & 0 \\ 0 & R(N_1+1) & \cdots & 0 \\ \vdots & \vdots & \ddots & \vdots \\ 0 & 0 & \cdots & R(N_u) \end{bmatrix} \tag{15}$$

Taking into account the general prediction form of a linear state-space model [35, 36], we can derive:

$$Y(k) = \Psi X(k) + \Upsilon u(k-1) + \Theta \Delta U(k) + \vartheta \tag{16}$$

$$\Psi = \begin{bmatrix} \tilde{C} \\ \tilde{C}\tilde{A} \\ \tilde{C}\tilde{A}^2 \\ \vdots \\ \tilde{C}\tilde{A}^{N_2-1} \end{bmatrix} \quad \Upsilon = \begin{bmatrix} \tilde{D} \\ \tilde{C}\tilde{B} + \tilde{D} \\ \tilde{C}\tilde{A}\tilde{B} + \tilde{C}\tilde{B} + \tilde{D} \\ \vdots \\ \tilde{C}\sum_{i=N_1}^{N_2} \tilde{A}^i \tilde{B} + \tilde{D} \end{bmatrix} \tag{17}$$

$$\Theta = \begin{bmatrix} \tilde{D} & 0 & \cdots & 0 \\ \tilde{C}\tilde{B} + \tilde{D} & \tilde{D} & & \vdots \\ \tilde{C}\tilde{A}\tilde{B} + \tilde{C}\tilde{B} + \tilde{D} & & \ddots & 0 \\ \vdots & & \ddots & \tilde{D} \\ \tilde{C}\sum_{i=1}^{N_2-2} \tilde{A}^i \tilde{B} + \tilde{D} & & \cdots & 0 \\ \vdots & & \ddots & \vdots \\ \tilde{C}\sum_{i=1}^{N_2-1} \tilde{A}^i \tilde{B} + \tilde{D} & & \cdots & \tilde{C}\sum_{i=1}^{N_2-N_u-1} \tilde{A}^i \tilde{B} + \tilde{D} \end{bmatrix} \tag{18}$$

Then, we can define: $E(k) = T(k) - \Psi X(k) - \Upsilon u(k-1) - \vartheta$. This expression is assumed as tracking error in sense of that it is the difference between the future target trajectory and the free response of the system that occurs over the prediction horizon if no input changes were made; if $\Delta U = 0$ is set. Using the last notation, we can write

$$J(k) = \Delta U^T H \Delta U + \Delta U^T \Phi + \mathrm{E}^T Q \mathrm{E},\ \Phi = -2\Theta^T Q \mathrm{E}(k), H = \Theta^T Q \Theta + R \tag{19}$$

Differentiating the gradient of J with respect to ΔU, gives the Hessian matrix: $\partial^2 J(k)/\partial \Delta U^2(k) = 2H = 2(\Theta^T Q \Theta + R)$. If $Q(i) \geq 0$ for each i (ensures that $\Theta^T Q \Theta \geq 0$) and if $R \geq 0$ then the Hessian is certainly positive-definite, which is enough to guarantee the reach of minimum.

To improve the robustness of the controller, an alternative formulation of the cost function is also considered

$$\begin{aligned} &J(k) = \textstyle\sum_{i=N_1}^{N_2} \|\hat{y}(k+i) - \omega(k+i)\|^2 Q + \sum_{i=N_1}^{N_u} \|\Delta u(k+i)\|^2 R \\ &\text{subject to } \Omega \Delta \mathrm{U} \leq \gamma \\ &\omega(k+i) = \alpha\omega(k+i-1) + (1-\alpha) r(k+i) \text{ and } \omega(k) = y(k) \end{aligned} \tag{20}$$

where ω defines a reference trajectory to follow, taking into account not only the desired reference values $r(k+i)$, but including the current value of the system output $y(k)$, as well. Thus, implementing the ω term and defining a value of α between 0 and 1, a more smooth approximation of the reference trajectory is achieved [37].

Constraints formulation

Since, $U(k)$ and $Y(k)$ are not explicitly included in the optimization problem, the constraints can be expressed in terms of ΔU signal:

$$\begin{bmatrix} F_1 \\ G\Theta \\ W \end{bmatrix} \Delta U \leq \begin{bmatrix} -F_2 u(k-1) + f \\ -G(\Psi X(k) + \Upsilon u(k-1)) + g \\ w \end{bmatrix} \tag{21}$$

The first row represents the constraints on the amplitude of the control signal, the second one the constraints on the output changes and the last the constraints on the rate change of the control.

Optimization procedures

Quadratic Programming-Active Set method

Using the active-set methods at each step of the algorithm must be defined a working set of constraints to be treated as the active set. The working set is a subset of the constraints that are actually active at the current point and the current point is feasible for the working set. The algorithm then proceeds to move on the surface defined by the working set of constraints to an improved point. At each step, an equality constraint problem is solved. If $\lambda_i \geq 0$ for all Lagrange multipliers, then the point is a local solution to the original problem. If, $\lambda_i < 0$ exists, the objective function value can be decreased by relaxing the constraint *i*. During the minimization process, it is necessary to monitor the values of the other constraints to be sure that they are not violated, since all points defined by the algorithm must be feasible. It often happens that while moving on the working surface, a new constraint boundary is encountered. It is necessary to add this constraint to the working set, then proceed to the redefined working surface [38]. Using the active-set notation, the problem can be formulated as:

$$\begin{gathered} \max_{\lambda > 0} \min_{\Delta U} \left[\Delta U^T H \Delta U + \Delta U \Phi + \mathrm{E}^T Q \mathrm{E} + \lambda^T (\Omega \Delta U - \gamma) \right] \\ \Delta U = -H^{-1}(\Phi + \Omega^T \lambda) \end{gathered} \tag{22}$$

Necessary conditions for optimization in presence of inequality constraints are the satisfaction of the KKT conditions:

$$\begin{aligned} -H^{-1}(\Phi + \Omega^T \lambda) &= 0 \\ \Omega \Delta U - \gamma &\leq 0 \\ \lambda^T (\Omega \Delta U - \gamma) &= 0 \\ \lambda &\geq 0 \end{aligned} \tag{23}$$

where the vector λ contains the Lagrange multipliers.

Hildreth Quadratic Programming

Using the active-set notation the optimal solutions are based on primal decision variables which have to be identified along with the current active constraints. A major problem is the dimension of the set of constraints which impacts the computational load. To overcome this problem, a dual method is introduced in order to be identified constraints that are not active, so they can be eliminated in the solution. In this purpose a very simple programming procedure can be adopted for finding optimal solutions of constrained minimization problems [38]. The dual problem to the original primal problem is derived below. Assuming feasibility, the primal problem is equivalent to (20). Substituting, the problem can be rewritten as:

$$\begin{gathered}\max_{\lambda>0}\left(\lambda^T M\lambda-\lambda^T K+\Phi^T M^{-1}\Phi\right)\\ M=\Omega H^{-1}\Omega^T, K=\gamma+\Omega H^{-1}\Phi\end{gathered} \tag{24}$$

The dual is also a quadratic programming problem with λ as the decision variable:

$$\begin{gathered}\Delta U=-H^{-1}\Phi-H^{-1}(\Omega_{act}\lambda_{act})\\ \min_{\lambda>0}\left(\lambda^T M\lambda+\lambda^T K+\gamma^T M^{-1}\gamma\right)\end{gathered} \tag{25}$$

The λ vector can be varied one component at a time. At a given step in the process, having obtained a vector $\lambda \geq 0$, our attention is fixed on a single component λ_i. The objective function may be regarded as a quadratic function in this single component. We adjust λ_i to minimize the objective function. If that requires $\lambda_i < 0$, we set $\lambda_i = 0$. In either case, the objective function is decreased. Then, we consider the next component $\lambda_i + 1$. If we consider one complete cycle through the components to be one iteration taking the vector λ^m to λ^{m+1}, the method can be expressed explicitly as:

$$\begin{aligned}\lambda_i^{m+1}&=\max(0,\alpha_i^{m+1})\\ \alpha_i^{m+1}&=-\frac{1}{h_{ii}}\left[k_i+\textstyle\sum_{j=1}^{i-1}h_{ij}\lambda_j^{m+1}+\sum_{j=i+1}^{n}h_{ij}\lambda_j^{m}\right]\end{aligned} \tag{26}$$

where the scalar *hij* is the *ij*th element in the matrix M, and k_i is the *i*th element in the vector K. Also, there are two sets of λ values in the computation: one involves λm and one involves the updated $\lambda m + 1$. Because the converged λ vector contains either zero or positive values of the Lagrange multipliers, we finally have

$$\begin{aligned}\Delta U&=-H^{-1}(\Phi+\Omega^T\lambda^*)\\ \lambda_{act}^*&=-(\Omega_{\text{act}}H^{-1}\Omega_{\text{act}}^T)^{-1}(\gamma_{\text{act}}+\Omega_{\text{act}}H^{-1}\Phi)\end{aligned} \tag{27}$$

Interior-point method

Since Karmarkar's breakthrough, many different interior-point methods have been developed. It is important to note that there exists in fact a whole collection of methods, sharing the same basic principles whose individual characteristics may vary. The fact that finding the optimal solution of a linear program is completely equivalent to solving the KKT conditions may suggest the use of a general method designed to solve systems of nonlinear equations. The most popular of these methods is the Newton's method, but a major problem in the resolution of the KKT conditions is the nonnegativity constraints on ΔU and γ which cannot directly be taken into account via mapping of Φ. One way of incorporating these constraints is to use a barrier term. Using such a barrier, it is possible to derive a parameterized family of unconstrained problems from an inequality-constrained problem [39].

A Newton-like interior-point convex algorithm has been implemented to calculate the optimal control sequence. Following the notation in (20), the algorithm requires the KKT conditions to be held as

$$\begin{aligned} -H^{-1}(\Phi+\Omega^T\lambda)&=0\\ \Omega\Delta U-\gamma-s&=0\\ s&\geq 0\\ \lambda&\geq 0\\ s_i\lambda_i&=0 \end{aligned} \tag{28}$$

where s is a vector of slacks of length m that convert inequality constraints to equalities depending on the number of linear inequalities and bounds. At first, the algorithm predicts a *Newton-Raphson* step, and then computes a corrector step. The corrector attempts to better enforce the nonlinear constraint $s_i\lambda_i = 0$. The predictor steps are formulated by the following residuals:

$$\begin{aligned} r_d&=H\Delta U-\Phi-\Omega^T\lambda\\ r_{ineq}&=\Omega\Delta U-\gamma-s\\ r_{s\lambda}&=S\lambda\\ r_c&=\frac{S^T\lambda}{m} \end{aligned} \tag{29}$$

Then, the algorithm iterates as:

$$\begin{pmatrix} H & 0 & -\Omega^T\\ \Omega & -I & 0\\ 0 & \Lambda & S \end{pmatrix}\begin{pmatrix}\Delta U\\ \Delta S\\ \Delta\Lambda\end{pmatrix}=\begin{pmatrix} r_d\\ r_{ineq}\\ r_{s\lambda}\end{pmatrix} \tag{30}$$

Additionally, to maintain the position in the interior, instead of trying to solve $s_i z_i = 0$, the algorithm takes a positive parameter σ, and tries to solve $s_i z_i = \sigma r_c$. Then, the algorithm replaces $r_{s\lambda}$ in the Newton step equation with $r_{s\lambda} + \Delta s\Delta\lambda - \sigma r_c\mathbf{1}$, where $\mathbf{1}$ is the vector of ones. Also, in order Newton equations to obtain a symmetric, more numerically stable system for the predictor step calculation are reordered. After calculating the corrected Newton step, the algorithm performs more calculations to get both a longer current step, and to prepare for better subsequent steps. These multiple correction calculations improve both performance and robustness of the algorithm.

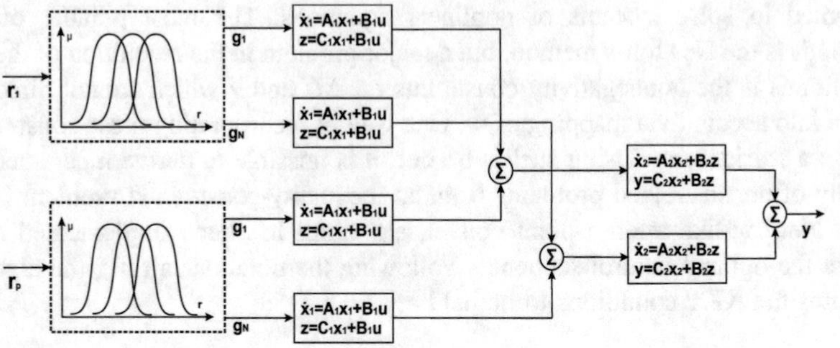

Fig. 1 Generalized scheme of the proposed state-space Fuzzy-Neural Hammerstein model

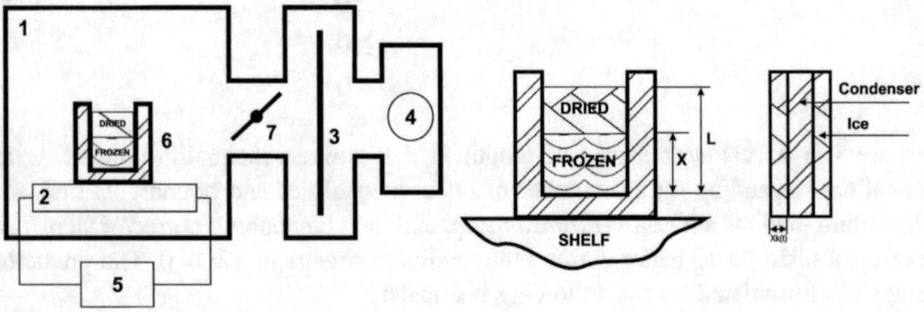

Fig. 2 Schematic diagram of a lyophilization plant

4 Results and Discussion

Reference plant model

To investigate the potentials of the proposed approaches, simulation experiments in MATLAB and Simulink environments, are performed. As reference for the proper controller operation is taken a validated first principle model of a small-scale freeze-drying plant. Referring to Fig. 2, a simplified diagram of the main components of the apparatus is shown. The plant consists of a drying chamber (1); temperature controlled shelves (2), a condenser (3) and a vacuum pump (4). The cooling and heating processed are supported by the shelves (6) heater, and refrigeration system (5). The chamber is isolated from the condenser by the valve (7) and the vacuum system is placed after the condenser.

The experiments assume that the product is a priori frozen and the chamber is evacuated in order to increase the partial vapor water pressure difference between the frozen ice zone and the chamber. Then the shelf heating system starts to provide enthalpy for the sublimation process and controller starts to operate. The used reference model accounts for a coupled heat and mass transfer governing the primary drying phase. The mass transfer is governed by simple diffusion and heat transfer is driven primarily through conduction from the shelves. Thus, the combined process forms a nonlinear fourth order system of differential equations.

In the conducted simulation experiments, the proposed reference model was used to simulate the nonlinear behavior of a freeze-drying plant in a theoretical way, without pre-generated real plant data. The experiments consider a freeze-drying batch for 50 vials filled with solution of amino acid. A detailed model description and the used process parameters are given in [40, 41].

Conditions and constraints

The following initial conditions for simulation experiments are assumed: $N_1 = 1$, $N_2 = 5$, $N_u = 3$; system reference $r = 255$ K; initial shelf temperature, before the start of the primary drying $Ts_{\text{in}} = 228$ K; initial thickness of the interface front $x = 0.0023$ m; thickness of the product $L = 0.003$ m. In the primary drying stage it is required to maintain the shelf temperature about 298 K, until the product will be dried. This circumstance requires of about 45 min of time for the primary drying stage of the process.

There are imposed the following constraints on the optimization problem: on the amplitude of the control signal—the heating shelves temperature 228 K $< T_s <$ 298 K; on the output changes—product temperature 238 K $< T_2 <$ 256 K; and on the rate change of the control signal 0.5 K $< \Delta T_s <$ 3 K.

Control system statement

As measured states used for model prediction are taken the moving ice front and the temperature inside the frozen layer. It is assumed, that both parameters are fully observable/estimable. The main idea behind the states selection is the gathering of valuable information for the process dynamics at each discrete sampling period of time by parameters which cannot be manipulated but they can indicate the current state of the process.

The first one of the selected states gives information about the evolution of the sublimation process, while the second one for the temperature gradient. Thus, the fuzzy-neural model predicts steps ahead along the prediction horizon the temperature profile of the product using estimates of the mentioned parameters. The optimization algorithms compute the optimal control actions using the model predictions. As optimal control, it is used the temperature of the heating shelves, which accounts for the evolution of the sublimation process during the freeze-drying cycle. The used optimization criterion represents a general cost function which includes terms of the system error, the control action increment and the assumed constraints. The underlying idea of using such function is to be minimized the error between the reference of the temperature of the product and its actual value estimated by the model by applying appropriate heating policy. The statement of the optimization problem ensures that the maximum allowed product temperature will be reached by satisfying the major system constraints, without overheating the product. The selection of the initial tuning parameters of the fuzzy-neural model is made heuristically, in order to be guaranteed random directions for the optimization procedures, while tuning of the cost function parameters is made following the well-known rule $N_u < N_2 > N_1$.

Discussion

To preserve the computational consistency, a final value of $x = 0.0001$ m for the interface front is chosen to stop the simulation experiments. Increasing the heating shelves temperature from 228 K indicates the start of the primary drying phase, confirmed by the initial drop of the product temperature, which represents the sudden loss of heat due to sublimation. The loss of heat due to sublimation vanishes after all of the unbound water has sublimed, then the enthalpy input from the shelf causes a sharp elevation of the product temperature.

The proposed FN Hammerstein model have a simplified structure based on the classical Takagi-Sugeno technique, which aims to ensure reliable and accurate modeling of the nonlinear dynamics of the lyophilization process, stating small number of parameters without additional need of computational power. The consequent parameters of the proposed fuzzy-neural rules are initialized at first with randomly selected coefficients in a normalized range. The learning procedure for the model parameters is executed *online* at each computational sampling. Thus, the model produces a predicted system output (the product temperature) in notion to current values of the input vector. The penalty terms/matrices into the objective functions are experimentally chosen.

Usually, the application of predictive control requires the development of an accurate process model and selection of an appropriate optimization approach. Many optimization policies are proposed in literature, but their real time use may be restricted by many factors, e.g., the nature of the plant process, the statement of the optimization problem the imposed constraints, etc. The described above HQP algorithm has been proved to be faster than classical active-set method, but its possible applicability in MPC control problems is less studied issue. For this purpose, its potentials was investigated in a simulation batch for two different values of the main diagonal of the penalty matrix R, using a standard cost function with constant reference signal.

The temperature versus time profile for the product and heating shelves temperatures for the representative vial are presented on Fig. 3. The prediction of the assumed states; x_1—interface position and x_2—temperature in the frozen region, is presented on Figs. 4 and 5. On Fig. 6 are demonstrated the squared errors of the model, during the controller operation. As can be seen for both cases the transient responses of the considered squared errors have a smooth nature and they are successfully minimized during the learning process for the model. This circumstance proves the proper operation of the model and ensures a well driven lyophilization cycle demonstrated by the transient responses of the moving ice front. The moving ice front is an important parameter which accounts for the reliable and optimal drying process.

For this purpose, it is used into the state-space model for a parameter being predicted by the model, along with the temperature in the frozen region. A slight error during the prediction of the moving ice front is observed, which is proved to vary by selecting the initial values of the respective rule consequent part coefficients or their learning rate.

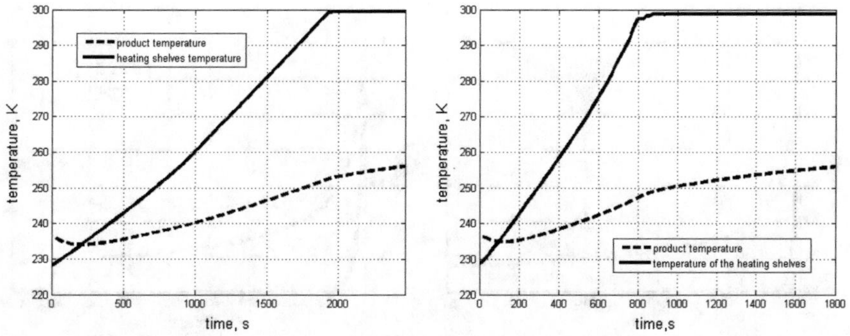

Fig. 3 Temperature profile when using HQP

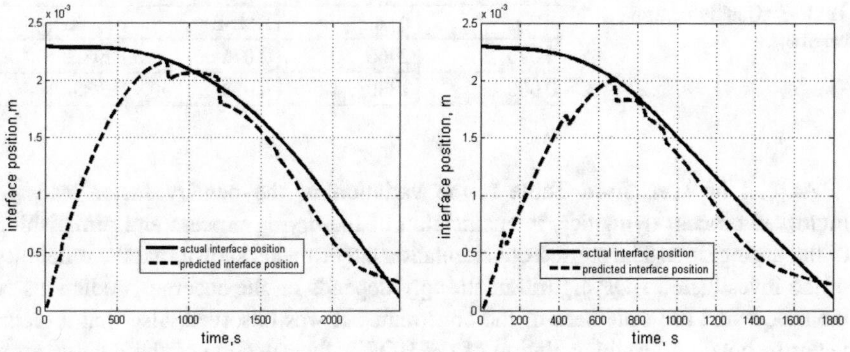

Fig. 4 State x_1 (interface position) prediction when using HQP

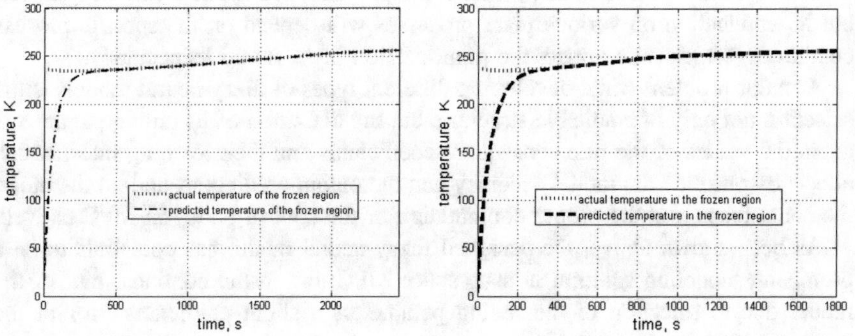

Fig. 5 State x_2 (frozen region temperature) prediction when using HQP

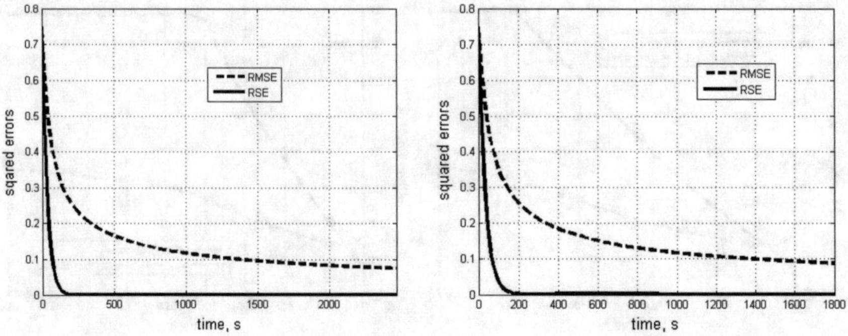

Fig. 6 Squared errors of the model during controller operation

Table 1 Quality control parameters

R(k)	t_p, s	RMSE	T_s, K
0.020	2460	0.076	298.3
0.008	1830	0.088	298.4

As can be seen from Table 1, the variation of the penalty factor leads to improved process dynamics, intensification of the drying process and diminishing of the drying time. The carried out simulation experiments shown that the reliability of the investigated HQP algorithm strongly depends on the accurate predictions of the model and the statement of the constraints. It was observed, also that a major factor to obtain a feasible solution of the HQP is the selection of the relative error tolerance to stop the algorithm, which impacts the number of the performed iterations for adjusting the Lagrange multipliers

The algorithm proves to be suitable for implementation in MPC control schemes, but its application on various plant processes will depend on the specific process conditions, which may impact the minimization process and its accuracy.

A major problem while developing different types of fuzzy-neural models is the selection not only of a suitable structure, but the definition of its tuning parameters as: initial values of the rule consequent coefficients and their learning rates, which affects by one side the model accuracy and the output predictions and on the other, the consistency and the proper computation of the control policy in MPC as well.

As been shown above, the designed fuzzy-neural model has potentials to be a promising modeling solution in state-space MPC, but in the nonlinear part of the model due to selection of the tuning parameters a slight prediction error of the predicted state is observed.

To overcome these problems, many approaches are described in literature, how to select good model tuning parameters instead of their heuristically selection. For that purpose, in this study, a potential approach to minimize the prediction error in the fuzzy-neural part of the model by introducing an adaptive learning rate scheduling has been investigated. As described above, the RSE is evaluated at each

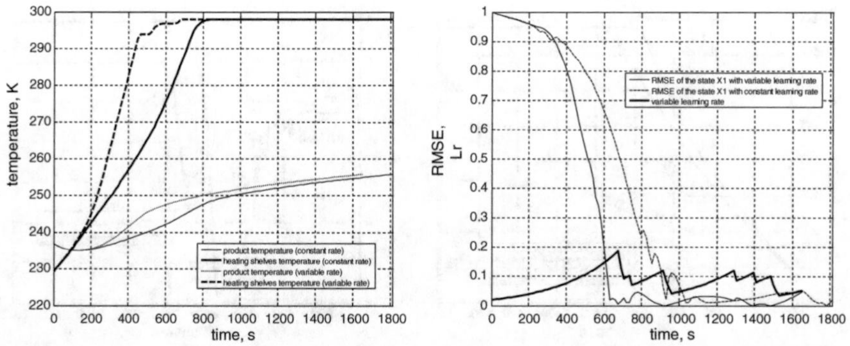

Fig. 7 Influence of the variable learning rate on RMSE of the predicted state

sampling period and compared to its previous value multiplied by a coefficient of admissible grow. Afterwards, the proposed approach at each sampling period computes a variable learning rate coefficient, depending on its previous value and multiplied by a grow factor, which depends on the RSE. To assess the efficiency of the proposed adaptive *online* learning rate approach, applied in calculation of the rule premise and consequent parameters, a simulation experiments with the classical active-set QP notation, have been performed.

In Fig. 7 are shown the obtained results in a MPC control system with constant and variable learning rates with initial value of 0.03 and equal initial conditions for the rule consequent parameters. As can be observed, in even random selection of the initials, the adopted adaptive learning rate approach, leads to significant improvement in the model prediction performance proved by the reduced instant value of the RMSE of the state prediction. A greater value of the learning rate is achieved inversely proportional when the error is high and it starts to decrease when it reaches values of the RMSE closer to zero. Also, an improved system performance is observed by reduction of the time of the transient response of the system, which proves again the positive effect of introduction of such adaptive learning rate approach. Additional studies on the selection of the error admissible grow coefficient and the learning rate scaling coefficients should be effectuated, since the proposed values may depend on the concrete plant process system dynamics.

Usually, when a MPC is designed, the reference signal is taken as constant, but in fact this may impact the controller performance leading to aggressive behavior. In this purpose arises the question how to be properly generated the reference signal and how this process affect the computation of the optimal control policy. A common recommendation in literature is to be used a smooth approximation of the output taking into account the previous value of the output, the desired reference scaled by an appropriate factor, leading to less or more smooth approximation. The simulation experiments using the cost functions (12) and (20) with $\alpha = 0.95$, on equal initial system conditions by using active-set QP are shown on Fig. 8. Since, the way of generation of the reference will impact the instant value of the system

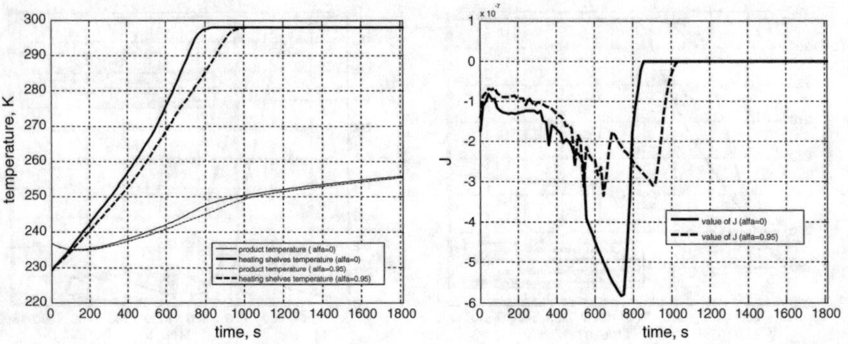

Fig. 8 MPC performance in case of smooth approximation of the system reference

error; this will affect also the performance of the optimization procedure and the calculated control trajectory. For that purpose, the instant value of the *J*, during the minimization process is investigated.

The value of the scaling factor α is taken as 0.95, in order to assess the smoothest case of reference approximation. The obtained results show that the approximation of the system reference leads to more smooth transient response of the process, with a smaller value of the applied control action for an equal system instant, which leads to less aggressive controller operation. On the other hand, in the sense of QP, the performance of the optimization procedure, which depends on the stated problem at and the assumed constraints, the current error value affects the way the control signal is being calculated. As can be seen from the results, the transient response of *J* is affected by the adaptive reference generation, especially when the velocity of the moving reference trajectory is changing significantly the value of *J* is decreasing.

Numerous methods for constrained optimization exist, but their application in various MPC strategies are less studied, because of the greater number of variances of the type of the predictive model being used coupled with the possible different optimization approaches, gives a rich algorithmic potential for predictive control. In this chapter, a comparative study between the widely used active-set QP and the interior-point QP optimization approaches is made. The interior-point method is a global optimization approach and a QP variant that uses a Newton prediction step and correction mechanism by a vector of slack variables. In Fig. 9 are shown the obtained simulation results with both optimization strategies, on equal conditions for the model and penalty terms and using a constant reference signal. As can be observed, the transient responses of the control signal and the controlled output slightly differ. Using the interior-point method leads a more active controller operation when the system error is high. A more smooth behavior of J is achieved when the change of system error is small. The transient response of the instant value of J shows also that greater values are reached when the velocity of the system error is changed. A major drawback of the interior-point algorithm is that the relative error term, to stop the algorithm is sensitive in notion to the optimization problem

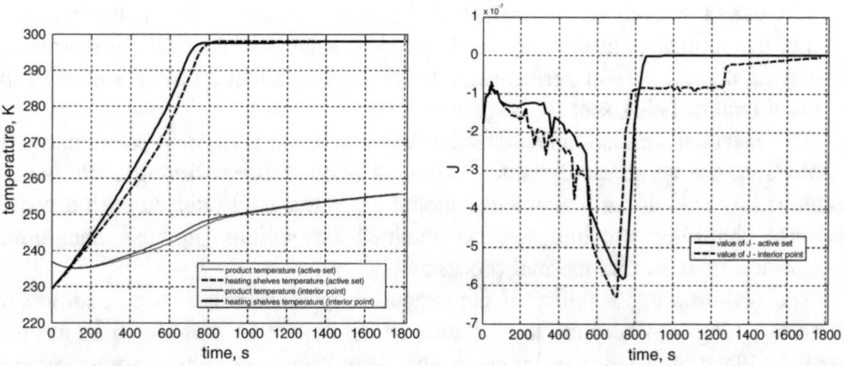

Fig. 9 Transient responses of the system outputs and the value of J when using active-set QP and interior-point QP

and the imposed constraints and may cause algorithm malfunctioning due to reaching infeasible point.

Nevertheless, the mentioned drawback is that the interior-point QP variant seems to be applicable in state-space predictive control, but its performance may be affected by the plant process-specific features and constraints which may impact the optimization process.

5 Conclusions

The performed investigations in this chapter have shown that the designed state-space Hammerstein modeling approach is an effective and reliable tool for modeling specific nonlinear system behavior by using a simple fuzzy-neural approach. The model is stated with small number of parameters, which effectively cover the nonlinear region of the process and ensure good system states and output predictions. The introduction of a variable learning rate for the rule consequent parameters of the model improves the model operation and diminishes the prediction error.

The proposed HQP optimization approach using a dual mechanism as variant of active-set QP may be a promising approach for MPC implementations. It manipulates the constraints which are not active and they are eliminated in the solution, which is very important in large-scale problems with a greater number of imposed constraints.

The considered interior-point QP variant optimization algorithm also has shown a good performance in the investigated state-space MPC. In spite of its more complicated computational performance, it may be applicable due to its potential being global optimization approach.

The use of smooth approximation of the system reference affects the computation of the optimal control policy, leading to less aggressive controller performance, achieving a good system performance by calculating smaller instant values of the optimal control being sent to the plant

The executed simulation results show also that a reliable system performance by controlling the nonlinear dynamics during a primary freeze-drying cycle can be achieved by applying a fuzzy-neural model predictive controller. In such a way, a flexible controller structure can be obtained for various products undergoing freeze-drying or similar thermal processing.

The real-time applicability of the proposed approaches has to be additionally studied taking into account some issues which cannot be considered in the presented simulation batches, as immeasurable disturbances, possible system operation faults, type of the product, etc., as well as the made assumptions. Nevertheless, in a theoretical aspect is proved that such control methodologies have major benefits, which can be considered for real-time process control.

Although, the state-space approach seems to be a promising solution for large scale drying plants, where the handling of the regime constraints is crucial and the system dynamic is relatively slow, which can accommodate with the computational procedures of the algorithms. A major advantage of the proposed control methodologies is the application of simple FN approach, which may impact the proper handling of some process uncertainties and disturbances.

References

1. Qin, S.J., Badgwell, T.A.: A survey of industrial model predictive control technology. Control Eng, Pract. **11**, 733–764 (2003)
2. Holkar, K.S., Waghmare L.M.: An overview of model predictive control. Int. J. Control Autom. **3**(4) (2010)
3. Pearson, R.K.: Selecting nonlinear model structures for computer control. J. Process Control **13** (2003)
4. Passino, K., Yourkovic, S.: Fuzzy Control. Adisson-Wesley (1998)
5. Dalhoumi, L.: Fuzzy predictive control based on Takagi-Sugeno model for nonlinear systems. In: Proceeding of 7th International Multi-Conference on Systems Signals and Devices, pp. 1–7 (2010)
6. Hadjili, M.: Modelling and control using Takagi-Sugeno fuzzy models. In: Proceeding of Electronics, Communications and Photonics Conference, pp. 1–6 (2011)
7. Mendes, J.: Adaptive fuzzy generalized predictive control based on Discrete-Time T-S fuzzy model. In: Proceeding of IEEE Conference of Emerging Technologies and Factory Automation, pp. 1–8 (2010)
8. Chadl,i M., Borne, P.: Multiple Models Approach in Automation: Takagi-Sugeno Fuzzy Systems, Wiley, (2012)
9. Bai, E.: An optimal two-stage identification algorithm for Hammerstein-Wiener nonlinear systems. Automatica **4**(3), 333–338 (1998)
10. Janczak, A.: Neural network approach for identification of Hammerstein systems. Int. J. Control **76**(17), 1749–1766 (2003)

11. Westwick, D., Kearney, R.: Identification of a Hammerstein model of the stretch reflex EMG using separable least squares. In: Proceedings of World Congress on Medical Physics and Biomedical Engineering. (2000)
12. Rizvi, S., Al-Duwaish, H.: A PSO-subspace algorithm for identification of Hammerstein models. In: Proceeding of IFAC Conference on Control Applications of Optimization (2009)
13. Ashooria, A., Moshiria, B., Khaki-Sedighb, A., Bakhtiari, M.: Optimal control of a nonlinear fed-batch fermentation process using model predictive approach. J. Process Control **19**(7), 1162–1173 (2009)
14. Li, Y., Kashiwa, H.: High-order volterra model predictive control and its application to a nonlinear polymerization process. Int. J. Autom. Comput. **2**, 208–214 (2006)
15. Franks, F.: Freeze-Drying of Pharmaceuticals and Biopharmaceuticals. The Royal Society of Chemistry (2007)
16. Oetjen, G., Hasley, P.: Freeze-Drying 2nd edn.. Willey (2004)
17. Prasad, K.: Downstream process technology: a new horizon in biotechnology. PHI Learning (2010)
18. Aghbashlo, M., Kianmehr, M.H., Nazghelichi, T., Rafiee, S.: Optimization of an artificial neural network topology for predicting drying kinetics of carrot cubes using combined response surface and genetic algorithm. Drying Technol. **29**(7), 770–779 (2011)
19. Trelea, I.C., Passot, S., Fonseca, F., Marin, M.: An interactive tool for the optimization of freeze-drying cycles based on quality criteria. Drying Technol. **25**(5), 741–751 (2010)
20. Fissore, D., Pisano, R., Barresi, A.: a model-based framework to optimize pharmaceuticals freeze drying. Drying Technol. **30**(9), 946–9589 (2012)
21. Pisano, R., Barresi, A., Fissore, D.: Innovation in monitoring food freeze drying. drying technology. Selected Papers Presented at the 17th International Drying Symposium, Part 2, vol. 29, issue (16), pp. 1920–1931 (2011)
22. Barresi, A., Pisano, R., Rasetto, V., Fissore, D., Marchisio, D.: Model-based monitoring and control of industrial freeze-drying processes: effect of batch non-uniformity. Drying Technol. **28**(5), 577–590 (2010)
23. Velardi, S., Hammouri, H., Barresi, A.: In-line monitoring of the primary drying phase of the freeze-drying process in vial by means of a Kalman filter based observer. Chem. Eng. Res. Des. **87**(10), 1409–1419 (2009)
24. Pisano, R., Fissore, D., Barresi, A.: Freeze-drying cycle optimization using model predictive control techniques. Ind. Eng. Chem. Res. **50**, 7363–7379 (2011)
25. Daraoui, N., Dufour, P., Hammouri, H., Hottot, A.: Model predictive control during the primary drying stage of lyophilisation. Control Eng. Pract. **18**(5), 483–494 (2010)
26. Cubillos, F., Vyhmeister, E., Acuña, G., Alvarez, P.: Rotary dryer control using a grey-box neural model scheme. Drying Technol. **29**(15), 1820–1827 (2011)
27. Polat, K., Kirmaci, V.: A novel data preprocessing method for the modeling and prediction of freeze-drying behavior of apples: multiple output-dependent data scaling. Drying Technol. **30** (2), 185–196 (2012)
28. Jumah, R., Mujumdar, A.: Modeling intermittent drying using an adaptive neuro-fuzzy inference system. Drying Technol. **23**(5), 1075–1092 (2005)
29. Terzyiska, M., Todorov, Y., Petrov, M.: Nonlinear model predictive controller using a fuzzy-neural Hammerstein model. In: Proceedings of international conference "Modern Trends in Control", pp. 299–308 (2006)
30. Todorov, Y., Petrov, M.: Model Predictive Control of a Lyophilization plant: a simplified approach using Wiener and Hammerstein systems. J. Control Intell. Syst. **39**(1), 23–32 (2011). Acta Press
31. Todorov, Y., Ahmed, S., Petrov, M.: Model predictive control of a Lyophilization plant: a newton method approach. J. Inform. Technol. Control **IX**(4), 9–15 (2011)
32. Todorov Y., Ahmed, S., Petrov M.: State-Space Predictive Control of a Lyophilization plant: A fuzzy-neural Hammerstein model approach. In: Proceedings of the 1st IFAC Workshop Dynamics and Control in Agriculture and Food Processing, pp. 181–186 (2012)

33. Todorov Y., Ahmed, S., Petrov, M., Chitanov, V.: implementations of a Hammerstein fuzzy-neural model for predictive control of a Lyophilization plant. In: Proceedings of the 6th IEEE Conference on "Intelligent Systems", vol. 2, pp. 315–319 (2012)
34. Osowski, S.: Sieci neuronowe do przetwarzania infromacji. Oficyna Wydawnycza Policehniki Warzawsikej (2000)
35. Maciejowski, J.: Predictive Control with Constraints. Prentiss Hall (2002)
36. Rossiter, A. Model Based Predictive Control: A Practical Approach. CRC Press (2003)
37. Camacho, E., Bordons, C.: Model Predictive Control, 2nd edn. (2004)
38. Wang, L.: Model Predictive Control System Design and Implementation Using MATLAB. Springer (2009)
39. Fletcher, R.: Practical Methods for Optimization, 2nd edn. Wiley (2006)
40. Schoen, M., Jefferis, R.: Simulation of a controlled freeze drying process. In: Proceedings of IASTED International Conference, pp. 65–68 (1993)
41. Schoen, M.: A Simulation model for primary drying phase of Freeze-drying. Int. J. Pharm. **114**, 159–170 (1995)

Free Search and Particle Swarm Optimisation Applied to Global Optimisation Numerical Tests from Two to Hundred Dimensions

Vesela Vasileva and Kalin Penev

Abstract This article presents an investigation on two real-value methods such as Free Search (FS) and Particle Swarm Optimisation (PSO) applied to global optimisation numerical tests. The objective is to identify how to facilitate assessment of heuristic, evolutionary, adaptive and other optimisation and search algorithms. Particular aim is to assess: (1) probability for success of given method; (2) abilities of given method for entire search space coverage; (3) dependence on initialisation; (4) abilities of given method to escape from trapping in local sub-optima; (5) abilities of explored methods to resolve multidimensional (one hundred dimensions) global optimisation tasks; (6) performance on two and hundred dimensional tasks; (7) minimal number of objective function calculation for resolving hundred dimensional tasks with acceptable level of precision. Achieved experimental results are presented and analysed. Discussion on FS and PSO essential characteristics concludes the article.

Keywords Global optimisation · Multidimensional optimisation · Numerical tests · Free search · Particle swarm optimisation · Heuristic methods

1 Introduction

One of the challenges of modern Computer Science is to cope with global optimisation tasks reaching acceptable level of precision for affordable period of time and with limited computational resources. According to the publications [1, 2] global optimisation refers to finding extremum (minimum or maximum) of a given

---

V. Vasileva (✉) · K. Penev
Technology School Maritime and Technology Faculty, Southampton Solent University, East Park Terrace, Southampton SO14 0YN, UK
e-mail: Vesela.Vasileva@solent.ac.uk

K. Penev
e-mail: Kalin.Penev@solent.ac.uk

V. Sgurev et al. (eds.), *Recent Contributions in Intelligent Systems*, Studies in Computational Intelligence 657,
DOI 10.1007/978-3-319-41438-6_18

nonconvex objective function. Scientists, engineers and practitioners often face in practice global optimisation problems and need reliable methods to resolve such tasks. For this purpose various search methods such as Genetic Algorithm [3], Particle Swarm Optimisation (PSO) [4, 5], Evolution Strategy [6], Differential Evolution [7] and Free Search (FS) [8] can be used. Majority of search and optimisation methods encounter difficulties when dealing with global optimisation problems. The main reasons of their failure are

- inability to generate non-zero probability for access to the whole search space;
- entrap in local sub-optimal solution;
- inability to escape from trapping;
- inability to abstract sufficient knowledge or use it effectively (if available) for a global task with multiple potential solutions.

In order to assess different algorithms many numerical tests are proposed and published [2, 9]. However, there is no enough evidence that available tests could provide sufficient assessment of given method and its applicability to complete global optimisation. This article aims to contribute to the knowledge using specific global tests.

Algorithms' abilities for global optimisation closely relates with their potential for highest level of adaptation to various tasks, [10] without retuning the search parameters, which is widely discussed in the literature [11–15]. In practice algorithms parameters retuning very often could be impossible due to luck or absence of knowledge for any particular optimisation tasks.

This study uses hard and introduces modified numerical tests for global optimisation supported by short and fast experimental methodology, which facilitates assessment of given optimisation algorithm.

Proposed methodology could measure the following:

- probability to resolve global and multidimensional tasks within finite (acceptable) period of time with limited computational resources;
- dependence of algorithms success on initial conditions;
- probability to resolve the problem, without initial knowledge;
- dependence on search parameters retuning for resolving various tasks;
- probability for algorithms' success if applied to similar real-world tasks.

2 Numerical Test

For the purposes of this investigation, all tests are modified for maximisation and designed in unified black box model, published in the literature [8]. If minimisation is required the objective functions should be with reversed sign.

2.1 Michalewicz Test Function

Objective function for Michalewicz test [2] is:

$$f(x_i) = \sum_{i=1}^{n} \sin(x_i)(\sin(ix_i^2/\pi))^{2m} \tag{1}$$

where the search space is $x_i \in [0.0, 3.0]$, $m = 10$, $i = 2$ and the optimum value is $f_{\text{opt}} = 1.8013$. [2] Global optimum is in the middle of the search space. Although this search space has local peak, previous investigation [16] suggests that it is relatively easy for resolving by modern powerful search algorithms. Therefore, for two-dimensional tests in this study modification with extension of the search space is proposed. For modified Michalewicz test search space boundaries are $x_i \in [-5.0, 5.0]$, $m = 10$, $i = 2$.

Two equal value $f_{\text{opt}} = 1.96785$ global optima are situated close to the search space boundaries. This makes the test difficult for methods, which rely on recombination due to the reduced probability for assess of the space between the optimal hills and adjoining boundary.

This test is suitable for assessment of algorithms' abilities to cover the whole search space with non-zero probability and to escape from trapping in sub-optimal areas. It is generalised for multidimensional space and used also for one hundred dimensional tests.

2.2 Five Hills Test

Five Hills numerical test is two-dimensional test specially designed for global optimisation. Its objective function is

$$\begin{aligned} f(x_i) = {} & \frac{11.4}{1+0.05\left((0-x_1)^2+(0-x_2)^2\right)} + \\ & \frac{9.9}{1+2\left((7-x_1)^2+(7-x_2)^2\right)} + \\ & \frac{9.9}{1+1.7\left((7+x_1)^2+(7+x_2)^2\right)} + \\ & \frac{10.0}{1+2.3\left((7-x_1)^2+(7+x_2)^2\right)} + \\ & \frac{9.7}{1+1.7\left((7+x_1)^2+(7-x_2)^2\right)} \end{aligned} \tag{2}$$

where $x_i \in [-10.0, 10.0]$. It has five local optima. A smooth hill in the middle dominates more than 60 % of the search space, but four corners' peaks are higher. The highest peak $f_{\max}$ (6.994, −6.994) = 12.0076 is situated in one of the four corners.

The highest area is less than 0.1 % of the search space only. Identification of the maximum requires good divergence, which could guarantee coverage of the whole search space, and then good convergence with small precise steps.

Five hills test could be transformed to time dependent test where the highest peak coordinates are variables, which change over the time.

2.3 Norwegian Test

Objective function for Norwegian test [17] is

$$f(x_i) = \prod_{i=1}^{n} \left(\cos(\pi x_i^3) \left(\frac{99 + x_i}{100} \right) \right) \tag{3}$$

In this study space boundaries are extended to $x_i \in [-1.1, 1.1]$. The highest peak is $f_{\max}$ (1.00011, 1.00011) = 1.00000113.

Search space is dominated by flat hill in the middle. This hill occupies more than 70 % of the space. It attracts and could easily entrap any search method. Higher peaks are located closely to the four corners. Identification of the highest peak could be very hard for majority of search methods. This test is generalised for multidimensional space and used for one hundred dimensional tests as well.

2.4 Sofia Test

Sofia numerical test is two dimensional, specially designed for testing and assessment of global optimisation algorithms.

$$\begin{aligned} f(x_i) = &\frac{1}{0.9\left(0.01 + x_1^2 + (x_2+1)^2\right)} + \\ &\frac{1}{0.9\left(0.01 + (x_1+3)^2 + x_2^2\right)} + \\ &\frac{1}{2.0\left(0.01 + (x_1-4.5)^2 + (x_2-4.5)^2\right)} + \\ &\frac{1}{3.0\left(0.01 + (x_1+4.5)^2 + (x_2-4.5)^2\right)} \end{aligned} \tag{4}$$

The search space is defined by $x_i \in [-5.0, 5.0]$.

It has seven local maxima—four in the corners and three peaks within the wide global valley. The global maximum is located close to the bottom of the valley in other words close to the global minimum. Gradient of the local correlation is in opposite direction to the global maximum for more than 90 % of the search space.

In order to discover this maximum, any search algorithm needs to guarantee non-zero probability for access to the whole search space and must be able to abstract sufficient knowledge about entire task and then use it effectively within limited (acceptable) period of time and with limited computational resources.

2.5 Bump Test

This is hard constrained global optimisation problem [3] transformed in this study for maximisation.

$$f(x_i) = \left| \sum_{i=1}^{n} \cos^4(x_i) - 2 \prod_{i=1}^{n} \cos^2(x_i) \right| \Big/ \sqrt{\sum_{i=1}^{n} i x_i^2} \tag{5}$$

$$\text{subject to: } \prod_{i=1}^{n} x_i > 0.75 \tag{6}$$

$$\text{and } \sum_{i=1}^{n} x_i < 15n/2 \tag{7}$$

where $x_i \in [0.0, 10.0]$, and start from $x_i = 5$, $i = 1, \ldots, n$. x_i are variables (expressed in radians) and n is the number of dimensions [14].

It has many peaks. It is very hard for most optimisation methods to cope with because its optimal value is defined by presence of constraint boundary formulated as a variables product (inequality 6) and because of initial condition—start from $x_i = 5$.

Start from the middle of the search space excludes from initial population locations, which could be accidentally near to the unknown optimal value.

Maximal values $f_{\max}$ and constraint boundaries defined as variables product (shortly noted as p) $p > 0.75$ are located on a steep slope so that very small changes of variables lead to substantial change of the objective function. In the same time product constraint (7) defines very uncertain boundary. Various variables combinations satisfy constraint (7) but produce different function values, which makes this test even harder. A given combination of variables produces highest function value when variables are in descending order [18]. It is generalised for multidimensional continuous search space and used for one hundred dimensional tests in this investigation.

3 Algorithms Essential Peculiarities

For this investigation two algorithms are used. These are Free Search [8] and Particle Swarm Optimisation [4, 5]. Other comparison between FS and PSO is published earlier [8].

3.1 Free Search

This section summarises and refines algorithms description. Free Search could be classified as adaptive heuristic. Adaptive is understood as an algorithm, which can adapt to various search spaces and can achieve optimal solutions without external changes of its parameters. According to the literature [8, 16, 10] several level of adaptation could be differentiated, namely:

- adaptation to various static known spaces;
- adaptation to time dependent (changeable) spaces;
- adaptation to unknown spaces where no prior knowledge about the problems exists and no proved right settings could be done.

Published earlier investigations show that FS can adapt to heterogeneous tasks [8], time-dependent tasks [19], and unknown constrained multidimensional tasks [18].

The description below aims to illustrate the manner in which a computational program can generate adaptive processes. FS process organisation differs from ordinary iterative and evolutionary processes. FS is organised as a harmony of short individual explorations with continuous self-control, task learning, knowledge update and behaviour improvement. FS explores the search space by heuristic trials. It generates a new solution as deviation of a current one:

$$x = x_0 + \Delta x \tag{8}$$

where x is a new solution, x_0 is a current solution and Δx is modification. Other interpretation of Δx is that this is simply individual's step.

Individuals in FS explore the search space walking step by step. Steps are described by x, x_0 and Δx, which are vectors of real numbers. The modification strategy is

$$\Delta x_{tji} = R_{ji} * (X\max_i - X\min_{i)}) * \text{random}_{tji}(0, 1) \tag{9}$$

where: i indicates dimension; $i = 1, \ldots, n$ for a multidimensional step; n is dimensions number; t is the current step $t = 1, \ldots, T$. T is the steps number limit per walk; R_{ji} indicates the maximal step size for individual j within dimension i. random$_{tji}$(0,1) generates random values between 0 and 1. Δx_{tji} indicates the actual step size for step t of individual j within dimension i. During the exploration an individual with abilities for large steps, which could exceed search space boundaries, can perform global exploration whereas another individual with abilities for small steps can do precise search around one location.

FS modification strategy is independent from the current or the best achievements. This independence highly supports adaptation and excludes well-known dilemma "exploration versus exploitation". Event exploration performs heuristic trials based on stochastic divergence from one location. The concrete value of the neighbourhood space for a particular exploration defines the extent of uncertainty of chosen individual. The exploration walk is followed by an individual assessment of the explored locations. The best location is marked. By analogy with nature the variable used for marking is called pheromone. The variable pheromone indicates the quality of the locations and may be interpreted also as result or cognition from previous activities. The assessment, during the exploration, is defined as follows:

$$f_{tj} = f\left(x_{tji}\right),\; f_j = \max\left(f_{tj}\right) \tag{10}$$

where f_{tj} is the value of the objective function achieved from individual j for step t. f_j is the quality of the location marked with pheromone from an individual after one exploration. The pheromone generation is generalised for the whole population:

$$P_j = f_j / \max\left(f_j\right) \tag{11}$$

where $\max(f_j)$ is the best achieved value from the population for the exploration.

This is a normalisation of the explored problem to an idealised qualitative (or cognitive) space, in which the algorithm operates. This idealised space is used to model the ideal space of notions in the mind of biological systems in which they generate decisions. Normalisation of any particular search space to one idealised space supports adaptation and successful performance across variety of problems without additional external adjustments. Decision-making policy in FS by analogy with biological systems depends on apprehension of the space. It is related with a variable called sense. The variable sense can be likened as a quantitative indicator of sensibility.

The algorithm tunes sensibility during the process of search as function of the explored problem. The same algorithm makes different regulations of the sense during the exploration of different problems. This is the model of adaptation in Free Search [8, 16]. Variable sense distinguishes individuals from solutions. The individuals are explorers differentiated from evaluated solutions and detached from the problem's search space. A solution in FS is a location from a continuous space marked with pheromone. Abstract individuals explore, select, evaluate, and mark these solutions.

The sensibility generation is

$$S_j = S\min + DS_j \tag{12}$$

$$\text{where } \Delta S_j = (S\text{max} - S\text{min}) * \text{random}_j(0, 1) \quad (13)$$

Smin and Smax are minimal and maximal possible values of the sensibility.

$$S\text{min} = P\text{min},\ S\text{max} = P\text{max}. \quad (14)$$

Pmin and Pmax are minimal and maximal possible values of the pheromone marks. The process continues with selection of a start location for a new exploratory walk. The ability for decision-making based on the achieved from the exploration (which can be in contradiction with the existing assumptions about the problem during the implementation of the algorithm) supports a good performance across variety of problems, adaptation and self-regulation without additional external adjustments. Selection for a start location x_{0j} for an exploration walk is

$$x_{0j} = x_k(P_k \geq S_j) \quad (15)$$

where: $j = 1, \ldots, m$, j is the number of the individuals; $k = 1, \ldots, m$, k is the number of the location marked with pheromone; x_{0j} is the start location selected from individual number j. After exploration the termination follows. Individual relation between sensibility and pheromone distributions affects decision-making policy of the whole population. A short discussion on three idealised general states of sensibility distribution can clarify FS self-regulation and how chaotic on first view accidental events can lead to purposeful behaviour. These are—uniform, enhanced and reduced sensibility.

In Fig. 1 (and in following Figs. 2 and 3) left side represents distribution of the variable sense within sensibility frame and across the animals from the population. Right side represents distribution of the pheromone marks within the pheromone frame and across the locations marked from previous generation. In case of uniformly distributed sensibility and pheromone (Fig. 1), individuals with low level of sensibility can select for start position any location marked with pheromone.

Individuals with high sensibility can select for start position locations marked with high level of pheromone and will ignore locations marked with low level of pheromone. It is assumed that during a stochastic process within a stochastic environment any deviation could lead to non-uniform changes of the process. The achieved results play role of deviator.

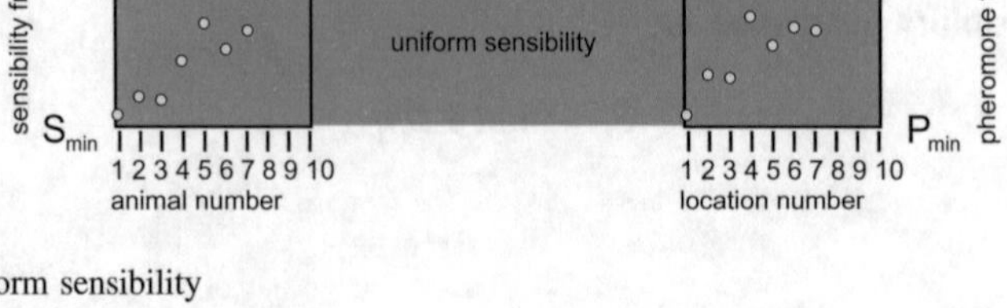

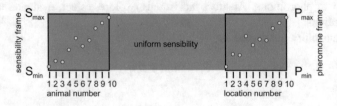

Fig. 1 Uniform sensibility

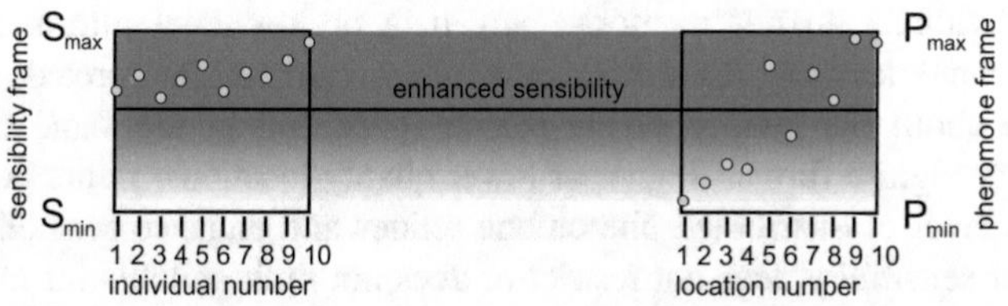

Fig. 2 Enhanced sensibility

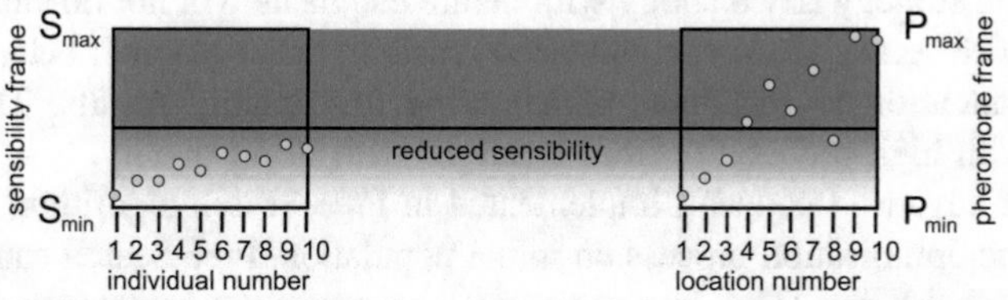

Fig. 3 Reduced sensibility

An enhancement of sensibility encourages animals to search around the area of the best-found solutions during previous exploration and marked with highest amount of pheromone. This situation appears naturally (within FS process) when pheromone marks are very different and stochastic generation of sensibility produces high values.

External adding of a constant or a variable to the sensibility of each animal could make forced sensibility enhancement. Individuals with enhanced sensibility will select and can differentiate more precisely locations marked with high level of pheromone and will ignore these indicated with lower level (Fig. 2).

By sensibility reduction an individual can be allowed to explore around locations marked with low level of pheromone. This situation naturally appears (within FS process) when the pheromone marks are very similar and randomly generated sensibility is low. In this case the individuals can select locations marked with low level of pheromone with high probability, which indirectly will decrease the probability for selection of locations marked with high level of pheromone. Subtracting of a constant or a variable from sensibility of each animal could make a forced reduction of sensibility frame (Fig. 3). Sensibility across all individuals varies. Different individuals can have different sensibility. It also varies during the optimisation process, and one individual can have different sensibility for different explorations.

Adaptive self-regulation of sense, action and pheromone marks is organised as follows. An achievement of better solutions increases the knowledge of the population for best possible solution. This knowledge clarifies pheromone and sensibility frames, which can be interpreted as an abstract approach for learning and sensibility can be described as high-level abstract knowledge about the explored space. This knowledge is acquired from the achieved and assessed results only.

The individuals do not memorise any data or low-level information, which consume computational resources. Sensibility can be interpreted as implicit knowledge about the quality of the search space and in the same time creates abilities to recognise further, higher or lower quality locations. Better achievements and higher level of distributed pheromone refines and enhance sensibility.

A higher sensibility does not restrict or does not limit abilities for movement. It implicitly regulates the action of the individuals in terms of selection of a start location for exploration [16]. Learning new knowledge does not change individual abilities for movement. Animals continue to do small or large steps according to the modification strategy (By analogy with nature elephants will not do small ants size steps and vice versa). However, enhanced sensibility changes their behaviour. They give less attention to locations, which bring low quality results. They can be attracted with high probability from locations with better quality.

Another advanced concept, implemented in Free Search algorithm, is independence of the optimisation process on initial population. Free Search can operate on any initial population. This is a conceptual improvement in comparison to other real-value methods for optimisation of non-discrete problems. Analysis of Genetic Algorithm [3], Particle Swarm Optimisation and Differential Evolution [6] suggests that these methods cannot operate when optimisation starts from one location. Genetic Algorithm starts effective work after the first mutation and DE and PSO cannot start at all.

Three types of initialisations presented below can illustrate FS abilities for start from stochastic initial locations, start from certain initial locations and start from one location.

Free Search can start from a stochastically selected set of initial solutions where all the initial locations x_{0ji} are random values:

$$x_{0ji} = \mathrm{Xmin}_i + (\mathrm{Xmax}_i - \mathrm{Xmin}_i) * \mathrm{random}_{ji}(0,1) \tag{16}$$

where Xmin_i and Xmax_i are the search space borders, $i = 1, \ldots, n$, n is the number of dimensions, $j = 1, \ldots, m$, m is the population size. random(0,1) is a random value between 0 and 1. A start from random locations guarantees non-zero probability for access to any location from the search space.

Free Search can start from certain initial population where all the initial locations x_{0ji} are prior-defined values a_{ji}:

$$x_{0ji} = a_{ji},\ a_{ji} \in [\mathrm{Xmin}_i, \mathrm{Xmax}_i] \tag{17}$$

where Xmin_i and Xmax_i are the search space borders, $i = 1, \ldots, n$, n is the number of dimensions, $j = 1, \ldots, m$, m is the population size and a_{ji} are constants, which belong to the search space.

A start from certain locations is a valuable ability for multi-start optimisation. It is useful from a practical point of view, when are available. The starting from certain locations can be used when some values are already achieved and the algorithm can continue from these values instead of repeating starts from random locations.

Free Search can start from an initial population where all the initial solutions x_{0ji} are in one location c:

$$x_{0ji} = c, c \in [X\min_i, X\max_i] \tag{18}$$

where $X\min_i$ and $X\max_i$ are the search space borders, $i = 1, \ldots, n$, n is the number of dimensions, $j = 1, \ldots, m$, m is the population size c is constant.

A start for one location is a difference from Particle Swarm Optimisation. Ability for start from one location closely relates with abilities to avoid stagnation and to escape for trapping in local optimal solutions. If during the optimisation process all individuals converge to one location, due to this ability FS quickly diverges and continues search process, in contrast PSO stagnates.

3.2 Particle Swarm Optimisation

This section discusses mainly PSO modification strategy in order to compare conceptual differences between FS and PSO. According to some publications, PSO intends to model a social behaviour of a group of individuals where it searches gradually for the optimum by changing the values of the set of solutions [4].

However, observation of PSO generated search process suggests that its behaviour could be liken as self-organised particles in a cloud systems.

Each particle (individual) shows a single intersection of all search dimensions and is defined as a potential solution to a test problem in multidimensional space.

At every iteration the particles appraise their position relative to an objective function (fitness) whether particles in a local neighbourhood allocate memories of their best positions then use those memories to accommodate their-own velocities, and thus positions [4].

The original PSO concept [4] is modified by adding inertia factor for velocities tuning [5]. This study uses modified PSO with variable inertia factor proposed earlier [5]. The velocity v is used to compute a new position for the particle as shown below:

$$x'_{\text{id}} = x_{\text{id}} + v_{\text{id}} \tag{19}$$

where x'_{id} is new position of particle i for dimension d, x_{id} is its current position and v_{id} is its velocity. The velocity vector v'_{id} for each particle is calculated using the best particles' achievement P_{id}, best for all population achievement g_{d} and inertia factor w according to the equation below:

$$v'_{\text{id}} = w * v_{\text{id}} + n_1 * \text{random}(0, 1) * (P_{\text{id}} - x_{\text{id}}) + n_2 * \text{random}(0, 1) * (g_{\text{d}} - x_{\text{id}}) \tag{20}$$

The constants n_1 (individual learning factor) and n_2 (social learning factor) are usually set with the equal values in terms of giving each component equal weight as the cognitive and social learning rate.

Both velocity component and inertia factor support adaptation to the explored test problem. PSO could be adjusted easily as it contains few parameters only.

4 Experimental Methodology

Proposed experimental methodology is simple and fast. All tests are evaluated for two dimensions. For all experiments for both FS and PSO, population is 10 individuals.

For two-dimensional tests PSO experiments are limited to 100 and 2000 iterations per each test. Total number of test function evaluations for 10 individuals are: $100 \times 10 = 1000$ and $2000 \times 10 = 20000$ accordingly. FS experiments are limited to 20 and 400 explorations with 5 steps per exploration. Total number of test function evaluations for 10 individuals are: $20 \times 5 \times 10 = 1000$ and $400 \times 5 \times 10 = 20000$.

For hundred dimensions PSO and FS are evaluated on Michalevicz and Norwegian tests and FS separately is evaluated on Bump test.

For all test PSO is applied with variable inertia factor, which enhances to some extent its ability for adaptation. Individual learning factor is 2.0. Group learning factor is 2.0. Inertia weight varies within the interval 0.5–1.5 with step 0.1. Initialisation is stochastic:

$$x_{0ji} = \mathrm{Xmin}_i + (\mathrm{Xmax}_i - X\mathrm{min}_i) * \mathrm{random}_{ji}(0, 1) \tag{21}$$

where $X\mathrm{max}_i$ – $X\mathrm{min}_i$ are search space boundaries; j is individual, i is dimension. For all tests FS is used with its standard set of parameters. Detailed description of these parameters is published [8]. In order to provide equal conditions for testing with PSO neighbour space varies within the interval 0.5–1.5 with step 0.1. Sensibility randomly varies within the interval 0.99999–1.0.

To evaluate probability for dependence on initialisation, 32 experiments with different initialisations per each inertia value for PSO and per each neighbour space value for FS are completed. This corresponds to 320 experiments per test per method in total. FS is evaluated additionally to the same number of experiments but with start for all individuals from a single location selected away from the global optimum. The single location for all tests is defined as

$$x_{0ji} = \mathrm{Xmin}_i + 0.1 * (\mathrm{Xmax}_i - X\mathrm{min}_i) \tag{22}$$

where $X\mathrm{max}_i$ – $X\mathrm{min}_i$ are search space boundaries; j is individual; i is dimension. For Sofia test FS is evaluated for extra hard initial conditions with start for all individuals from a single location close to the local sub-optimal solutions, which purposefully increases probability for trapping. The single location for these tests is defined as

$$x_{0ji} = X\max_i - 0.25 * (X\max_i - X\min_i) \tag{23}$$

where $X\max_i$ – $X\min_i$ are search space boundaries; j is individual; i is dimension.

In order to compare algorithms performance on multidimensional global test, FS and PSO are tested on Michalewicz, Norwegian test and FS only on Bump test for 100 dimensions.

For 100 dimensions, Michalewicz test is implemented for search space $x_i \in [0.0, 3.0]$.

Experiments for Michalewicz and Norwegian tests are limited to 1000000 iterations for PSO and 200000 explorations with 5 steps for FS. A criterion for termination for the experiments on Bump test is reaching particular optimal value [4, 20].

5 Experimental Results for 2-Dimensional Tests

PSO completed four tests with start from randomly generated initial solutions. Due to its modification strategy, PSO is unable to process from single initial solution.

FS completed all tests starting from single and random locations. For each experiment 320 results are obtained, analysed and summarised. Achieved experimental results for two-dimensional tests are assessed by the following criteria:

- Michalewicz test for $x_i \in [-5.0, 5.0]$ has two equal value maxima $f_{\max}(-4.96599769, -4.71238898) = 1.96785066$ and $f_{\max}(-4.96599769, 1.57079633) = 1.96785066$. Second local maximum is 1.80130341. As successful are accepted results above 1.96.
- Norwegian test for $x_i \in [-1.1, 1.1]$ has maximum at $f_{\max}(1.00011, 1.00011) = 1.00000113$. Second local maximum is 0.98344. As successful are accepted results above 0.99.
- Five hills test for $x_i \in [-10.0, 10.0]$ has maximum at $f_{\max}(6.99344, -6.9942) = 12.0076$. Second local maximum is 11.9076. As successful are accepted results above 11.91.
- Sofia test for $x_i \in [-5.0, 5.0]$ has maximum at $f_{\max}((-3.0003, 0.000100313)) = 122.256$. Second local maximum is 106.148. For this test, as successful are accepted results above 110.
- Bump test for $x_i \in [-10.0, 10.0]$ has maximum at $f_{\max}(1.601116247$ $0.468424618) = 0.36497$. As successful are accepted results above 0.36.

Table 1 presents the number of successful results from 320 experiments reached by FS with start from single location and FS and PSO with start from random locations for 100 and 2000 iterations (for FS 20 and 400 explorations 5 steps each).

Table 1 FS and PSO successful results on two-dimensional tests

Successful results	Iterations					
	One location		Random locations			
	FS 100	FS 2000	FS 100	FS 2000	PSO 100	PSO 2000
Michalewicz	93	320	138	313	16	16
Norwegian	11	173	9	160	9	118
Five Hills	41	260	21	242	4	52
Sofia	21	74	24	65	2	2
Bump	30	312	56	316	126	275
Total	196	1139	248	1096	157	463

Table 2 FS on Sofia test start form extra hard location

FS starting from extra hard single location	Iterations	
	100	2000
Sofia	7	41

Fig. 4 FS and PSO successful results

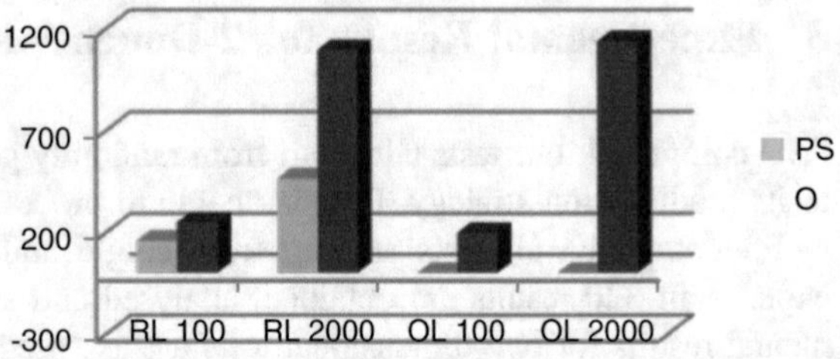

Table 2 presents the number of successful results reached by FS on Sofia test from 320 experiments with start from extra hard singe location for 100 iterations and for 2000 iterations.

On the Fig. 4, RL indicates start form random locations and OL indicates start form one location. Figure 4 summarises the total number of successful results achieved by FS on five tests for 100 and 2000 iterations with start from one location and random locations and PSO on five tests with start from random locations.

On the Fig. 5, RL indicates start form random locations and OL indicates start form one location. Figure 5 shows successful results achieved on Sofia test by PSO for 100 and 2000 iterations with start from random locations and by FS for 100 and 2000 iterations with start from random locations, single location and extra hard single location. Results presented in Tables I and II suggest that probability for success on explored tests:—within 100 iterations is low for both methods. For majority of tests (except Bump test and FS on Michalewicz test) probability for success is below 10 %;—within 2000 iterations for PSO is moderate on Norwegian (16 %) and Five hills (37 %) and low for Michalewicz (5 %) and Sofia (0.6 %); for FS, which does not depend on initialisation, for start from random locations is 61 % and for start from single location is 64 %. FS can resolve with 60 % probability

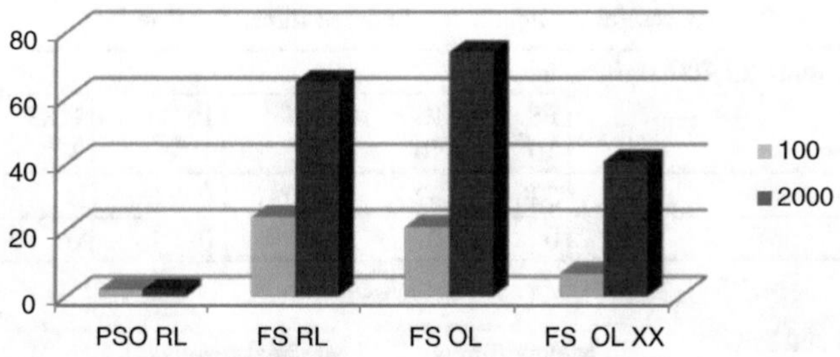

Fig. 5 FS and PSO on Sofia test

global optimisation tests. FS is independent on initialisation. FS would be reliable with high probability for global optimisation real-world tasks. On two-dimensional Bump test PSO demonstrates high convergence, for 100 iterations reaches 39 % successful results and outperforms FS, which reaches 17.5 %. However for 2000 iterations FS has 98.75 % success versus 85.94 % for PSO.

6 Experimental Results for 100-Dimensional Tests

Experimental results for 100 dimensional tests are assessed by the following criteria:

On 100 dimensional Michalewicz test for $x_i \in [0.0, 3.0]$ maximal achieved result by Free Search on experiments limited to two hundred thousand explorations of five steps, which corresponds to then million iterations is $f100_{max} = 99.6191$. Firm confirmation whether this is the global optimum could be a subject of further research. For statistical purposes, as successful are accepted results above 99.4.

On 100-dimensional Norwegian test for $x_i \in [-1.1, 1.1]$ maximal achieved result by Free Search on experiments limited to two hundred thousand explorations of five steps, which corresponds to then million iterations is $f100_{max} = 1.00004$. Firm confirmation whether this is the global optimum could be a subject of further research. For statistical purposes, as successful are accepted results above 1.0.

For 100 dimension FS and PSO are tested on 100 dimensional Michalewicz and Norwegian tests. Completed are three series of experiments limited to 10^5, 10^6 and 10^7 iterations for PSO and 20000, 200000, 2000000 exploration 5 steps for FS. These limitations guarantee identical number of objective functions calculations. Although they perform well on two-dimensional Bump test they need more iterations to cope with 100 dimension, therefore only FS is tested separately.

Table 3 presents the number of successful results on 100 dimensional tests from 320 experiments reached by FS and PSO with start from random locations for 10^5, 10^6 and 10^7 iterations limits. Used implementation of PSO although performed well on two-dimensional tests was trapped at sub-optimal solutions for booth 100 dimensional tests. FS resolved with acceptable level of precision 100 dimensional

Table 3 FS and PSO successful results on one hundred dimensional tests

Successful results for 100 dimensions	Iterations					
	FS 10^5	FS 10^6	FS 10^7	PSO 10^5	PSO 10^6	PSO 10^7
Michalewicz	320	320	320	0	0	0
Norwegian	0	0	117	0	0	0

Table 4 FS and PSO maximal results on Michalevicz and Norwegian tests

Method/limits	Maximal results	
	Michalewicz 100D	Norwegian 100D
FS/20000 * 5	99.5808	0.750627
FS/200000 * 5	99.6157	0.967082
FS/2000000 * 5	99.6191	1.00004
PSO/100000	79.2948	0.220553
PSO/1000000	79.2948	0.224411
PSO/10000000	79.2948	0.225525

Table 5 FS and PSO mean results on Michalevicz and Norwegian tests

Method/limits	Mean results	
	Michalewicz 100D	Norwegian 100D
FS/20000 * 5	99.5021065	0.69120580
FS/200000 * 5	99.6109537	0.92401155
FS/2000000 * 5	99.618175	0.98937421
PSO/100000	31.9071906	0.00747217
PSO/1000000	33.0173021	0.00798572
PSO/10000000	34.28708729	0.008360065

Table 6 FS and PSO standard deviation on Michalevicz and Norwegian tests

Method/limits	Standard deviation	
	Michalewicz 100D	Norwegian 100D
FS/20000 * 5	0.033529264	0.02712148
FS/200000 * 5	0.002655801	0.01853559
FS/2000000 * 5	0.000480039	0.00843936
PSO/100000	0.03091758	20.51973245
PSO/1000000	0.03013784	20.98219184
PSO/10000000	?	?

Michalewicz test for all tests. More explorations just add to the precision. For 100-dimensional Norwegian tests, however, FS reached global maximum only for 117 from 320 experiments limited to 2000000 explorations * 5 step (corresponding to 10^7 iterations).

Table 4 presents maximal results on 100 dimensional tests from 320 experiments reached by FS and PSO with start from random locations for 10^5, 10^6 and 10^7 iterations limits.

Table 5 presents mean results on 100-dimensional test from 320 experiments reached by FS and PSO with start from random locations for 10^5, 10^6 and 10^7 iterations limits.

Table 6 presents standard deviation on 100-dimensional tests from 320 experiments reached by FS and PSO with start from random locations for 10^5, 10^6 and 10^7 iterations limits. Experiments on 100 dimensional tests confirm published earlier [8, 16] Free Search abilities to:

- avoid stagnation;
- escape from trapping in sub-optimal solutions;
- avoid dependence on initial population;
- adapt to unknown space.

Used implementation of PSO, although resolved two-dimensional tests, has difficulties to cope with one-hundred dimensional versions of these tests. Tables 4 and 5 suggest that cannot achieve global optimum within defined iterations limits. Table 6 shows that increasing the number of iterations lead to an increase of standard deviation. Reasons for this behaviour and an increase of the PSO convergence for global multidimensional tasks could be a subject of further research.

Although PSO and FS show good results on two-dimensional Bump test on one hundred dimensional version of this test used PSO implementation could not reach acceptable results. After substantial number of sequential multi-start experiments Free Search reaches maximal value of $F\max_{100} = 0.84568545610035528$, constraint perimeter for this value is $p_{100} = 0.75000000000000466$.

Variables for this value are presented in Table 7 and could be used for comparison to other methods or for start location for further refining of this value.

7 Discussion

This section discusses FS performance and how FS avoids fundamental acceptances, on which are based attempts to prove inability of the modern heuristic methods to perform well on heterogeneous global optimisation problems.

In the same time FS does not contradict the existing conclusion that there is no algorithm, which performs best across all possible range of problems [12, 13]. FS is based on several advanced concepts, whose harmony contributes to the adaptation to various problems and to the good performance on unknown tasks. In some extent, these concepts differ from existing assumptions, published in the literature.

Table 7 Variables for Bump test for $F_{\max 100} = 0.84568545610035528$

×0	9.4220110174916201	×34	2.9539993057768896	×68	0.45441962950622516
×1	6.2826269503345156	×35	2.946700290458983	×69	0.45356136749912962
×2	6.268383174968319	×36	2.9393411869342341	×70	0.45270186591116279
×3	3.1685544120244855	×37	2.9319048348438392	×71	0.45185207098772395
×4	3.1614801076277121	×38	2.9243754868990863	×72	0.45101725657683717
×5	3.1544518696911341	×39	2.9167792641902861	×73	0.45017531809106459
×6	3.1474495832289167	×40	0.48215961508911009	×74	0.44936319426877097
×7	3.1404950479045404	×41	0.48103824313351567	×75	0.4485377329920564
×8	3.1335624856086999	×42	0.47990261875141216	×76	0.4477274215608178
×9	3.1266689455963657	×43	0.47878287205581088	×77	0.44690913523308484
×10	3.1197829709276546	×44	0.47768430094929853	×78	0.44610180306836622
×11	3.1129330728191893	×45	0.47661284860125969	×79	0.44530695939065651
×12	3.1060985093274232	×46	0.47553401880423046	×80	0.44452775570335706
×13	3.0992700218084379	×47	0.47446785125492774	×81	0.44374601157744475
×14	3.0924554965695812	×48	0.47341194393636354	×82	0.44296210126059304
×15	3.0856546345558531	×49	0.47238534512975816	×83	0.44218165327387443
×16	3.0788600779867954	×50	0.471363407442003	×84	0.44144485059511346
×17	3.0720627258530895	×51	0.47032875754599018	×85	0.44065200218379885
×18	3.065265700036762	×52	0.46930945558813297	×86	0.43991686744809411
×19	3.0584746707061194	×53	0.46831837263410242	×87	0.4391561530937918
×20	3.0516637032956146	×54	0.4673564573538872	×88	0.43841830158401307
×21	3.0448584948795405	×55	0.46635873320297838	×89	0.43768312099339146
×22	3.0380286239783816	×56	0.4653959047174871	×90	0.43694766056794315
×23	3.0311926837052807	×57	0.46442273745998447	×91	0.43620589320100195
×24	3.0243298573367188	×58	0.4634765652864925	×92	0.43550809718429934
×25	3.0174564446916823	×59	0.46251430644316693	×93	0.43477896752008077
×26	3.0105548474617962	×60	0.46159995229591577	×94	0.43406590079301738
×27	3.0036206348180787	×61	0.46065790486205743	×95	0.43335315526735246
×28	2.9966580744764033	×62	0.45977027595464537	×96	0.43265449240042753
×29	2.9896625571719468	×63	0.45882918141692586	×97	0.43194929446147168
×30	2.9826197373857943	×64	0.45794414020951224	×98	0.43126789176092595
×31	2.9755470498458614	×65	0.45703884159185892	×99	0.430563651
×32	2.9684103339517844	×66	0.45616083798158447		
×33	2.9612203185135004	×67	0.45530543288455649		

7.1 Adaptation Versus Preliminary Parameters Settings

According to the literature "There is much recent evidence to support the use of advanced search algorithms in wide range scientific, technological and social fields. Advantages have been demonstrated in design, engineering, transport, business, diagnosis, planning and control. However, when objectives are incomplete and when search environments are uncertain and infinite, which is common for real-world tasks, performance may fall drastically. An approach to cope with complex problems is by the use of prior knowledge about the problem and preliminary setting of the optimisation parameters" [21].

Assumptions for existence of prior knowledge about the explored problem as a premise, as a requirement, for adaptation, such as: "Systems that include prior case knowledge provide opportunities for systems to adapt and learn in uncertain environment" [21] are opposed to the concept for black box model, which excludes requirements for prior knowledge about the search problem.

It is accepted that Free Search has to be able to adapt and to adjust its behaviour to the explored problem without prior knowledge. FS has to be able to cope with the problem by utilisation of the knowledge achieved during the process of search, only. FS does not require prior knowledge and preliminary settings related to the particular problem.

At the same time, FS does not restrict a use of prior knowledge if it exists and if it supports the search process [8]. In the literature on Evolutionary Computation, the notion adaptation is interpreted as: "The process of generating a set of behaviours that more closely match or predict a specific environmental regime. An increased ecological-physiological efficiency of an individual relative to others in the population". [22] In FS the term adaptation is understood not only as an adjustment of the behaviour to the different environments or as prediction of the changes of a time-dependent data space.

The attribute adaptive is understood as ability, as a potential, for variations and changes in the behaviour. On conceptual level, Free Search attempts to avoid any rules, acceptances or settings, which may restrict the ability for adaptation.

According to some publications [21] one system could be able to adapt to different environments or to changes of the current environment if it performs prediction, control and feedback and action. FS is based on the concept that if the individuals of a population are set in advance and suffers restrictions to do changes because of prior acceptances or prior knowledge, then precise prediction of the possible changes and a full control on, and feedback from, the environment become meaningless. These individuals would be unable to adapt due to the imposed restrictions. Free Search is not set to perform well on a certain class of problems. It has rich potential for: local search; global exploration; sharp convergence; wide divergence. The behaviour of Free Search is not previously set. The individuals possess these abilities and the algorithm varies their behaviour and exhibits their potential during the optimisation process. FS attempts to adapt to the explored problem. It can be formulated that during the search process the individuals are exploring and learning the optimisation problem and then decide how to behave. Furthermore, the individuals can change their behaviour during this process.

7.2 *Unrestricted Step Size Versus Regulation by Step Size*

The idea for variation of the parameters of evolutionary algorithms has been discussed since earlier stages of iterative computation. For an iterative scheme to succeed, multiple elements had to be altered simultaneously, leading to a combinatorial explosion of possibilities [20, 23].

For global search problems, uncertainty of the behaviour of the individuals establishes two problems—(1) to overcome stagnation of the search process and (2) to cope with a large number of possibilities as result of given probability for change of the individuals behaviour. The idea for alteration of the search parameters affects the step size. Earlier published results indicate that: “Depending upon the step size it can be reached a point of stagnation, which may or may not be near to the solution. For well-conditioned systems and with some knowledge about the problem the method achieves quickly area of the best solution. Badly conditioned system usually gets trapped at a point of stagnation. A change of the step size may result in further progress, but after a while the search gets trapped again”. [23]

Free Search attempts to cope with stagnation or trapping in a local sub-optimum by: elimination of any restrictions of the step size, which is limited by the space borders, only, (Precisely, the step size is not restricted, even, by the space borders. The individuals can step outside of the borders. However, the results achieved outside of the borders are ignored.); use of a modification strategy that highly encourages escaping from trapping; individual decision-making policy, which allows individuals to explore remote locations. In FS the term configuration of the optimisation parameters is understood as a frame for variation of the parameters. In FS parameters are not set to a particular value. They can be changed and adapted flexibly during the process of search within a defined frame.

In PSO increasing the divergence for wide global exploration of the search space requires a large step. Large steps encourage escaping from trapping. However, increasing the convergence speed to an exact optimal location requires small precise steps. Small steps lead to trapping in local sub-optima. This principle for regulation of the convergence and divergence has the following contradictions: Large steps—low probability for trapping but low convergence to the optimum; Small steps—high convergence to the optimum but high probability for trapping.

Free Search uses adaptive self-regulation of the convergence and divergence: Increasing the difference between achieved results increases the number of possible variations but decreases the probability for exploration of locations with a low level of pheromone.

Decreasing the probability to explore locations with a low level of pheromone increases the convergence speed to better locations.

Decreasing difference between achieved results, which indicates a high possibility of trapping, leads to an increasing of the probability for exploration of locations marked with a low level of pheromone. Increasing the probability to explore locations with a low level of pheromone, combined with the ability for large steps, increases the divergence and encourages escaping from trapping in local sub-optima. This manner of regulation is possible, because of the new concept for sense. It has no contradictions and harmonises the behaviour of the individuals:

- A large tolerance of the results and a low probability for trapping—increases the convergence;
- A small tolerance of the results and a high probability for trapping—increases the divergence.

Regulation of the process by sensibility:—allows regulation of the convergence and divergence without determination or discrimination of the search space;—allows use of large and small steps during the whole search process.

In some cases, for example, when the random initial population generates accidentally solution within the area of the global optimum Particle Swarm Optimisation may reach global optimum. Two successful results on optimisation of Sofia test illustrate this case.

7.3 *Creativity Versus Constant Principles*

Creativity is understood as a capability to discover an original successful approach to different and changeable environments, which leads to the ability to adapt and to solve different and time-dependent problems.

It is based on the definition for intelligence and intelligent system: "Intelligence may be defined as the capability of a system to adapt its behaviour to meet its goals in a range of environments. For natural species, survival is a necessary goal in any given or changeable environment. For human created systems, as machines and computers, both goals and environments may be imbued by the machine's creator" [20].

Constant principles are understood as acceptance based principles, which are applicable and valid for a large period and a large number of cases. The applicability for large period leads to the acceptance that these principles are not changeable.

It is accepted that for an infinite period or for an infinite space constant principles do not exist. Acceptance of any principle can be only temporary and must not restrict the capability for adaptation.

This study is based on a clear understanding that all knowledge including all logical and mathematical rules are grounded on acceptances under certain conditions. These rules and that knowledge may not be applicable to uncertain, infinite and unknown problems and environments, where required for the acceptances conditions may not exist.

An infinite variety of real-world problems can be difficult to solve for many reasons such as those detailed below:

"The problem is so complicated that just to facilitate any answer at all, which requires a use of simplified models of the problem that any result is essentially useless" [15]. In the thesis these types of difficulties are defined as: obstacles caused from the continuous search space. The simplified model of the continuity can be considered as a model with a low level of precision, where complex details can be ignored.

"The number of possible solutions in the search space is so large and it forbids an exhaustive search for the best answer" [15]. In the thesis formulation of these types of difficulties is—obstacles caused from the infinite search space.

"The objective function that describes the quality of any proposed solution is noisy or time dependent, thereby requiring not just a single solution but an entire series of solutions" [15]. In the thesis, the description, of such types of difficulties, is—obstacles caused from time-dependent or dynamic search space.

"The possible solutions are so heavily constrained that constructing even one feasible answer is difficult, let alone searching for an optimal solution" [15]. In the thesis these types of difficulties are described as obstacles caused from the constrained search space. "The person solving the problem is inadequately prepared or imagines some psychological barrier that prevents him from discovering a solution" [15]. In the thesis this is described as obstacles caused from the constant principles, un-creativity or inability to adapt.

"Systems often contain fixed conceptual models of the engineering task based on concrete assumptions, which are often impossible to change. Even when adaptive methodologies act on higher-level representations such as in composition modelling, there are a finite number of ways to learn with new information". [21].

Free Search attempts to avoid constant assumptions and principles on conceptual level. FS aims to eliminate restrictions for a number of ways to learn new information and a number of ways to behave. During the process of search FS temporary accepts the way of learning and behaving, however, it is always able to change them. Therefore, Free Search can be applied successfully to infinite, continuous, changeable and constraint space. If the current experience suggests a change of the behaviour FS can change it. The algorithm does not rely on prior constant principles or rules. How it is implemented in Free Search?

In Free Search the concepts for pheromone and sensibility are a dynamic cognitive abstraction of the process of search and of the search space. Before initialisation sensibility is undefined and pheromone does not exist. After the initial exploration walk, the animals assess their-own achievements and indicate with pheromone their-own best-found location. The amount of deposited pheromone is proportional to the quality of the location according to the explored objective function. The amount of pheromone defines a frame of acquired knowledge. The pheromone can be considered as abstracted cognition about the search space. The frame of the knowledge defines the way of learning, which is implemented by the frame of sensibility for each animal. The ability to refine the sense can be considered as ability for learning. The two frames of the pheromone and of the sensibility are equal. So that the full abstracted cognition about the space can be learned. The amount of pheromone depends on the quality of the location according to the objective function. However, the sensibility does not depend upon the quality of a particular location or particular pheromone. The sensibility does not depend on the explored problem and on the objective function, as well. The sense makes the animals independent from the problem. The sense, also, differentiates the animals between each other as different independent individuals. The animals can explore the same problem. They can be in the same location but that is no reason to be equal. That is no reason for all the animals to continue in the same way. The animals in FS are different. They have different sensibility. They can take different, even contrary, decisions for further explorations. Consequently they can proceed in different ways. This is understood as creative, intelligent individualism. That individualism differentiates a team from a flock.

The cognition, which animals abstract from exploration walk has two aspects. The first aspect is acquaintance of the search space. Selected locations are marked and can be used for further exploration. The search space is not unknown any more after the initial exploration. The second aspect is improvement of their-own skills. Each animal refines and perfects its own level of sensibility. So that it can be better prepared to select the appropriate location marked with pheromone, which is most suitable for its sensibility. Sensibility varies for a variety of animals. They can have either different or equal sensibility. The pheromone amount varies for a variety of selected locations. The locations can be marked either with different with or equal amounts of pheromone. A relation of these two variables is an act of selection.

Selection of further directions for search can be classified as a decision-making event. The event of decision-making is, in fact, an act of assessment of the individuals' knowledge and skills. If the animals' sensibility is appropriate for selection of locations, which lead to the desired optimum, it follows, that the animals have good way of learning, good cognition and successful behaviour. If the animals cannot select a successful location it follows that the way of learning can be improved.

A creation of the sensibility can be considered as a high-level cognitive abstraction about the explored problem, in comparison to memorizing of the explored path, marked with pheromone, which can be considered as a data level abstraction. This is a conceptual difference with ant algorithms.

Abstraction, as a form of cognition, based on separation in thought of essential characteristics and relationships, it is one of the fundamental ways that humans can cope with complexity. In real life high-level abstractions can cover a large amount of data and information. Operation with large sets of information leads to high-level thinking and behaviour. Respectively, in computational modelling, high-level abstractions can lead to high performance and high quality results, which the experiments presented in this article confirm.

8 Conclusion

This article presents an evaluation of FS and PSO on global and multidimensional optimisation numerical tests. New and modified test together with experimental methodology are presented. From the achieved results, it can be concluded that the implemented in unified Black Box model [8] tests and simple multi-start experimental methodology are suitable to measure and assess:—search algorithms performance;—probability for success of explored methods;—abilities of explored methods for entire search space coverage;—dependence on initialisation; abilities of explored methods to escape from trapping in local sub-optima;—probability for success on global optimisation of other real-value coded methods.

Experimental results could be summarised as follows: FS demonstrates good overall performance and abilities to escape from trapping on both two and hundred dimensional tests. With high probability PSO cannot escape from trapping. In particular on two-dimensional Sofia tests for start from extra hard single location FS still can resolve the test within 2000 iterations with probability 13 %. Due to its specific modification strategy, which requires difference between individuals, PSO cannot start and proceed from single location. Experiments confirm previous investigation [8] and experiments on one hundred dimensions extend the knowledge about FS abilities to avoid stagnation in sub-optimal solutions and to resolve with high probability global multidimensional optimisation test. FS is able to explore entire search space with non-zero probability for access to any arbitrary location. FS is independent on initialisation. PSO depends on initialisation. PSO could be very fast if some of the initial locations are close to the global optimum. FS could be reliable for real-world global multidimensional optimisation.

Future research could focus on evaluation of other methods and test and application to real-world problems in communications, science and industry.

Acknowledgements Preparation of this article is supported by Southampton Solent University Research & Enterprise Fund, grant 516/17062011.

I would like to thank also to my students Asim Al Nashwan, Dimitrios Kalfas, Georgius Haritonidis and Michael Borg for the design, implementation and overclocking of desktop PC used for completion of the experiments presented in this article.

References

1. Weisstein, E.: Global Optimisation, WolframMathWorld. http://mathworld.wolfram.com/GlobalOptimisation.html. Accessed 27 Sept 2013
2. Hedar, A.-R.: Global Optimisation, Global Optimisation Methods and Codes, Kyoto University, Japan. http://www-optima.amp.i.kyoto-u.ac.jp/member/student/hedar/Hedar_files/go.htm. Accessed 27 Sept 2013
3. Holland, J.: Adaptation in Natural and Artificial Systems. University of Michigan Press (1975)
4. Eberhart, R., Kennedy, J.: Particle swarm optimisation. In: Proceedings of the 1995 IEEE International Conference on Neural Networks, vol. 4, pp. 1942–1948. IEEE Press (1995)
5. Shi, Y., Eberhart, R.: A modified PSO optimizer. In: Proceedings of the IEEE, International Conference on Evolutionary Computation, pp. pp. 69–73. IEEE Press, Piscataway (1998)
6. Rechenberg, I.: The evolutionary strategy: a mathematical model of Darwinian evolution. In: Frehlend, E. (ed.) Synergetics: from Microscopic to Macroscopic Order. Springer Series in Synergetics, vol. 22, pp. 122–132 (1984)

7. Price, K., Storn, R.: Differential Evolution—A Simple and Efficient Adaptive Scheme for Global Optimisation over Continuous Spaces. TR-95-012, International Computer Science Institute, 1947 Center Street, Berkeley, CA 94704-1198, Suite 600 (1995)
8. Penev, K.: Free Search of Real Value or How to Make Computers Think, Gegov, A. (ed.). St. Qu pbl., ISBN 978-0955894800, UK (2008)
9. Liang, J.J., Suganthan, P.N., Deb, K.: Novel composition test functions for numerical global optimisation. In: Proceedings of Swarm Intelligence Symposium, 2005 pp. 68–75 (2005)
10. Gabrys, B., Leiviska, K., Strackeljan, J. (eds.): Do smart adaptive systems exist?—Best Practice for Selection and Combination of Intelligent Methods. Springer series on Studies in Fuzziness and Soft Computing. vol. 173. Springer (2005)
11. Angeline, P.: Evolutionary optimisation versus particle swarm optimisation: philosophy and performance difference. In: The 7th Annual Conference on Evolutionary Programming, San Diego, USA (1998)
12. Igel, C., Toussaint, M.: A no-free-lunch theorem for non-uniform distribution of target functions. J. Math. Modell. Algorithms **3**, 313–322 (2004) (© 2004 Kluwer Publications)
13. Wolpert, D.H., Macready, W.G.: No free lunch theorems for optimisation. IEEE Trans. Evol. Comput. **1**(1), 67–82 (1997)
14. Keane, A.J.: A brief comparison of some evolutionary optimisation methods. In: Rayward-Smith, V., Osman, I., Reeves, C., Smith, G.D. (eds.) Modern Heuristic Search Methods. Wiley, New York, pp 255–272 (1996). ISBN 0471962805
15. Michalewicz, Z., Fogel, D.: How to Solve It: Modern Heuristics. Springer (2002). ISBN 3-540-66061-5
16. Penev, K., Ruzhekov, A.: Adaptive intelligence applied to numerical optimisation. In: Numerical Methods and Applications. Lecture Notes in Computer Science, vol. 6046, issue no. 1, pp. 280–288. Springer (2010)
17. Brekke, E.F.: Complex Behaviour in Dynamical Systems, The Norwegian University of Science and Technology (2005). http://www.academia.edu/545835/COMPLEX_BEHAVIOR_IN_DYNAMICAL_SYSTEMS pp. 37–38. Accessed 27 Sept 2013
18. Penev, K.: Free Search—a model of adaptive intelligence. In: Proceedings of the 2009 International Conference on Adaptive and Intelligent Systems. IEEE Computer Society Conference Publishing Service, pp. 92–97 (2009). ISBN 278-0-7695-3827-3/09
19. Penev, K.: Free search in tracking time dependent optima. In: Advanced Topics on Evolutionary Computing. WSEAS Press, Bulgaria, pp. 127–132 (2008). ISBN 978 960 6766 58 9
20. Fogel, D.B.: Evolutionary Computation—Toward a New Philosophy of Machine Intelligence. IEEE Press (2000). ISBN 0-7803-5379-X
21. Smith, I.F.C.: Mixing Maths with Cases. In: Parmee, I. (ed.) Adaptive Computing in Design and Manufacturing, UK, pp. 13–21 (2004)
22. Eiben, A.E., Smith, J.E.: Introduction to Evolutionary Computing, pp. 15–35. Springer (2003). ISBN 3-540-40184-9
23. Bremermann, H.: Optimisation through evolution and recombination. In: Yovits, M., Jacobi, G., Goldstein, G. (ed.) Self-Organizing Systems, pp. 93–106. Spartan Books, New York (1962)

Intuitionistic Fuzzy Sets Generated by Archimedean Metrics and Ultrametrics

Peter Vassilev

Abstract For a nonempty universe E it is shown that the standard intutitionistic fuzzy sets (*IFSs*) over E are generated by Manhattan metric. For several other types of intuitionistic fuzzy sets the metrics, generating them, are found. As a result a general metric approach is developed. For a given abstract metric d, the corresponding objects are called d-intuitionistic fuzzy sets. Special attention is given to the case when d is a metric generated by a subnorm. If d is generated by an absolute normalized norm (the Archimedean case), an important result is established: the class of all d-intuitionistic fuzzy sets over E is isomorphic (in the sense of bijection) to the class of all *IFSs* over E. In § 4, instead of $\mathbb{R}^2$, the Cartesian product $\mathbb{Q}^2$, of the rational number field $\mathbb{Q}$ with itself, is considered. It is shown that $\mathbb{Q}^2$ may be transformed in infinitely many ways (depending on family of primes p) into a field with non-Archimedean field norm Φ_p generated by p-adic norm. Using the corresponding ultrametric d_{Φ_p} on $\mathbb{Q}^2$, objects called d_{Φ_p}-intuitionistic fuzzy sets over E are defined (the non-Archimedean case). Thus, for the first time intuitionistic fuzzy sets depending on ultrametric are introduced.

1 Preliminary Definitions

Below we remind some important definitions to be used in the investigation.

Definition 1 Let M be a nonempty set. A map $d : M \times M \to [0, +\infty)$ is called metric (or distance) on M (M-metric) if the[1] following conditions are fulfilled:

(1) For $x, y \in M$ $d(x, y) = 0$ iff $x = y$ (coincidence axiom);
(2) For $x, y \in M$ $d(x, y) = d(y, x)$ (symmetry);

[1]Here and further iff means "if and only if".

P. Vassilev (✉)
Bioinformatics and Mathematical Modelling Department,
Institute of Biophysics and Biomedical Engineering,
Bulgarian Academy of Sciences, 105 Acad. G. Bonchev Str., Sofia 1113, Bulgaria
e-mail: peter.vassilev@gmail.com

V. Sgurev et al. (eds.), *Recent Contributions in Intelligent Systems*,
Studies in Computational Intelligence 657,
DOI 10.1007/978-3-319-41438-6_19

(3) For $x, y, z \in M$ $d(x, y) \leq d(x, z) + d(z, y)$ (triangle inequality).

The ordered couple (M, d) is called metric space.

Definition 2 *M*-metric *d* is said to be ultrametric (non-Archimedean metric, if 3) is replaced by the stronger condition

(4) For $x, y, z \in M$ $d(x, y) \leq \max(d(x, z), d(z, y))$.

If *d* is not an ultrametric, then *d* is called Archimedean metric.

Definition 3 Let *F* be Abelian group with group operation $\oplus$ and zero element 0_F. We call a map $\varphi : F \to [0, +\infty)$ *subnorm*[2] on *F* (*F*-subnorm) if the following three conditions are fulfilled:

(5) For $x \in F$ $\varphi(x) = 0$ iff $x = 0_F$;
(6) For $x \in F$ $\varphi(x) = \varphi(-x)$ (where $-x$ is the inverse of *x* with respect to $\oplus$);
(7) For $x, y \in F$ $\varphi(x \oplus y) \leq \varphi(x) + \varphi(y)$.

Definition 4 *F*-subnorm φ is said to be non-Archimedean if (7) is replaced by the stronger condition

(8) For $x, y \in F$ $\varphi(x \oplus y) \leq \max(\varphi(x), \varphi(y))$.

If φ is not non-Archimedean subnorm, then φ is called Archimedean subnorm.

Any *F*-subnorm φ generates *F*-metric $d = d_\varphi$ that for $x, y \in F$ is given by

$$d_\varphi(x, y) = \varphi(x \oplus (-y)), \tag{1}$$

where $-y$ denotes the inverse element of *y* (with respect to operation $\oplus$) in *F*.

Thus, φ transforms the Abelian group *F* into a metric space (F, d_φ).

Using Definition 3, the notions norm on vector space and norm on field may be easily defined. Namely, let *F* be a vector space over the number field *K*. Then *F* is Abelian group with respect to an additive operation $\oplus$ and a second operation $\odot$ (multiplication between the elements of *F* and *K*) is introduced, satisfying the basic axioms for the vector space.

Definition 5 Let *F* be a vector space over *K* with additive operation $\oplus$ and multiplicative operation $\odot$. An *F*-subnorm φ is said to be norm (vector norm) if the condition:

$$\varphi(\lambda \odot x) = |\lambda| \varphi(x) \tag{2}$$

for $x \in F$ and $\lambda \in K$ is satisfied.

Let *F* be a field with additive operation $\oplus$ and multiplicative operation $\odot$. Then *F* is Abelian group with respect to $\oplus$.

---

[2] Some authors use the term "group norm" [8, p. 89], [5].

Definition 6 F-subnorm φ is said to be norm (field norm) if for $x, y \in F$

$$\varphi(x \odot y) = \varphi(x)\varphi(y) \,. \tag{3}$$

Remark 1 A vector (or a field) F-norm φ is called Archimedean or non-Archimedean if φ is Archimedean or non-Archimedean, considered as a subnorm.

It is clear that every vector norm φ on a vector space F transforms F into a metric space (F, d_φ), because of (1). The same is true when φ is a field norm on F. The corresponding metric d_φ is Archimedean or non-Archimedean if the subnorm φ is Archimedean or non-Archimedean, respectively.

Let for $n \geq 1$ $\mathbb{R}^n$ is the set of all n-tuples with real components and $\mathbb{R}^n_+$ is the set of all n-tuples with real nonnegative components. $\mathbb{R} \stackrel{\text{def}}{=} \mathbb{R}^1$ is the real number field. Below we give examples of vector norms on $F = \mathbb{R}^n$, when $\mathbb{R}^n$ is considered as a vector space over $K = \mathbb{R}$ with additive and multiplicative operations:

$$x \oplus y = (x_1 + y_1, \dots, x_n + y_n);\ \lambda \odot x = (\lambda x_1, \dots, \lambda x_n) \tag{4}$$

with: $x = (x_1, \dots, x_n)$, $y = (y_1, \dots, y_n)$, $\lambda \in K$.

Example 1 For a fixed real $p \in [1, +\infty)$ the $\mathbb{R}^n$-norm $\varphi_n^{(p)}$ is introduced by

$$\varphi_n^{(p)}(x) = \left(\sum_{i=1}^{n} |x_i|^p \right)^{\frac{1}{p}}, \tag{5}$$

for $x = (x_1, \dots, x_n)$.

Let us note (see [12, 13]) that the norm $\varphi_n^{(1)}$ is called Manhattan norm (sometimes, wrongly,[3] Hamming norm—see [8]) and the corresponding distance $d_{\varphi_n^{(1)}}$ is called Manhattan metric, Manhattan distance (sometimes, wrongly, Hamming distance), taxicab metric or rectilinear distance. The norm $\varphi_n^{(2)}$ is called Euclidean norm and the corresponding metric $d_{\varphi_n^{(2)}}$ is the widely used Euclidean metric.

Example 2 The limit case $p = \infty$ (in (5)) yields

$$\varphi_n^{(\infty)}(x) \stackrel{\text{def}}{=} \lim_{p \to \infty} \varphi_n^{(p)}(x) = \max(|x_1|, \dots, |x_n|) \,.$$

The last norm is known as Chebyshev norm, supremum norm, uniform norm, or infinity norm and the corresponding metric as Chebyshev distance, maximum metric (for $n = 2$ also chessboard distance) [9, p. 143].

[3]Manhattan norm must be called Hamming norm only when the components of x are binary.

Since all $\mathbb{R}^n$-norms are equivalent to Euclidean norm, which is Archimedean, any $\mathbb{R}^n$-norm is Archimedean too. Therefore, the above norms $\varphi = \varphi_n^{(p)}$ are also Archimedean. Moreover, they are absolute norms, i.e., satisfying the condition

$$\varphi(x) = \varphi(x^*) , \tag{6}$$

where $x^* \stackrel{\text{def}}{=} (|x_1|, \dots, |x_n|)$.

The norms $\varphi_n^{(p)}$ are also known as: $p, L^p, L_p, \ell^p, \ell_p$-norms or as Minkowski norms, since for them (7) represents the well-known Minkowski inequality (cf. [7, p. 189, Theorem 9]).

It is necessary to note that for $p < 1$, $\varphi_n^{(p)}$ is not a norm, since in this case Minkowksi's inequality is violated. This is the reason to consider subnorms, instead of norms, to make it possible for p to take values in the entire interval $(0, +\infty]$.

Definition 7 For a fixed real $p \in (0, +\infty]$ the Archimedean $\mathbb{R}^n$-subnorm $\tilde{\varphi}_n^{(p)}$ is introduced by

$$\tilde{\varphi}_n^{(p)}(x) = \begin{cases} \sum_{i=1}^{n} |x_i|^p, \text{ if } p \in (0, 1) \\ \varphi_n^{(p)}(x), \text{ if } p \in [1, +\infty] , \end{cases} \tag{7}$$

with $x = (x_1, \dots, x_n)$.

Although for $p \in (0, 1)$ $\tilde{\varphi}_n^{(p)}$ satisfies (6) and Minkowski inequality, it is not norm ((2) is violated). But for $p \in (0, +\infty]$ the metric space $\left(\mathbb{R}^n, d_{\tilde{\varphi}_n^{(p)}}\right)$ exists due to (1).

2 Introduction

In 1965, with his groundbreaking paper Zadeh [21], generalized the idea of characteristic function of a set introducing the notions membership function and fuzzy set (*FS*). This started a new direction of research, now widely known as Fuzzy set Theory. In 1983, with his pioneering report, Atanassov [1] generalized the concept of *FS* by defining the intuitionistic fuzzy sets (*IFSs*) over a nonempty universe E. The crucial idea was the introduction of a second generalized characteristic function, called non-membership function, related to the complement of a set $A \subset E$ to E. Thus the theory of *IFS* appeared. The concept of *IFS* contains the concept of *FS* as a particular case. Following the numerous publications of K. Atanassov on *IFSs*, addressing different areas of application and his two monographs devoted to *IFS* from 1999 [3] and 2012 [4], recently many authors, recognizing his enormous contributions to the field, have begun to refer to *IFSs* as Atanassov sets or Atanassov's intuitionistic fuzzy sets.

For $A \subset E$ and mappings $\mu_A : E \to [0, 1]$, $\nu_A : E \to [0, 1]$, such that

$$\mu_A(x) + \nu_A(x) \leq 1 \, (x \in E) \, , \tag{8}$$

Atanassov introduced the object called intuitionistic fuzzy set $\tilde{A}$ as

$$\tilde{A} = \left\{ \langle x, \mu_A(x), \nu_A(x) \rangle \, | x \in E \right\} \, .$$

He named the mappings μ_A and ν_A membership and non-membership functions and the map $\pi_A : E \to [0, 1]$, that is given by

$$\pi_A(x) = 1 - \mu_A(x) - \nu_A(x) \, (x \in E) \, ,$$

is called hesitancy function. So

$$\mu_A(x) + \nu_A(x) + \pi_A(x) = 1 \, (x \in E)$$

and the numbers $\mu_A(x), \nu_A(x), \pi_A(x)$ are said to be degree of membership, degree of non-membership and hesitancy degree/margin of the element $x \in E$ to the set A.

Further we shall denote by IFS(E) the class of all *IFSs* over E.

Later Atanassov in [2] considered (besides the mentioned above standard *IFSs*) other types of intuitionistic fuzzy sets for which the condition (8) is replaced by

$$\mu_A^2(x) + \nu_A^2(x) \leq 1 \, (x \in E) \, . \tag{9}$$

He called these sets 2-intuitionistic fuzzy sets (or briefly 2-*IFSs*).

More generally, for a fixed real $p \in (0, +\infty)$, it is possible to consider intuitionistic fuzzy sets for which condition (8) is replaced by

$$\mu_A^p(x) + \nu_A^p(x) \leq 1 \, (x \in E) \, . \tag{10}$$

By analogy these sets are called p-intuitionistic fuzzy sets (or briefly *p-IFSs*) and for fixed p the class of all *p-IFSs* over E is denoted by p-IFS(E). Such kind of sets have been studied by several authors. For example see [14, 17].

All these attempts to extend and generalize the notion *intuitionistic fuzzy set* are interesting and admit appropriate applications, but unfortunately share one and the same disadvantage—they are not a part of a general scheme that is able to explain what is common between all of them. We observe that conditions (8)–(10) may be rewritten as follows:

$$\varphi_2^{(1)}((\mu_A(x), \nu_A(x))) \leq 1 \, (x \in E) \, ; \tag{11}$$

$$\varphi_2^{(2)}((\mu_A(x), \nu_A(x))) \leq 1 \, (x \in E) \, ; \tag{12}$$

$$\tilde{\varphi}_2^{(p)}((\mu_A(x), \nu_A(x))) \leq 1 \, (x \in E) \, . \tag{13}$$

In this form they are obviously connected with the corresponding metrics in $\mathbb{R}^2$. For (11) it is the Manhattan metric, generated by the Manhattan norm $\varphi_2^{(1)}$. For (12) it is the Euclidean metric, generated by the Euclidean norm $\varphi_2^{(2)}$. And for (13) we have two cases. When $p \geq 1$, it is the L^p metric, that is generated by $\varphi_2^{(p)}$ norms. And for p satisfying $0 < p < 1$, it is the metric that is generated by the subnorm $\tilde{\varphi}_2^{(p)}$ from (7).

The mentioned above suggests the idea for a unified metric approach to the notion *intuitionistic fuzzy set*. Under this approach, the standard *IFSs* (Atanassov sets) correspond to Manhattan metric; 2-*IFSs* correspond to Euclidean metric; and generally *p-IFSs* correspond to the metric generated by the subnorm $\tilde{\varphi}_2^{(p)}$ (which is a norm for the mentioned cases $p = 1$ and $p = 2$ and moreover only for real $p \geq 1$).

3 Intuitionistic Fuzzy Sets Depending on $\mathbb{R}^2$-Metrics. Abstract Approach

Definition 8 An $\mathbb{R}^2$-metric d is called normalized if

$$d((1,0),(0,0)) = d((0,1),(0,0)) = 1 \, .$$

Definition 9 An $\mathbb{R}^2$-subnorm (norm) φ is called normalized if

$$\varphi((1,0)) = \varphi((0,1)) = 1 \, .$$

The class of all normalized $\mathbb{R}^2$-norms is denoted by N_2 and the class of all absolute normalized $\mathbb{R}^2$-norms is denoted by AN_2.

We note that $AN_2 \subset N_2$ and the inclusion is strict, since the $\mathbb{R}^2$-norm

$$\varphi((u,v)) \stackrel{\text{def}}{=} \sup_{t \in [0,1]} |u - t\,v|$$

satisfies $\varphi \in N_2$, but $\varphi \notin AN_2$ because $\varphi((1,-1)) = 2 \neq 1 = \varphi((|1|, |-1|))$.

It is clear that any normalized subnorm (norm) φ generates a normalized metric $d = d_\varphi$, given by (1), where $\oplus$ is given by (4) with $n = 2$. When φ is a norm, the corresponding metric d_φ is an Archimedean metric.

Definition 10 Let E be a universe, $A \subset E$, $\mu_A : E \to [0,1]$, $\nu_A : E \to [0,1]$ are mappings and d is a normalized metric. We say that the set

$$\tilde{A} = \{\langle x, \mu_A(x), \nu_A(x)\rangle | x \in E\}$$

is d-intuitionistic fuzzy set (briefly d-*IFS*[4]) over E generated by A (through the metric d), if the relation

$$d((\mu_A(x), \nu_A(x)), (0,0)) \leq 1 \,(x \in E) \tag{14}$$

holds. The mappings μ_A and ν_A are called membership and non-membership functions. The mapping $\pi_A : E \to [0,1]$, given by

$$\pi_A(x) \stackrel{\text{def}}{=} 1 - d((\mu_A(x), \nu_A(x)), (0,0)), \tag{15}$$

is called hesitancy function.

For $x \in E$ the numbers $\mu_A(x)$, $\nu_A(x)$ and $\pi_A(x)$ are called degree of membership, degree of non-membership and hesitancy degree/margin of the element x to the set A. The class of all d-*IFSs* over E is denoted by d-IFS(E).

Let B_d be the closed disc in $\mathbb{R}^2$ (centered at $(0,0)$) with respect to the metric d, i.e.,

$$B_d = \{(u,v) | (u,v) \in \mathbb{R}^2 \;\&\; d((u,v),(0,0)) \leq 1\}.$$

Further ∂B_d denotes the contour of B_d.

Definition 11 The set

$$K_d \stackrel{\text{def}}{=} B_d \cap \mathbb{R}^2_+$$

is called interpretation domain for the class d-IFS(E).

Let φ be a normalized subnorm and $d = d_\varphi$ is defined by (1). Then (14) becomes

$$\varphi((\mu_A(x), \nu_A(x))) \leq 1$$

and (15) may be rewritten as:

$$\pi_A(x) = 1 - \varphi((\mu_A(x), \nu_A(x))).$$

When $d = d_\varphi$, with $\varphi \in N_2$, the sets B_{d_φ} and K_{d_φ} are always convex but for an arbitrary normalized subnorm φ the convexity of B_{d_φ} and K_{d_φ} is not guaranteed.

In particular $K_{d_{\varphi_2^{(1)}}}$ is the so-called interpretation triangle (with vertexes $(0,0)$, $(1,0)$, $(0,1)$) for the standard *IFSs*. The set $K_{d_{\varphi_2^{(2)}}}$ is a quarter of the Euclidean unit disc in $\mathbb{R}^2_+$ (centered at $(0,0)$). It is the interpretation domain for the 2-*IFSs*. The set $K_{d_{\varphi_2^{(\infty)}}}$ coincides with the Cartesian product $[0,1] \times [0,1]$ and it is the interpretation domain for the class $d_{\varphi_2^{(\infty)}}$-IFS(E).

[4]A more precise denotation would be $^{(\mathbb{R}^2)}d$-*IFS* but we will omit it since there is no danger of misunderstanding.

The well-known facts: $\varphi \in N_2$ implies $\varphi \le \varphi_2^{(1)}$; $\varphi \in AN_2$ implies $\varphi_2^{(\infty)} \le \varphi$, yield

Proposition 1 *Let $\varphi', \varphi'' \in AN_2$ and $\varphi_2^{(\infty)} < \varphi'' < \varphi' < \varphi_2^{(1)}$. Then for the convex sets*

$$B_{d_{\varphi_2^{(\infty)}}}, B_{d_{\varphi''}}, B_{d_{\varphi'}}, B_{d_{\varphi_2^{(1)}}}; K_{d_{\varphi_2^{(\infty)}}}, K_{d_{\varphi}}, K_{d_{\varphi'}}, K_{d_{\varphi_2^{(1)}}}$$

the strict inclusions hold

$$B_{d_{\varphi_2^{(1)}}} \subset B_{d_{\varphi'}} \subset B_{d_{\varphi''}} \subset B_{d_{\varphi_2^{(\infty)}}}; K_{d_{\varphi_2^{(1)}}} \subset K_{d_{\varphi'}} \subset B_{d_{\varphi''}} \subset K_{d_{\varphi_2^{(\infty)}}} = [0,1] \times [0,1] .$$

If $\varphi \in N_2$ the curve $\delta_{d_\varphi} \stackrel{\text{def}}{=} \partial B_{d_\varphi} \cap \mathbb{R}_+^2$ is a continuous (because of the continuity of the $\mathbb{R}^2$-norms) and concave function passing through the points $(1,0)$ and $(0,1)$ and if $\varphi \in AN_2$ this curve is contained in $[0,1] \times [0,1]$.

Let $\tilde{\varphi}_2^{(p)}$ be the subnorm introduced with (7) (when $n = 2$). Then the following assertion is true.

Proposition 2 *Let p run over $(0, +\infty]$. Then $K_{d_{\tilde{\varphi}_2^{(p)}}}$ are closed sets in the topology of $\mathbb{R}^2$. They grow with p and tend to $K_{d_{\varphi_2^{(\infty)}}} = [0,1] \times [0,1]$.*

For $p \ge 1$ these sets are convex, coincide with $K_{d_{\varphi_2^{(p)}}}$ (in particular, the curves $\delta_{d_{\tilde{\varphi}_2^{(p)}}} \stackrel{\text{def}}{=} \partial B_{d_{\tilde{\varphi}_2^{(p)}}} \cap \mathbb{R}_+^2$ coincide with $\delta_{d_{\varphi_2^{(p)}}}$) and contain the interpretation triangle $K_{d_{\varphi_2^{(1)}}}$ for the standard IFSs, tending to it when p tends to $1 + 0$.

For $0 < p < 1$, the sets $K_{d_{\tilde{\varphi}_2^{(p)}}}$ are not convex. They are contained in $K_{d_{\varphi_2^{(1)}}}$, tending to it when p tends to $1 - 0$.

Corollary 1 *Let $p, q \in (0, +\infty]$ and $p < q$. Then $d_{\varphi_2^{(p)}}$-IFS$(E) \subset d_{\varphi_2^{(q)}}$-IFS$(E)$ and if $0 < p < q < +\infty$, then p-IFS$(E) \subset q$-IFS(E).*

Some important properties of the class AN_2 are based on

Definition 12 With Ψ_2 is denoted the class of all continuous convex functions ψ in $[0,1]$ which for $t \in [0,1]$ satisfy the condition

$$\max(1-t, t) \le \psi(t) \le 1 .$$

The fundamental result for the class AN_2 is given by Bonsall and Duncan [6, p. 37, Lemma 3]. Below we represent it in the following form

Theorem 1 *There exists a bijection between AN_2 and Ψ_2. Namely, if $\psi \in \Psi_2$, then there exists a unique $\varphi \in AN_2$ such that for $t \in [0,1]$*

$$\psi(t) = \varphi((1-t, t)) \tag{16}$$

and if $\varphi \in AN_2$, *then there exists a unique* $\psi \in \Psi_2$, *such that for* $(\mu, \nu) \in \mathbb{R}^2$

$$\varphi((\mu, \nu)) = \begin{cases} (|\mu| + |\nu|)\,\psi\left(\frac{|\nu|}{|\mu|+|\nu|}\right), & \textit{if } (\mu, \nu) \neq (0, 0) \\ 0, \ \textit{if } (\mu, \nu) = (0, 0) \,. \end{cases} \tag{17}$$

Since for $p \geq 1$ $\varphi_2^{(p)} \in AN_2$, the associated function $\psi = \psi_p \in \Psi_2$ (from Bonsall–Duncan's bijection) is given by

$$\psi_p(t) = ((1-t)^p + t^p)^{\frac{1}{p}} \,.$$

Let p run over $[1, +\infty]$ and L^* be the class of all L^p-norms on $\mathbb{R}^2$. Since for $(\mu, \nu) \in \mathbb{R}^2$

$$\varphi_2^{(p)}((\mu, \nu)) = \varphi_2^{(p)}((\nu, \mu)) \,,$$

we obtain

$$L^* \subset SYMAN_2 \,, \tag{18}$$

where $SYMAN_2$ denotes the class of all symmetric norms $\varphi \in AN_2$, i.e., such that

$$\varphi((\mu, \nu)) = \varphi((\nu, \mu))$$

for all $(\mu, \nu) \in \mathbb{R}^2$.

The inclusion (18) is strict, since if $\lambda \in (0, 1)$, $p, q \geq 1$ and $p \neq q$, then the norm

$$\varphi_{\lambda,p,q} \stackrel{\text{def}}{=} (1-\lambda)\varphi_2^{(p)} + \lambda\varphi_2^{(q)} \,,$$

satisfies $\varphi_{\lambda,p,q} \in SYMAN_2$ and $\varphi_{\lambda,p,q} \notin L^*$. On the other hand, if $\varphi \in SYMAN_2$ and $\psi \in \Psi_2$ is its associate function, then for $t \in [0, 1]$ the relation

$$\psi(1-t) = \psi(t)$$

holds. Hence, the strict inclusion $SYMAN_2 \subset AN_2$ is true. Therefore, the following result is valid:

Proposition 3 *The strict inclusions*

$$L^* \subset SYMAN_2 \subset AN_2 \subset N_2$$

hold.

Our main result in this paragraph is the following theorem, which gives the connection between the standard *IFS* and the introduced here d_φ-*IFS*, when $\varphi \in AN_2$ (here we remind that the metric d_φ is Archimedean)

Theorem 2 *Let E be a universe and $\varphi \in AN_2$ be fixed norm. Then there exists a bijection between the classes d_φ-IFS(E) and IFS(E).*

Proof Let $\psi \in \Psi_2$ be the associated function (from (16)) for φ and T_φ be the mapping which juxtaposes to

$$A \stackrel{\text{def}}{=} \{\langle x, \mu(x), \nu(x)\rangle | x \in E\} \in d_\varphi\text{-IFS}(E)$$

the set

$$B \stackrel{\text{def}}{=} \{\langle x, \mu^*(x), \nu^*(x)\rangle | x \in E\},$$

where

$$\mu^*(x) = \begin{cases} \mu(x)\psi\left(\frac{\nu(x)}{\mu(x)+\nu(x)}\right), & \text{if } \mu(x)+\nu(x) > 0 \\ 0, \text{ if } \mu(x)+\nu(x) = 0 \text{ ;} \end{cases} \tag{19}$$

$$\nu^*(x) = \begin{cases} \nu(x)\psi\left(\frac{\nu(x)}{\mu(x)+\nu(x)}\right), & \text{if } \mu(x)+\nu(x) > 0 \\ 0, \text{ if } \mu(x)+\nu(x) = 0 \text{ .} \end{cases} \tag{20}$$

We will prove that T_φ is a bijection between d_φ-IFS(E) and IFS(E).

First we have to show that $B \in$ IFS(E).

From $A \in d_\varphi$-IFS(E) we obtain

$$\varphi((\mu(x), \nu(x))) \leq 1 \; (x \in E) \text{ .} \tag{21}$$

Now (17), (19) and (20) yield

$$\mu^*(x) + \nu^*(x) = \begin{cases} (\mu(x)+\nu(x))\psi\left(\frac{\nu(x)}{\mu(x)+\nu(x)}\right), & \text{if } \mu(x)+\nu(x) > 0 \\ 0, \text{ if } \mu(x)+\nu(x) = 0 \end{cases} =$$

$$\varphi((\mu(x), \nu(x))) \leq 1 \text{ .}$$

Hence (from (21)) $B \in$ IFS(E).

Second, we have to show that T_φ is injection.

Let us assume the opposite. Then there would exist mappings $\mu_i : E \to [0, 1]$, $\nu_i : E \to [0, 1]$, $i = 1, 2$, such that

$$(\mu_1, \nu_1) \neq (\mu_2, \nu_2) \text{ ;} \tag{22}$$

$$(\mu_1^*, \nu_1^*) = (\mu_2^*, \nu_2^*) \text{ .} \tag{23}$$

Obviously, (22) means that the following condition holds:

(i_1) There exists $x_0 \in E$, such that at least one of the equalities

$$\mu_1(x_0) = \mu_2(x_0);\ \nu_1(x_0) = \nu_2(x_0)$$

is violated.

On the other hand, (23) means that for $x \in E$ it is fulfilled:

$$\mu_1^*(x) = \mu_2^*(x);\ \nu_1^*(x) = \nu_2^*(x)\ .$$

In particular

$$\mu_1^*(x_0) = \mu_2^*(x_0);\ \nu_1^*(x_0) = \nu_2^*(x_0)\ . \tag{24}$$

For x_0 we have

(i_2) At least one of the equalities

$$\mu_1(x_0) + \nu_1(x_0) = 0\ ;\ \mu_2(x_0) + \nu_2(x_0) = 0\ ,$$

is violated.

The assumption that (i_2) is not true, yields

$$\mu_1(x_0) = \nu_1(x_0) = \mu_2(x_0) = \nu_2(x_0) = 0\ ,$$

which contradicts to (i_1).

Therefore, because of (i_2), there are only three possible cases

(A) $\mu_1(x_0) + \nu_1(x_0) > 0\ \&\ \mu_2(x_0) + \nu_2(x_0) = 0$;
(B) $\mu_1(x_0) + \nu_1(x_0) = 0\ \&\ \mu_2(x_0) + \nu_2(x_0) > 0$;
(C) $\mu_1(x_0) + \nu_1(x_0) > 0\ \&\ \mu_2(x_0) + \nu_2(x_0) > 0$.

Let (A) hold. Then

$$\mu_2(x_0) = \nu_2(x_0) = 0\ .$$

From (19) and (20) with: $\mu = \mu_2$; $\mu^* = \mu_2^*$; $\nu = \nu_2$; $\nu^* = \nu_2^*$; $x = x_0$, it follows:

$$\mu_2^*(x_0) = \nu_2^*(x_0) = 0.$$

The last equalities and (24) yield

$$\mu_1^*(x_0) = \mu_2^*(x_0) = 0;\ \nu_1^*(x_0) = \nu_2^*(x_0)\ . \tag{25}$$

Definition 12 provides for $t \in [0,1]$

(i_3) $\psi(t) > 0$.

Putting $\mu = \mu_1$; $\mu^* = \mu_1^*$; $\nu = \nu_1$; $\nu^* = \nu_1^*$; $x = x_0$ (in (19) and (20)), from (25) and (i_3) we obtain

$$\mu_1(x_0) = \nu_1(x_0) = 0 .$$

But the above contradicts to (A).

In the same manner, the case (B) also leads us to contradiction.

Let (C) hold. We put

$$\psi\left(\frac{\nu_1(x_0)}{\mu_1(x_0) + \nu_1(x_0)}\right) = z;\ \psi\left(\frac{\nu_2(x_0)}{\mu_2(x_0) + \nu_2(x_0)}\right) = -w . \tag{26}$$

From (19), for: $\mu = \mu_1$; $\nu = \nu_1$; $\mu^* = \mu_1^*$; $x = x_0$, we obtain

$$\mu_1^*(x_0) = \mu_1(x_0)z \tag{27}$$

and for: $\mu = \mu_2$; $\nu = \nu_2$; $\mu^* = \mu_2^*$; $x = x_0$, we obtain

$$\mu_2^*(x_0) = -\mu_2(x_0)w . \tag{28}$$

From (20), for: $\mu = \mu_1$; $\nu = \nu_1$; $\nu^* = \nu_1^*$; $x = x_0$, we obtain

$$\nu_1^*(x_0) = \nu_1(x_0)z \tag{29}$$

and for: $\mu = \mu_2$; $\nu = \nu_2$; $\nu^* = \nu_2^*$; $x = x_0$, we obtain

$$\nu_2^*(x_0) = -\nu_2(x_0)w . \tag{30}$$

Then, because of (24), we get the linear homogeneous system with unknowns z and w :

$$\begin{cases} \mu_1(x_0)z + \mu_2(x_0)w = 0 \\ \nu_1(x_0)z + \nu_2(x_0)w = 0 . \end{cases}$$

Now (i_3) and (26) imply $z \neq 0$ and $w \neq 0$, i.e. this system has a nontrivial solution. Then, because of the well known result of the linear algebra, the determinant

$$\Delta = \begin{vmatrix} \mu_1(x_0) & \mu_2(x_0) \\ \nu_1(x_0) & \nu_2(x_0) \end{vmatrix}$$

equals to 0.

This means that the vector-columns of Δ are linearly dependent. Then, due to (C), these vectors are different from the zero-vector. Hence, there exists a real number $k \neq 0$, such that

$$\mu_2(x_0) = k\mu_1(x_0);\ \nu_2(x_0) = k\nu_1(x_0) .$$

The last two equalities imply

$$\psi\left(\frac{\nu_2(x_0)}{\mu_2(x_0)+\nu_2(x_0)}\right) = \psi\left(\frac{k\nu_1(x_0)}{k\mu_1(x_0)+k\nu_1(x_0)}\right) = \psi\left(\frac{\nu_1(x_0)}{\mu_1(x_0)+\nu_1(x_0)}\right).$$

The above equalities and (26) yield $-w = z$. Hence, from (27)–(30), we obtain

$$\mu_1^*(x_0) = z\mu_1(x_0);\ \mu_2^*(x_0) = z\mu_2(x_0);\ \nu_1^*(x_0) = z\nu_1(x_0);\ \nu_2^*(x_0) = z\nu_2(x_0) .$$

The last equalities and (24) yield

$$z\mu_1(x_0) = z\mu_2(x_0);\ z\nu_1(x_0) = z\nu_2(x_0) .$$

Hence

$$\mu_1(x_0) = \mu_2(x_0);\ \nu_1(x_0) = \nu_2(x_0) ,$$

since $z \neq 0$. But the last contradicts to (i_1), and therefore to (22).

Thus, we proved that T_φ is injection.

Third, we have to show that T_φ is surjection.

Let $B \stackrel{\text{def}}{=} \{\langle x, \mu^*(x), \nu^*(x)\rangle | x \in E\} \in \text{IFS}(E)$. For any $x \in E$ we put

$$\mu(x) = \begin{cases} \dfrac{\mu^*(x)}{\psi\left(\frac{\nu^*(x)}{\mu^*(x)+\nu^*(x)}\right)}, \text{ if } \mu^*(x)+\nu^*(x) > 0 \\ 0, \text{ if } \mu^*(x)+\nu^*(x) = 0 ; \end{cases} \tag{31}$$

$$\nu(x) = \begin{cases} \dfrac{\nu^*(x)}{\psi\left(\frac{\nu^*(x)}{\mu^*(x)+\nu^*(x)}\right)}, \text{ if } \mu^*(x)+\nu^*(x) > 0 \\ 0, \text{ if } \mu^*(x)+\nu^*(x) = 0 . \end{cases} \tag{32}$$

We will show that

$$\mu : E \to [0, 1];\ \nu : E \to [0, 1] . \tag{33}$$

Let $x \in E$ is such that $\mu^*(x) + \nu^*(x) = 0$. Then (31) and (32) imply: $\mu(x) = 0$ and $\nu(x) = 0$, i.e. $\mu(x), \nu(x) \in [0, 1]$.

Let $x \in E$ is such that $\mu^*(x) + \nu^*(x) > 0$. Then (31) and (32) yield

$$\mu(x) = \frac{\mu^*(x)}{\psi\left(\frac{\nu^*(x)}{\mu^*(x)+\nu^*(x)}\right)}; \; \nu(x) = \frac{\nu^*(x)}{\psi\left(\frac{\nu^*(x)}{\mu^*(x)+\nu^*(x)}\right)} \; . \tag{34}$$

Let us put

$$t = \frac{\nu^*(x)}{\mu^*(x) + \nu^*(x)} \, .$$

Since $\psi \in \Psi_2$, then Definition 12 implies

$$\psi\left(\frac{\nu^*(x)}{\mu^*(x) + \nu^*(x)}\right) = \psi(t) \geq \max(t, 1-t) = \max\left(\frac{\nu^*(x)}{\mu^*(x) + \nu^*(x)}, \frac{\mu^*(x)}{\mu^*(x) + \nu^*(x)}\right).$$

The last and (34) provide that (33) will be proved if the following inequalities hold

$$\mu^*(x) \leq \max\left(\frac{\nu^*(x)}{\mu^*(x) + \nu^*(x)}, \frac{\mu^*(x)}{\mu^*(x) + \nu^*(x)}\right)$$

$$\nu^*(x) \leq \max\left(\frac{\nu^*(x)}{\mu^*(x) + \nu^*(x)}, \frac{\mu^*(x)}{\mu^*(x) + \nu^*(x)}\right).$$

But these inequalities follow from the inequality

$$\mu^*(x) + \nu^*(x) \leq 1 \, , \tag{35}$$

which is true, since $B \in \text{IFS}(E)$.

We will prove that $\mu(x)$ and $\nu(x)$, given by (31) and (32), satisfy (21).

According to (17) we have

$$\varphi(\mu(x), \nu(x)) = \begin{cases} (\mu(x) + \nu(x))\psi\left(\frac{\nu(x)}{\mu(x)+\nu(x)}\right), \text{ if } \mu(x) + \nu(x) > 0 \\ 0, \text{ if } \mu(x) + \nu(x) = 0 \, . \end{cases} \tag{36}$$

Equalities (31), (32) and (i_3) imply that

$$\mu(x) + \nu(x) = 0 \text{ iff } \mu^*(x) + \nu^*(x) = 0.$$

From the last it follows that (36) may be rewritten as

$$\varphi(\mu(x), \nu(x)) = \begin{cases} (\mu(x) + \nu(x))\psi\left(\frac{\nu(x)}{\mu(x)+\nu(x)}\right), \text{ if } \mu^*(x) + \nu^*(x) > 0 \\ 0, \text{ if } \mu^*(x) + \nu^*(x) = 0 \, . \end{cases} \tag{37}$$

Let $x \in E$ is such that $\mu^*(x) + \nu^*(x) = 0$. Then $\mu(x) + \nu(x) = 0$. Hence: $\mu(x) = 0$; $\nu(x) = 0$ and $\varphi(\mu(x), \nu(x)) = 0$, i.e. (21) holds.

Let $x \in E$ is such that $\mu^*(x) + \nu^*(x) > 0$. Then (31), (32) and (37) yield

$$\varphi(\mu(x), \nu(x)) = \frac{\mu^*(x) + \nu^*(x)}{\psi\left(\frac{\nu^*(x)}{\mu^*(x)+\nu^*(x)}\right)} \psi\left(\frac{\nu(x)}{\mu(x) + \nu(x)}\right) . \tag{38}$$

Equalities (31) and (32) imply

$$\psi\left(\frac{\nu(x)}{\mu(x) + \nu(x)}\right) = \psi\left(\frac{\frac{\nu^*(x)}{\psi\left(\frac{\nu^*(x)}{\mu^*(x) + \nu^*(x)}\right)}}{\frac{\mu^*(x)}{\psi\left(\frac{\nu^*(x)}{\mu^*(x) + \nu^*(x)}\right)} + \frac{\nu^*(x)}{\psi\left(\frac{\nu^*(x)}{\mu^*(x) + \nu^*(x)}\right)}}\right) .$$

Hence (because of (i_3))

$$\psi\left(\frac{\nu(x)}{\mu(x) + \nu(x)}\right) = \psi\left(\frac{\nu^*(x)}{\mu^*(x) + \nu^*(x)}\right) . \tag{39}$$

Equalities (38) and (39) yield

$$\varphi(\mu(x), \nu(x)) = \mu^*(x) + \nu^*(x) .$$

The last equality and (35) immediately prove (21).

Let $A \overset{\text{def}}{=} \{\langle x, \mu(x), \nu(x)\rangle | x \in E\}$. From the proved (21) and (33) it follows: $A \in d_\varphi$-IFS(E).

Equalities (19), (20) and (39) immediately yield

$$T_\varphi(A) = B.$$

Hence T_φ is surjection. Therefore T_φ is a bijection.

Theorem 2 is proved. □

Remark 2 From the proof of Theorem 2 it is seen that T_φ is an injection also for the case: $\varphi \in N_2 \setminus AN_2$. But in this case it is not guaranteed that T_φ is surjection. The last means that for $\varphi \in N_2 \setminus AN_2$ it is not true (in the general case) that T_φ is a bijection.

Let T_φ^{-1} denote the inverse mapping of T_φ. From the proof of Theorem 2 we obtain the following

Corollary 2 *The mappings T_φ and T_φ^{-1} admit the representations*

$$T_\varphi\langle\mu(x),\nu(x)\rangle = \begin{cases} \langle\frac{\mu(x)\varphi((\mu(x),\nu(x)))}{\mu(x)+\nu(x)}, \frac{\nu(x)\varphi((\mu(x),\nu(x)))}{\mu(x)+\nu(x)}\rangle, \text{ if } \mu(x)+\nu(x)>0 \\ \langle 0,0\rangle, \text{ if } \mu(x)=\nu(x)=0\,, \end{cases}$$

*where μ and ν are the membership and non-membership functions of an element from the class d_φ-*IFS*(E);*

$$T_\varphi^{-1}\langle\mu(x),\nu(x)\rangle = \begin{cases} \langle\mu(x)\frac{\mu(x)+\nu(x)}{\varphi((\mu(x),\nu(x)))}, \nu(x)\frac{\mu(x)+\nu(x)}{\varphi((\mu(x),\nu(x)))}\rangle, \text{ if } \mu(x)+\nu(x)>0 \\ \langle 0,0\rangle, \text{ if } \mu(x)=\nu(x)=0\,, \end{cases}$$

where μ and ν are the membership and non-membership functions of an element from the class IFS*(E).*

Another Corollary from Theorem 2 is

Theorem 3 *Let $\varphi',\varphi''\in AN_2$. Then the mapping $T_{\varphi',\varphi''}: d_{\varphi'}$-IFS$(E)\to d_{\varphi''}$-IFS$(E)$, given by*

$$T_{\varphi',\varphi''} \stackrel{\text{def}}{=} T_{\varphi''}^{-1}T_{\varphi'}\,,$$

is a bijection between $d_{\varphi'}$-IFS(E) and $d_{\varphi''}$-IFS(E).

Corollary 3 *Let $p\in[1,+\infty]$. Then the mapping $T_p: d_{\varphi_2^{(p)}}$-IFS$(E)\to$ IFS(E), that for $\tilde{A}=\{\langle x,\mu(x),\nu(x)\rangle | x\in E\}\in d_{\varphi_2^{(p)}}$-IFS$(E)$ is given by*

$$T_p\langle\mu(x),\nu(x)\rangle = \begin{cases} \langle\mu(x)\frac{(\mu^p(x)+\nu^p(x))^{\frac{1}{p}}}{\mu(x)+\nu(x)}, \nu(x)\frac{(\mu^p(x)+\nu^p(x))^{\frac{1}{p}}}{\mu(x)+\nu(x)}\rangle, \text{ if } \mu(x)+\nu(x)>0 \\ \langle 0,0\rangle, \text{ if } \mu(x)=\nu(x)=0\,, \end{cases}$$

is a bijection between $d_{\varphi_2^{(p)}}$-IFS(E) and IFS(E) and for $p\neq+\infty$ T_p is a bijection between p-IFS(E) and IFS(E).

Corollary 4 *Let $p,q\in[1,+\infty]$ and $T_{p,q}\stackrel{\text{def}}{=}T_q^{-1}T_p$. Then $T_{p,q}$ is a bijection between $d_{\varphi_2^{(p)}}$-IFS(E) and $d_{\varphi_2^{(q)}}$-IFS(E) and when $p,q\neq+\infty$ $T_{p,q}$ is a bijection between p-IFS(E) and q-IFS(E).*

The validity of the following proposition is a matter of direct check.

Proposition 4 (see [15]) *Let $p, q \in (0, +\infty)$. Then there exists a bijection $T^*_{p,q}$ between p-IFS(E) and q-IFS(E), that for $\tilde{A} = \{\langle x, \mu(x), \nu(x)\rangle | x \in E\} \in p$-IFS($E$) is given by*

$$T^*_{p,q}(A) \stackrel{\text{def}}{=} \{\langle x, \mu(x)^{\frac{p}{q}}, \nu(x)^{\frac{p}{q}}\rangle | x \in E\} .$$

All preceding related to d-IFS(E), represents the entire apparatus of d-intuitionistic fuzzy sets, considered for an abstract normalized metric d and especially for the case when $d = d_\varphi$ is generated by a normalized subnorm (norm) φ. The corresponding *d-IFS* operators are considered and studied by us in [19].

4 Intuitionistic Fuzzy Sets Based on $\mathbb{Q}^2$-Ultrametrics

Let E be a nonempty universe, $\mathbb{Q}$ be the rational number field and $\mathbb{Q}^2$ be the Cartesian product $\mathbb{Q} \times \mathbb{Q}$. In the preceding paragraph we discussed the class d-IFS(E). The elements of this class are the so-called d-intuitionistic fuzzy sets (*d-IFS*) over E. They are generated by an arbitrary normalized $\mathbb{R}^2$-metric d. In Definition 10 there is no restriction for d to be Archimedean metric. Therefore, it is possible, to consider the class d-IFS(E) for a normalized ultrametrics d too. Unfortunately, this possibility does not exist for metrics $d = d_\varphi$, generated by $\mathbb{R}^2$-norms φ, since every $\mathbb{R}^2$-norm is Archimedean. So to consider ultrametrics $d = d_\varphi$, generated by norms, we must forgo $\mathbb{R}^2$ and replace it by $\mathbb{Q}^2$. Further we consider only field norms (which again, for brevity, we call norms).

$\mathbb{Q}$ is a classical example of a field with characteristic 0, in which there exist non-Archimedean norms. According to Ostrowski's theorem (see Theorem 4), each one of these norms is equivalent to the so-called p-adic norm φ_p, for an appropriate $p \in \mathbb{P}$, where $\mathbb{P}$ denotes the set of all primes. The main problem is to show that $\mathbb{Q}^2$ can be turned in infinitely many ways into a field in which there exist infinitely many non-Archimedean norms Φ_p, generated by the $\mathbb{Q}$-norms φ_p. Solving successfully this problem, we can apply the general scheme for d-IFS(E) to introduce ${}^{(\mathbb{Q}^2)}d_{\Phi_p}$-intuitionistic fuzzy sets over E. Examples of such sets, depending on ultrametric d_{Φ_p}, are presented here for the first time. Similar considerations with extension of the norm φ_p to the p-adic number field $\mathbb{Q}_p$ are potentially viable. They correspond to the possibility for introducing ${}^{(\mathbb{Q}_p^2)}d_{\Phi_p}$-intuitionistic fuzzy sets with the extended norm Φ_p, acting on $\mathbb{Q}_p^2 \stackrel{\text{def}}{=} \mathbb{Q}_p \times \mathbb{Q}_p$.

4.1 $\mathbb{Q}$-*Norms*

Here we recall basic facts for $\mathbb{Q}$-norms.

Let $\mathbb{Z}$ be the set of all integers. For $u, v, w \in \mathbb{Z}$ and $w > 0$, we write

$$u \equiv v \pmod{w},$$

if w divides $u - v$, and write

$$u \not\equiv v \pmod{w},$$

if w does not divide $u - v$.

Definition 13 Let $a \in \mathbb{Z} \setminus \{0\}$ and $p \in \mathbb{P}$. By $ord_p a$ we denote the greatest nonnegative $m \in \mathbb{Z}$ such that

$$a \equiv 0 \pmod{p^m}.$$

For instance: $ord_2 12 = 2; ord_3 162 = 4; ord_7 15 = 0$.

It is easy to see that if $a, b \in \mathbb{Z} \setminus \{0\}$, then

$$ord_p(a\,b) = ord_p a + ord_p b \,. \tag{40}$$

Also we agree that $ord_p 0 \stackrel{\text{def}}{=} +\infty$. Hence (40) is true for all $a, b \in \mathbb{Z}$.

In other words $ord_p x$ and $log_p x$ share the same additive property (namely, (40)).

Definition 14 Let $x \in \mathbb{Q}$ and $x = \frac{a}{b}$ be an arbitrary representation of x with $a, b \in \mathbb{Z}$. For any $p \in \mathbb{P}$ we set

$$ord_p x = ord_p a - ord_p b$$

Then it is easy to verify that $ord_p x$ does not depend on the representation of x. Indeed, if

$$x = \frac{a\,c}{b\,c}$$

with $c \in \mathbb{Z} \setminus \{0\}$, then due to (40) we have:

$$ord_p x = ord_p(a\,c) - ord_p(b\,c) = ord_p a + ord_p c - ord_p b - ord_p c = ord_p a - ord_p b,$$

which shows that Definition 14 is correct.

Let $x \in \mathbb{Q}$ and $p \in \mathbb{P}$. We define

$$\varphi_p(x) = \begin{cases} \left(\frac{1}{p}\right)^{ord_p x}, & \text{if } x \neq 0; \\ 0, & \text{if } x = 0\,. \end{cases} \tag{41}$$

The following fact is well known ([11, p. 2]):

Lemma 1 *For any $p \in \mathbb{P}$, φ_p is $\mathbb{Q}$-norm, satisfying the inequality*

$$\varphi_p(x + y) \leq \max(\varphi_p(x), \varphi_p(y))\ x, y \in \mathbb{Q}\,. \tag{42}$$

From (42) and Remark 1 it follows that φ_p is non-Archimedean $\mathbb{Q}$-norm. It is known as p-adic norm.

The result below is well known ([11, p. 7]):

Lemma 2 *Let F be a field, and f and g are F-norms. Then f is equivalent with g on F iff there exists positive real number α, such that for every $x \in F$ it is fulfilled*

$$f(x) = (g(x))^{\alpha}\,.$$

For an arbitrary real $\alpha \in (0, 1]$ we introduce

$$\varphi_{\infty}^{\{\alpha\}}(x) \stackrel{\text{def}}{=} (|x|)^{\alpha}\,.$$

Then $\varphi_{\infty}^{\{\alpha\}}$ is Archimedean $\mathbb{Q}$-norm.

The following two assertions are well known ([11, p. 7]):

Lemma 3 *The norms $\varphi_{\infty}^{\{\alpha\}}, \alpha \in (0, 1]$ are equivalent $\mathbb{Q}$-norms.*

Lemma 4 *If $p, q \in \mathbb{P}$ and $p \neq q$, then φ_p and φ_q are not equivalent $\mathbb{Q}$-norms.*

Definition 15 $\mathbb{Q}$-norm φ is called trivial,[5] if for $x \in \mathbb{Q} \setminus \{0\}$

$$\varphi(x) = 1\,.$$

All norms which are not trivial are called nontrivial.

The following result of Ostrowski ([11, p. 3]) gives complete description of all nontrivial $\mathbb{Q}$-norms:

Theorem 4 *(Ostrowski's theorem) Let φ be a nontrivial $\mathbb{Q}$-norm.*

If φ is non-Archimedean, then there exists $p \in \mathbb{P}$, such that φ is equivalent with φ_p on $\mathbb{Q}$.

If φ is Archimedean, then φ is equivalent with $\varphi_{\infty}^{\{1\}}$ on $\mathbb{Q}$.

4.2 Examples of $\mathbb{Q}^2$-Metrics Generated by Nontrivial $\mathbb{Q}$-Norms

Let φ be an arbitrary nontrivial $\mathbb{Q}$-norm. The above theorem (together with Lemma 2) completely describes φ. Since φ transforms $\mathbb{Q}$ into metric space $(\mathbb{Q}, d_{\varphi})$ where

[5]This definition remains valid if $\mathbb{Q}$ is replaced by arbitrary field.

$$d_\varphi(x,y) = \varphi(x-y)\ (x,y \in \mathbb{Q}),$$

the question for introducing $\mathbb{Q}^2$-metric generated by φ arises. The same question for a finite Cartesian product of arbitrary metric spaces is answered in [20, Theorem 1, p. 125]. Since $\mathbb{Q}^2 = \mathbb{Q} \times \mathbb{Q}$ is a particular case of the mentioned result, we obtain the following.

Let $\hat{\varphi}$ be an arbitrary $\mathbb{Q}^2$-subnorm generated by the $\mathbb{Q}$-norm φ. For example, we may put

$$\hat{\varphi}((x,y)) = \varphi(x) + \varphi(y),\ x,y \in \mathbb{Q}$$

or

$$\hat{\varphi}((x,y)) = \max(\varphi(x), \varphi(y)),\ x,y \in \mathbb{Q},$$

or to use another appropriate formula.

Since $\hat{\varphi}$ generates $\mathbb{Q}^2$-metric $d_{\hat{\varphi}}$ given by

$$d_{\hat{\varphi}}((x_1,y_1),(x_2,y_2)) = \hat{\varphi}((x_1-x_2, y_1-y_2)), (x_i,y_i) \in \mathbb{Q}^2, (i=1,2),$$

then the metric space $(\mathbb{Q}^2, d_{\hat{\varphi}})$ is obtained.

4.3 $\mathbb{Q}^2$ Considered as a Field

In § 4.2 we showed that the set $\mathbb{Q}^2$ may be transformed into metric space in infinitely many ways. Here we will show that there are infinitely many ways to turn the set $\mathbb{Q}^2$ into field. To this end in $\mathbb{Q}^2$ is introduced the additive operation $\oplus$ (addition) defined for $(a_1,b_1), (a_2,b_2) \in \mathbb{Q}^2$ by

$$(a_1,b_1) \oplus (a_2,b_2) = (a_1+a_2, b_1+b_2). \tag{43}$$

Obviously the operation $\oplus$ turns $\mathbb{Q}^2$ in Abelian group with zero element $(0,0)$, and each element $(a,b) \in \mathbb{Q}^2$ has a unique inverse element: $(-a,-b)$. Also we may introduce in $\mathbb{Q}^2$ multiplicative operation (multiplication) in infinitely many ways so $\mathbb{Q}^2$ is transformed into a field (with respect to $\oplus$ and the chosen multiplication). Namely, let $\odot_D$ be given by

$$(a_1,b_1) \odot_D (a_2,b_2) = (a_1a_2 + Db_1b_2, a_1b_2 + a_2b_1), \tag{44}$$

where $D \neq 1$ is a nonzero rational number and $\sqrt{|D|}$ is an irrational number (if $D \neq -1$).

It is easy to see, that if $D \neq -1$, then $\mathbb{Q}^2$, together with operations (43) and (44), is a field isomorphic to the field $\mathbb{Q}(\sqrt{D})$ with elements $x + \sqrt{D}y$ $(x, y \in \mathbb{Q})$. When $D = -1$, $\mathbb{Q}^2$, together with the operations (43) and (44), is a field isomorphic to the field with elements $x + i\,y$ $(x, y \in \mathbb{Q}, i = \sqrt{-1})$, which is a subfield of the complex number field. This field is called Gaussian rational field.

Further the mentioned above fields (for fixed D) will be denoted by $\mathbb{Q}^2(D)$.

Obviously

$$(0, 1) \odot_D (0, 1) = D(1, 0)\ .$$

Therefore $(0, 1) \in \mathbb{Q}^2(D)$ represents $\sqrt{D}$.

4.4 $\mathbb{Q}^2(D)$-Norms Generated by $\mathbb{Q}$-Norms

Here we will show how any nontrivial $\mathbb{Q}$-norm generates a nontrivial $\mathbb{Q}^2(D)$-norm. In particular any p-adic $\mathbb{Q}$-norm generates non-Archimedean $\mathbb{Q}^2(D)$-norm. In the next assertion the norms are required to satisfy only (5) and (3) but not necessarily (7).

Lemma 5 *Let φ be $\mathbb{Q}$-norm. Then the function*

$$\Phi((a, b)) \stackrel{\text{def}}{=} \sqrt{\varphi(a^2 - Db^2)} \tag{45}$$

is $\mathbb{Q}^2(D)$-norm.

Proof Since φ is $\mathbb{Q}$-norm, we have (5) and (3). Let for $(a, b) \in \mathbb{Q}^2$ $\Phi((a, b)) = 0$.

Then (45) yields $\varphi(a^2 - Db^2) = 0$. Hence $a^2 - Db^2 = 0$ since φ is norm.

If $D < 0$, then the last equality yields: $a = 0, b = 0$.

If $D > 0$ and $b \neq 0$, then $\sqrt{D} = \frac{a}{b} \in \mathbb{Q}$. But the last contradicts to $\sqrt{D}$ being an irrational number. Therefore $b = 0$. Hence $a = 0$.

From (45) $\Phi((a, b)) \geq 0$ $(a, b) \in \mathbb{Q}^2$.

Thus we proved that $\Phi : \mathbb{Q}^2 \to [0, +\infty)$ and (5) is satisfied.

Let $(a_1, b_1) \in \mathbb{Q}^2, (a_2, b_2) \in \mathbb{Q}^2$. It remains to prove (3) for Φ, i.e.,

$$\Phi((a_1, b_1) \odot_D (a_2, b_2)) = \Phi((a_1, b_1))\,\Phi((a_2, b_2))\ .$$

But:

$$\Phi((a_1, b_1) \odot_D (a_2, b_2)) = \Phi((a_1a_2 + Db_1b_2, a_1b_2 + a_2b_1)) =$$
$$\sqrt{\varphi([(a_1a_2 + Db_1b_2)^2 - D(a_1b_2 + a_2b_1)^2])} =$$
$$\sqrt{\varphi([(a_1^2a_2^2 + D^2b_1^2b_2^2 - Da_1^2b_2^2 - Da_2^2b_1^2])} =$$
$$\sqrt{\varphi([(a_1^2 - Db_1^2)\,(a_2^2 - Db_2^2)])} = \text{(applying (3) for } \varphi) =$$

$$\sqrt{\varphi((a_1^2 - Db_1^2)}\sqrt{\varphi((a_2^2 - Db_2^2)} = \Phi((a_1, b_1))\Phi((a_2, b_2)),$$

and (3)is proved. Thus Φ is $\mathbb{Q}^2(D)$-norm.

Lemma 5 is proved. □

Corollary 5 *If φ is a nontrivial $\mathbb{Q}$-norm then Φ is a nontrivial $\mathbb{Q}^2(D)$-norm.*

Corollary 6 *If $\varphi(1) = \varphi(-D) = 1$, then Φ is a normalized norm.*[6]

Further instead of D we take arbitrary fixed $q \in \mathbb{P}$.

We need the following definition (cf. [10, p. 50])

Definition 16 Let $r \in \mathbb{P}$. Then the number $t \in \mathbb{Z}, t \not\equiv 0 \pmod{r}$ is called quadratic residue modulo r, if there exists $x \in \mathbb{Z}$, such that $x^2 \equiv t \pmod{r}$. Otherwise (i.e., if there does not exist such x) t is called quadratic nonresidue modulo r.

Let $p \in \mathbb{P}$ be fixed, $\varphi = \varphi_p$ is put in (45) (see (41)), and Φ is denoted by Φ_p.

Using only elementary techniques we will prove the following

Theorem 5 *Let $p, q \in \mathbb{P}$ and q be quadratic nonresidue modulo p. Then Φ_p is a non-Archimedean $\mathbb{Q}^2(q)$-norm.*

Before giving the proof of Theorem 5 we need the following

Lemma 6 *Let $p, q \in \mathbb{P}$ and q be quadratic nonresidue modulo p. If $x, y \in \mathbb{Z}$, then*

$$ord_p(x^2 - qy^2) = 2\min(ord_p x, ord_p y). \tag{46}$$

Proof Since q is quadratic nonresidue modulo p, we have $q \neq p$. Indeed, $q = p$ yields

$$p^2 \equiv q \pmod{p},$$

i.e., q is quadratic residue modulo p, which is a contradiction.

If $x = y = 0$, then $+\infty = ord_p x = ord_p y = \min(ord_p x, ord_p y)$ and the assertion of Lemma 6 is true. If exactly one of the numbers x and y is different from 0, then exactly one of the numbers $ord_p x, ord_p y$ is less than $+\infty$. Therefore, the assertion of Lemma 6 is also true.

It remains only to consider the case $x \neq 0, y \neq 0$.

Let $ord_p x = \alpha, ord_p y = \beta$. We put:

$$x^* = \frac{x}{p^\alpha};\ y^* = \frac{y}{p^\beta}.$$

Let $\alpha < \beta$. Then

$$x^2 - qy^2 = p^{2\alpha}.((x^*)^2 - q.p^{2(\beta-\alpha)}.(y^*)^2).$$

[6] We remind that Φ is a normalized norm, if $\Phi((1, 0)) = \Phi((0, 1)) = 1$.

Hence

$$ord_p(x^2 - qy^2) = 2\alpha = 2ord_p x = 2\min(ord_p x, ord_p y)\ ,$$

since

$$x^* \not\equiv 0 \pmod{p}\ .$$

Let $\beta < \alpha$. Then

$$x^2 - qy^2 = p^{2\beta}.(p^{2(\alpha-\beta)}.(x^*)^2 - q.(y^*)^2)\ ,$$

i.e.,

$$ord_p(x^2 - qy^2) = 2\beta = 2ord_p y = 2\min(ord_p x, ord_p y)\ ,$$

since

$$y^* \not\equiv 0 \pmod{p}\ .$$

Let

$$ord_p x = ord_p y = \gamma\ .$$

Then

$$x^2 - qy^2 = p^{2\gamma}.((x^*)^2 - q.(y^*)^2)$$

and

$$x^* \not\equiv 0 \pmod{p}, y^* \not\equiv 0 \pmod{p}\ .$$

Hence

$$ord_p(x^2 - qy^2) \geq 2\gamma\ .$$

If we assume that

$$ord_p(x^2 - qy^2) > 2\gamma\ ,$$

then

$$(x^*)^2 - q(y^*)^2 \equiv 0 \pmod{p}\ .$$

The last contradicts to the fact that q is quadratic nonresidue modulo p. Therefore

$$ord_p(x^2 - qy^2) = 2\gamma = 2\min(ord_p x, ord_p y)\ .$$

Lemma 6 is proved. □

Definition 17 Let $u, v \in \mathbb{Z}$. By $\gcd(u, v)$ is denoted the greatest common divisor of u and v.

Proof (of Theorem 5) Lemma 5 provides that Φ_p is a $\mathbb{Q}^2(q)$-norm. It remains to show that Φ_p is non-Archimedean norm, i.e., for $(a_1, b_1), (a_2, b_2) \in \mathbb{Q}^2$ the inequality

$$\Phi_p((a_1, b_1) \oplus (a_2, b_2)) \leq \max(\Phi_p((a_1, b_1)), \Phi_p((a_2, b_2))) . \tag{47}$$

holds.

Using the definition of $\oplus$ we rewrite (47) in the form

$$\Phi_p(a_1 + a_2, b_1 + b_2) \leq \max(\Phi_p((a_1, b_1)), \Phi_p((a_2, b_2))) . \tag{48}$$

Let

$$a_1 = \frac{A_1}{E_1}, a_2 = \frac{A_2}{E_2}, b_1 = \frac{B_1}{G_1}, b_2 = \frac{B_2}{G_2} , \tag{49}$$

where

$$\gcd(A_1, E_1) = \gcd(A_2, E_2) = \gcd(B_1, G_1) = \gcd(B_2, G_2) = 1 . \tag{50}$$

Then

$$a_1^2 - qb_1^2 = \frac{A_1^2G_1^2 - qB_1^2E_1^2}{E_1^2G_1^2} ; \tag{51}$$

$$a_2^2 - qb_2^2 = \frac{A_2^2G_2^2 - qB_2^2E_2^2}{E_2^2G_2^2} ; \tag{52}$$

$$(a_1 + a_2)^2 - q(b_1 + b_2)^2 = \frac{G_1^2G_2^2(A_2E_1 + A_1E_2)^2 - qE_1^2E_2^2(B_2G_1 + B_1G_2)^2}{E_1^2E_2^2G_1^2G_2^2} . \tag{53}$$

Applying (46) for $x = A_1G_1$ and $y = B_1E_1$ and using (51), and the definitions of Φ_p and φ_p, we obtain

$$\Phi_p((a_1, b_1)) = p^{ord_pE_1 + ord_pG_1 - \min(ord_pA_1G_1, ord_pB_1E_1)} . \tag{54}$$

Applying (46) for $x = A_2G_2$ and $y = B_2E_2$ and using (52), and the definitions of Φ_p and φ_p, we obtain

$$\Phi_p((a_2, b_2)) = p^{ord_pE_2 + ord_pG_2 - \min(ord_pA_2G_2, ord_pB_2E_2)} . \tag{55}$$

Applying (46) for $x = G_1G_2(A_2E_1 + A_1E_2)$ and $y = E_1E_2(B_2G_1 + B_1G_2)$ and using (53), and the definitions of Φ_p and φ_p, we obtain

$$\Phi_p((a_1 + a_2, b_1 + b_2)) = p^{ord_pE_1 + ord_pE_2 + ord_pG_1 + ord_pG_2 - \min(ord_px, ord_py)} , \tag{56}$$

where

$$ord_p x = ord_p G_1 + ord_p G_2 + ord_p(A_2E_1 + A_1E_2)\ ; \tag{57}$$

$$ord_p y = ord_p E_1 + ord_p E_2 + ord_p(B_2G_1 + B_1G_2)\ . \tag{58}$$

Below we consider the following two cases separately

$$ord_p y \leq ord_p x \tag{I}$$

$$ord_p x \leq ord_p y\ . \tag{II}$$

In order to prove Theorem 5, we must prove (48) for each of these cases.

Let (I) hold. Then

$$\min(ord_p x, ord_p y) = ord_p y$$

and (56) and (58) yield

$$\Phi_p((a_1 + a_2, b_1 + b_2)) = p^{ord_p G_1 + ord_p G_2 - ord_p(B_2G_1 + B_1G_2)}\ . \tag{59}$$

We will consider the following three subcases

$$G_1 \equiv 0 \pmod{p}\ \&\ G_2 \equiv 0 \pmod{p} \tag{I_1}$$

$$G_1 \not\equiv 0 \pmod{p}\ \&\ G_2 \not\equiv 0 \pmod{p} \tag{I_2}$$

$$G_1 \equiv 0 \pmod{p}\ \&\ G_2 \not\equiv 0 \pmod{p} \tag{I_3}$$

The case $G_1 \not\equiv 0 \pmod{p}$ & $G_2 \equiv 0 \pmod{p}$ is analogous to (I_3) and we will omit it.

Let (I_1) be fulfilled. Then

$$B_1 \not\equiv 0 \pmod{p}, B_2 \not\equiv 0 \pmod{p}\ ,$$

because of (50). Therefore

$$ord_p(B_2G_1 + B_1G_2) \geq \min(ord_p B_2G_1, ord_p B_1G_2)\ .$$

But $ord_p B_1 = ord_p B_2 = 0$. Therefore:

$$ord_p B_2G_1 = ord_p B_2 + ord_p G_1 = ord_p G_1\ ;$$

$$ord_p B_1G_2 = ord_p B_1 + ord_p G_2 = ord_p G_2\ .$$

Hence

$$ord_p(B_2G_1 + B_1G_2) \geq \min(ord_pG_1, ord_pG_2)$$

and (59) yields

$$\Phi_p((a_1 + a_2, b_1 + b_2)) \leq p^{ord_pG_1+ord_pG_2-\min(ord_pG_1, ord_pG_2)}. \quad (60)$$

Let

$$ord_pG_1 \leq ord_pG_2 .$$

Then (60) yields

$$\Phi_p((a_1 + a_2, b_1 + b_2)) \leq p^{ord_pG_2} . \quad (61)$$

Let us consider (55). Since

$$ord_pB_2E_2 = ord_pB_2 + ord_pE_2 = ord_pE_2 ,$$

we have

$$\Phi_p((a_2, b_2)) = p^{ord_pG_2+ord_pE_2-\min(ord_pA_2G_2, ord_pE_2)} .$$

But obviously

$$ord_pE_2 - \min(ord_pA_2G_2, ord_pE_2) \geq 0 .$$

From the last inequality and the above equalities, we conclude that

$$p^{ord_pG_2} \leq \Phi_p((a_2, b_2)) .$$

The last inequality and (61) yield

$$\Phi_p((a_1 + a_2, b_1 + b_2)) \leq \Phi_p((a_2, b_2)).$$

Therefore (48) directly follows. If

$$ord_pG_2 \leq ord_pG_1 ,$$

then the considerations are analogous. In this case we reach the inequality

$$\Phi_p((a_1 + a_2, b_1 + b_2)) \leq \Phi_p((a_1, b_1)) ,$$

from where we obtain again (48).

Thus we proved that if (I_1) holds, then (48) holds too.

Let (I_2) hold. Then:

$$ord_pG_1 = ord_pG_2 = 0$$

and (59) yields

$$\Phi_p((a_1 + a_2, b_1 + b_2)) = p^{-ord_p(B_2G_1+B_1G_2)} . \tag{62}$$

If

$$B_2G_1 + B_1G_2 = 0 , \tag{63}$$

then from (58) and (I) we conclude that $y = 0$ and $+\infty = ord_p y \leq ord_p x$. Therefore $x = 0$ and due to (57)

$$A_2E_1 + A_1E_2 = 0 . \tag{64}$$

From (63) and (64) it follows

$$(a_1 + a_2, b_1 + b_2) = (0, 0) .$$

Therefore

$$\Phi_p((a_1 + a_2, b_1 + b_2)) = 0$$

and since Φ_p takes only nonnegative values, then (48) holds.

Let

$$B_2G_1 + B_1G_2 \neq 0 .$$

Then (62) yields

$$\Phi_p((a_1 + a_2, b_1 + b_2)) \leq 1 . \tag{65}$$

From (54) and (55) we find

$$\Phi_p((a_1, b_1)) = p^{ord_pE_1 - \min(ord_pA_1, ord_pB_1E_1)} ; \tag{66}$$

$$\Phi_p((a_2, b_2)) = p^{ord_pE_2 - \min(ord_pA_2, ord_pB_2E_2)} , \tag{67}$$

since

$$ord_pA_1G_1 = ord_pA_1 + ord_pG_1 = ord_pA_1 ;$$

$$ord_pA_2G_2 = ord_pA_2 + ord_pG_2 = ord_pA_2 .$$

If at least one of the congruences

$$E_i \equiv 0 \pmod{p}, i = 1, 2$$

holds, e.g.,

$$E_1 \equiv 0 \pmod{p} ,$$

then $\gcd(A_1, E_1) = 1$ yields

$$A_1 \not\equiv 0 \pmod{p}$$

and therefore

$$ord_p A_1 = 0 .$$

In this case (65) and (66) yield

$$\Phi_p((a_1, b_1)) = p^{ord_p E_1} > 1 \geq \Phi_p((a_1 + a_2, b_1 + b_2)) .$$

The last inequality yields (48).

Analogously in case that

$$E_2 \equiv 0 \pmod{p} ,$$

from (67), we obtain the inequality

$$\Phi_p((a_2, b_2)) = p^{ord_p E_2} > 1 \geq \Phi_p((a_1 + a_2, b_1 + b_2)) ,$$

which immediately yields (48).

Let us have

$$E_1 \not\equiv 0 \pmod{p} \;\&\; E_2 \not\equiv 0 \pmod{p} .$$

Then from (62) and the equalities

$$ord_p E_1 = ord_p E_2 = ord_p G_1 = ord_p G_2 = 0$$

we find

$$\Phi_p((a_1, b_1)) = p^{-\min(ord_p A_1, ord_p B_1)} ; \tag{68}$$

$$\Phi_p((a_2, b_2)) = p^{-\min(ord_p A_2, ord_p B_2)} ; \tag{69}$$

$$\Phi_p((a_1 + a_2, b_1 + b_2)) = p^{-ord_p(B_1 G_2 + ord_p B_2 G_1)} . \tag{70}$$

But obviously

$$ord_p(B_1 G_2 + B_2 G_1) \geq \min(ord_p B_1 G_2, ord_p B_2 G_1) .$$

Hence

$$ord_p(B_1 G_2 + B_2 G_1) \geq \min(ord_p B_1, ord_p B_2) .$$

The last inequality and (70) yield

$$\Phi_p((a_1 + a_2, b_1 + b_2)) \le p^{-\min(ord_p B_1, ord_p B_2)} . \tag{71}$$

If at least for one $i \in \{1, 2\}$ we have

$$B_i \not\equiv 0 \pmod p ,$$

then

$$ord_p B_i = 0$$

and hence

$$\min(ord_p B_1, ord_p B_2) = 0 .$$

Then (71) yields

$$\Phi_p((a_1 + a_2, b_1 + b_2)) \le p^0 = 1 . \tag{72}$$

But (68) and (69) yield:

$$\Phi_p((a_i, b_i)) = p^{-\min(ord_p A_i, ord_p B_i)} = p^0 = 1 . \tag{73}$$

From (72) and (73) immediately follows (48).

Let us have

$$B_1 \equiv 0 \pmod p \;\&\; B_2 \equiv 0 \pmod p .$$

If $i \in \{1, 2\}$ is the index for which we have

$$\min(ord_p B_1, ord_p B_2) = ord_p B_i ,$$

then (71) yields

$$\Phi_p((a_1 + a_2, b_1 + b_2)) \le p^{-ord_p B_i} .$$

From (68) or (69) it follows

$$\Phi_p((a_i, b_i)) = p^{-\min(ord_p A_i, ord_p B_i)} \ge p^{-ord_p B_i} . \tag{74}$$

The last two inequalities give

$$\Phi_p((a_1 + a_2, b_1 + b_2)) \le \Phi_p((a_i, b_i))$$

and thus (48) is fulfilled.

Let (I_3) hold. Then (59) yields

$$\Phi_p((a_1 + a_2, b_1 + b_2)) = p^{ord_p G_1 - ord_p \{B_2 G_1 + B_1 G_2\}} \tag{75}$$

and from (54) we obtain

$$\Phi_p((a_1, b_1)) = p^{ord_p E_1 + ord_p G_1 - \min(ord_p A_1 G_1, ord_p E_1)} . \tag{76}$$

We note that (77) follows from (54), since

$$G_1 \equiv 0 \pmod{p} \ \& \ \gcd(B_1, G_1) = 1$$

imply

$$B_1 \not\equiv 0 \pmod{p} .$$

Therefore

$$ord_p B_1 = 0 .$$

Hence

$$ord_p B_1 E_1 = ord_p B_1 + ord_p E_1 = ord_p E_1 .$$

Since

$$B_1 \not\equiv 0 \pmod{p} \ \& \ G_2 \not\equiv 0 \pmod{p} ,$$

then

$$B_1 G_2 \not\equiv 0 \pmod{p}$$

and

$$B_2 G_1 \equiv 0 \pmod{p} .$$

Therefore

$$ord_p\{B_2 G_1 + B_1 G_2\} = 0$$

and (76) yields

$$\Phi_p((a_1 + a_2, b_1 + b_2)) = p^{ord_p G_1} . \tag{77}$$

But obviously

$$ord_p E_1 - \min(ord_p A_1 G_1, ord_p E_1) \geq 0 .$$

Then (76) yields

$$\Phi_p((a_1, b_1)) \geq p^{ord_p G_1} . \tag{78}$$

From (77) and (78) we obtain

$$\Phi_p((a_1 + a_2, b_1 + b_2)) \leq \Phi_p((a_1, b_1)) .$$

The last yields (48).

Thus we proved that if (I) holds, then (48) holds too. Completely analogously one may prove that if (II) holds, then (48) holds too. To accomplish the proof the following replacements in the above considerations are needed: E_1 with G_1, E_2 with G_2, G_1 with E_1, G_2 with E_2, A_1 with B_1, A_2 with B_2, B_1 with A_1 and B_2 with A_2.

Theorem 5 is proved. □

If we consider the field $\mathbb{Q}^2(-q)$, where $q = 1$ or $q \in \mathbb{P}$, then in analogous manner the following two theorems may be proved.

Theorem 6 *Let $p \in \mathbb{P}$ be such that -1 is a quadratic nonresidue modulo p. If for $(a, b) \in \mathbb{Q}^2$ $\tilde{\Phi}_p$ is given by*

$$\tilde{\Phi}_p((a,b)) = \sqrt{\varphi_p(a^2 + b^2)}\,,$$

then $\tilde{\Phi}_p$ is non-Archimedean $\mathbb{Q}^2(-1)$-norm.

Theorem 7 *Let $p, q \in \mathbb{P}$ and $-q$ be quadratic nonresidue modulo p. If for $(a, b) \in \mathbb{Q}^2$ Φ^*_p is given by*

$$\Phi^*_p((a,b)) = \sqrt{\varphi_p(a^2 + qb^2)}\,,$$

then Φ^*_p is non-Archimedean $\mathbb{Q}^2(-q)$-norm.

As a corollary from Theorems 5–7 and Corollary 6, we obtain

Theorem 8 *Let Φ_p, $\tilde{\Phi}_p$, and Φ^*_p be the norms from Theorems 5–7. Then each one of these norms is a normalized norm.*

Proof Let $p \in \mathbb{P}$ be fixed and $q \in \mathbb{P}$ satisfies the conditions of Theorem 5 or Theorem 7. Then $q \not\equiv 0 \pmod{p}$ and according to Definition 13

$$ord_p q = ord_p(-q) = 0\,.$$

But we also have

$$ord_p 1 = ord_p(-1) = 0\,.$$

Hence

$$\varphi_p(1) = \varphi_p(-1) = \varphi_p(q) = \varphi_p(-q) = 1.$$

Now, for $D = -1, D = q$ and $D = -q$, from Corollary 6 (for $\varphi = \varphi_p$) Theorem 8 follows. □

4.5 An Infinite Class of Prime Numbers Generating Non-Archimedean $\mathbb{Q}^2(D)$-Norms

Using some well-known facts from number theory (concerning the theory of quadratic residues) we will show that there are infinitely many examples of prime numbers, satisfying the conditions of Theorems 5–7. In such way, with the help of these theorems, we will introduce infinitely many non-Archimedean norms on $\mathbb{Q}^2$ (considered as a field), that are generated by appropriate p-adic $\mathbb{Q}$-norms.

Let us begin with the following well known result

Theorem 9 *(Dirichlet's Theorem (see [10, pp. 25, 249])) Let* $A, B \in \mathbb{Z} \setminus \{0\}$ *and* $\gcd(A, B) = 1$. *The infinite arithmetic progression*

$$A + kB,\ k = 0, 1, 2, \ldots ,$$

contains infinitely many prime numbers.

The same Theorem allows elegant and unexpected formulation (observed by us), namely,

Theorem 10 *Let* $A, B \in \mathbb{Z}$. *The infinite arithmetic progression*

$$A + kB,\ k = 0, 1, 2, \ldots ,$$

contains infinitely many prime numbers iff it contains at least two different prime numbers.

We also need the following

Definition 18 Let $a \in \mathbb{Z}$ and $p \in \mathbb{P} \setminus \{2\}$. The Legendre symbol $\left(\frac{a}{p}\right)$ is introduced by

$$\left(\frac{a}{p}\right) \stackrel{\text{def}}{=} \begin{cases} 1, & \text{if } a \text{ is quadratic residue modulo } p \\ -1, & \text{if } a \text{ is quadratic nonresidue modulo } p \\ 0, & \text{if } a \equiv 0 \pmod{p} \end{cases}$$

Some very important properties of this symbol needed for the further considerations are given as the following

Theorem 11 *Let* $p, q \in \mathbb{P} \setminus \{2\}$ *and* $a, b \in \mathbb{Z} \setminus \{0\}$. *Then it is fulfilled ([10, pp. 51, 53])*

$$\left(\frac{1}{p}\right) = 1; \left(\frac{-1}{p}\right) = (-1)^{\frac{p-1}{2}};$$

$$\left(\frac{2}{p}\right) = (-1)^{\frac{p^2-1}{8}};$$

$$\left(\frac{a}{p}\right) = \left(\frac{b}{p}\right) \quad \text{if } a \equiv b \pmod{p};$$

$$\left(\frac{ab}{p}\right) = \left(\frac{a}{p}\right)\left(\frac{b}{p}\right) \quad \textit{(multiplicativity of Legendre symbol)};$$

$$\left(\frac{p}{q}\right) = (-1)^{\frac{p-1}{2}\frac{q-1}{2}}\left(\frac{q}{p}\right) \quad \textit{(Gauss quadratic reciprocity law)}.$$

With the help of Theorem 11 it can be directly checked that the following assertions are true:

Lemma 7 *Let $p \in \mathbb{P}$. Then -1 is quadratic nonresidue modulo p iff p belongs to the infinite arithmetic progression*

$$3 + k4,\ k = 0, 1, 2, \dots\ .$$

Lemma 8 *Let $q \in \mathbb{P}$ and $q \equiv 3 \pmod 8$. If $p \in \mathbb{P}$ belongs to the union of the following two infinite arithmetic progressions:*

$$2q + 1 + k(4q),\ k = 0, 1, 2, \dots\ ;$$

$$q + 2 + k(4q),\ k = 0, 1, 2, \dots\ ,$$

then q is quadratic nonresidue modulo p.

Lemma 9 *Let $q \in \mathbb{P}$ and $q \equiv 5 \pmod 8$. If $p \in \mathbb{P}$ belongs to the infinite arithmetic progression*

$$q + 2 + k(2q),\ k = 0, 1, 2, \dots\ ,$$

then q is quadratic nonresidue modulo p.

Lemma 10 *Let $q \in \mathbb{P}$ and $q \equiv 7 \pmod 8$. If $p \in \mathbb{P}$ belongs to the infinite arithmetic progression*

$$2q + 1 + k(4q),\ k = 0, 1, 2, \dots\ ,$$

then q is quadratic nonresidue modulo p.

Lemma 11 *Let $q \in \mathbb{P}$ and $q \equiv 1 \pmod 8$. If $p \in \mathbb{P}$ belongs to the infinite arithmetic progression*

$$2q + 1 + k(4q),\ k = 0, 1, 2, \dots\ ,$$

then $-q$ is quadratic nonresidue modulo p.

Lemma 12 *Let $q \in \mathbb{P}$ and $q \equiv 3 \pmod 8$. If $p \in \mathbb{P}$ belongs to the infinite arithmetic progression*

$$q + 2 + k(4q),\ k = 0, 1, 2, \dots\ ,$$

then $-q$ is quadratic nonresidue modulo p.

Lemma 13 *Let $q \in \mathbb{P}$ and $q \equiv 5 \pmod 8$. If $p \in \mathbb{P}$ belongs to the infinite arithmetic progression*

$$2q + 1 + k(4q),\ k = 0, 1, 2, \dots\ ,$$

then $-q$ is quadratic nonresidue modulo p.

Lemma 14 *Let $q \in \mathbb{P}$ and $q \equiv 7 \pmod 8$. If $p \in \mathbb{P}$ belongs to the infinite arithmetic progression*

$$2q - 1 + k\,(4q),\; k = 0, 1, 2, \dots\;,$$

then $-q$ is quadratic nonresidue modulo p.

From Theorems 6 and 8, according to Lemma 7, we obtain

Theorem 12 *Let $p \in \mathbb{P}$ and $p \equiv 3 \pmod 4$. Then $\tilde{\Phi}_p$, defined for $(a, b) \in \mathbb{Q}^2$, by*

$$\tilde{\Phi}_p((a, b)) = \sqrt{\varphi_p(a^2 + b^2)}\,,$$

is a normalized non-Archimedean $\mathbb{Q}^2(-1)$-norm.

From Theorems 5 and 8, according to Lemma 8, we obtain

Theorem 13 *Let $q \in \mathbb{P}$ and $q \equiv 3 \pmod 8$. If $p \in \mathbb{P}$ belongs to the union of the following two infinite arithmetic progressions:*

$$2q + 1 + k\,(4q),\; k = 0, 1, 2, \dots\;;$$

$$q + 2 + k\,(4q),\; k = 0, 1, 2, \dots,$$

then Φ_p, defined for $(a, b) \in \mathbb{Q}^2$ by

$$\Phi_p((a, b)) = \sqrt{\varphi_p(a^2 - qb^2)}\,,$$

is a normalized non-Archimedean $\mathbb{Q}^2(q)$-norm.

From Theorems 5 and 8, according to Lemma 9, we obtain

Theorem 14 *Let $q \in \mathbb{P}$ and $q \equiv 5 \pmod 8$. If $p \in \mathbb{P}$ belongs to the infinite arithmetic progression*

$$q + 2 + k\,(2q),\; k = 0, 1, 2, \dots\;,$$

then Φ_p, defined for $(a, b) \in \mathbb{Q}^2$ by

$$\Phi_p((a, b)) = \sqrt{\varphi_p(a^2 - qb^2)}\,,$$

is a normalized non-Archimedean $\mathbb{Q}^2(q)$-norm.

From Theorems 5 and 8, according to Lemma 10, we obtain

Theorem 15 *Let $q \in \mathbb{P}$ and $q \equiv 7 \pmod 8$. If $p \in \mathbb{P}$ belongs to the infinite arithmetic progression*

$$2q + 1 + k\,(4q),\; k = 0, 1, 2, \dots\;,$$

then Φ_p, *defined for* $(a, b) \in \mathbb{Q}^2$ *by*

$$\Phi_p((a, b)) = \sqrt{\varphi_p(a^2 - qb^2)}\,,$$

is a normalized non-Archimedean $\mathbb{Q}^2(q)$*-norm.*

From Theorems 7 and 8, according to Lemma 11, we obtain

Theorem 16 *Let* $q \in \mathbb{P}$ *and* $q \equiv 1 \pmod 8$*. If* $p \in \mathbb{P}$ *belongs to the infinite arithmetic progression*

$$2q + 1 + k\,(4q),\ k = 0, 1, 2, \dots\ ,$$

then Φ_p^*, *defined for* $(a, b) \in \mathbb{Q}^2$ *by*

$$\Phi_p^*((a, b)) = \sqrt{\varphi_p(a^2 + qb^2)}\,,$$

is normalized non-Archimedean $\mathbb{Q}^2(-q)$*-norm.*

From Theorems 7 and 8, according to Lemma 12, we obtain

Theorem 17 *Let* $q \in \mathbb{P}$ *and* $q \equiv 3 \pmod 8$*. If* $p \in \mathbb{P}$ *belongs to the infinite arithmetic progression*

$$q + 2 + k\,(4q),\ k = 0, 1, 2, \dots\ ,$$

then Φ_p^*, *defined for* $(a, b) \in \mathbb{Q}^2$ *by*

$$\Phi_p^*((a, b)) = \sqrt{\varphi_p(a^2 + qb^2)}\,,$$

is a normalized non-Archimedean $\mathbb{Q}^2(-q)$*-norm.*

From Theorems 7 and 8, according to Lemma 13, we obtain

Theorem 18 *Let* $q \in \mathbb{P}$ *and* $q \equiv 5 \pmod 8$*. If* $p \in \mathbb{P}$ *belongs to the infinite arithmetic progression*

$$2q + 1 + k\,(4q),\ k = 0, 1, 2, \dots\ ,$$

then Φ_p^*, *defined for* $(a, b) \in \mathbb{Q}^2$ *by*

$$\Phi_p^*((a, b)) = \sqrt{\varphi_p(a^2 + qb^2)}\,,$$

is a normalized non-Archimedean $\mathbb{Q}^2(-q)$*-norm.*

From Theorems 7 and 8, according to Lemma 14, we obtain

Theorem 19 *Let $q \in \mathbb{P}$ and $q \equiv 7 \pmod 8$. If $p \in \mathbb{P}$ belongs to the infinite arithmetic progression*

$$2q - 1 + k(2q),\ k = 0, 1, 2, \dots ,$$

then Φ_p^*, defined for $(a, b) \in \mathbb{Q}^2$ by

$$\Phi_p^*((a, b)) = \sqrt{\varphi_p(a^2 + qb^2)}\,,$$

is a normalized non-Archimedean $\mathbb{Q}^2(-q)$-norm.

Remark 3 According to Theorem 9, for each one of Theorems 12–19, there exist infinitely many $p \in \mathbb{P}$, satisfying its respective conditions.

4.6 $^{(\mathbb{Q}^2)}d_\Phi$-Intuitionistic Fuzzy Sets Depending on Normalized Non-Archimedean $\mathbb{Q}^2$-Norms Φ

In § 4.3 we introduced the field $\mathbb{Q}^2(D)$(where $D \neq 1$ is nonzero rational number and if $D \neq -1$, then $\sqrt{|D|}$ is irrational number), which consists of ordered pairs of rational numbers, that are added component wise and multiplied by the rule (44).

Definition 19 Let $A \subset E$ and $\mu_A : E \to [0, 1] \cap \mathbb{Q}$, $\nu_A : E \to [0, 1] \cap \mathbb{Q}$ are mappings. Let Φ be $\mathbb{Q}^2(D)$-norm and d_Φ be $\mathbb{Q}^2$- metric that for $(a_1, b_1), (a_2, b_2) \in \mathbb{Q}^2$ is given by

$$d_\Phi((a_1, b_1), (a_2, b_2)) = \Phi((a_1 - a_2, b_1 - b_2))\,.$$

Then the set

$$\tilde{A} = \{\langle x, \mu_A(x), \nu_A(x)\rangle | x \in E\}$$

is called $^{(\mathbb{Q}^2)}d_\Phi$-intuitionistic fuzzy set ($^{(\mathbb{Q}^2)}d_\Phi$*-IFS*) over E, generated by A (through the $\mathbb{Q}^2$-metric d_Φ) if the relation

$$\Phi((\mu_A(x), \nu_A(x))) \leq 1$$

holds.

The mappings μ_A, ν_A are called membership and non-membership function and the mapping $\pi_A : E \to [0, 1]$, that for $x \in E$ is given by

$$\pi_A(x) = 1 - \Phi((\mu_A(x), \nu_A(x))),$$

is called hesitancy function.

For $x \in E$ the numbers $\mu_A(x)$, $\nu_A(x)$ and $\pi_A(x)$ are called degree of membership, degree of non-membership and hesitancy degree/margin of the element x to the set A.

The class of all ${}^{(\mathbb{Q}^2)}d_{\Phi}$*-IFSs* over E is denoted by ${}^{(\mathbb{Q}^2)}d_{\Phi}$-IFS($E$).

Let $q \in \mathbb{P}$, $p \in \mathbb{P}$, $p \neq q$ and φ_p is p-adic $\mathbb{Q}$-norm defined by (41). In § 4.4 the following results were established:

(a) If $D = q$ and q is quadratic nonresidue modulo p, then the mapping $\boldsymbol{\Phi}_p$, given for $(a, b) \in \mathbb{Q}^2$ by

$$\boldsymbol{\Phi}_p((a, b)) = \sqrt{\varphi_p(a^2 - qb^2)}\,,$$

is a normalized non-Archimedean $\mathbb{Q}^2(D)$-norm (Theorems 5 and 8);

(b) If $D = -q$ and $-q$ is quadratic nonresidue modulo p, then the mapping $\boldsymbol{\Phi}^*_p$, given for $(a, b) \in \mathbb{Q}^2$ by

$$\boldsymbol{\Phi}^*_p((a, b)) = \sqrt{\varphi_p(a^2 + qb^2)}\,,$$

is a normalized non-Archimedean $\mathbb{Q}^2(D)$-norm (Theorems 7 and 8);

(c) If $D = -1$ and -1 is quadratic nonresidue modulo p, then the mapping $\tilde{\boldsymbol{\Phi}}_p$, given for $(a, b) \in \mathbb{Q}^2$ by

$$\tilde{\boldsymbol{\Phi}}_p((a, b)) = \sqrt{\varphi_p(a^2 + b^2)}\,,$$

is a normalized non-Archimedean $\mathbb{Q}^2(D)$-norm (Theorems 6 and 8).

In § 4.5 we proved that there exist infinitely many prime numbers q and p, satisfying the conditions in (a) and (b). The main results, corresponding to the case (a), are Theorems 13–15. The main results corresponding to the case (b), are Theorems 16–19.

There also exist infinitely many prime numbers p, satisfying (c). The main result corresponding to the case (c), is Theorem 12.

Let q and p satisfy the conditions in (a). Then the normalized non-Archimedean norm $\boldsymbol{\Phi} = \boldsymbol{\Phi}_p$ generates a $\mathbb{Q}^2$-ultrametric $d_{\boldsymbol{\Phi}} = d_{\boldsymbol{\Phi}_p}$ given for $(a_1, b_1), (a_2, b_2) \in \mathbb{Q}^2$ by the formula:

$$d_{\boldsymbol{\Phi}_p}((a_1, b_1), (a_2, b_2)) \stackrel{\text{def}}{=} \boldsymbol{\Phi}_p((a_1 - a_2, b_1 - b_2)) = \sqrt{\varphi_p((a_1 - a_2)^2 - q(b_1 - b_2)^2)}\,.$$

The metric $d_{\boldsymbol{\Phi}_p}$ generates the class ${}^{(\mathbb{Q}^2)}d_{\boldsymbol{\Phi}_p}$-IFS($E$). For

$$\{\langle x, \mu(x), \nu(x)\rangle | x \in E\} \in {}^{(\mathbb{Q}^2)}d_{\boldsymbol{\Phi}_p}\text{-IFS}(E)$$

we have

$$\mu : E \to [0, 1] \cap \mathbb{Q};\ \nu : E \to [0, 1] \cap \mathbb{Q}$$

and

$$\Phi_p((\mu(x), \nu(x))) \leq 1 ,$$

i.e.,

$$\varphi_p((\mu(x))^2 - q(\nu(x))^2) \leq 1 .$$

The hesitancy function $\pi_{d_{\Phi_p}} : E \to [0, 1]$ now is given by

$$\pi_{d_{\Phi_p}}(x) = 1 - \Phi_p((\mu(x), \nu(x)) ,$$

i.e., by

$$\pi_{d_{\Phi_p}}(x) = 1 - \sqrt{\varphi_p((\mu(x))^2 - q(\nu(x))^2)} .$$

Let q and p satisfy the conditions in (b). Then the normalized non-Archimedean norm $\Phi = \Phi_p^*$ generates a $\mathbb{Q}^2$-ultrametric $d_\Phi = d_{\Phi_p^*}$ given for $(a_1, b_1), (a_2, b_2) \in \mathbb{Q}^2$ by the formula

$$d_{\Phi_p^*}((a_1, b_1), (a_2, b_2)) \overset{\text{def}}{=} \Phi_p^*((a_1 - a_2, b_1 - b_2)) = \sqrt{\varphi_p((a_1 - a_2)^2 + q(b_1 - b_2)^2)} .$$

The metric $d_{\Phi_p^*}$ generates the class ${}^{(\mathbb{Q}^2)}d_{\Phi_p^*}$-IFS($E$). For

$$\{\langle x, \mu_A(x), \nu_A(x)\rangle | x \in E\} \in {}^{(\mathbb{Q}^2)}d_{\Phi_p^*}\text{-IFS}(E)$$

we have

$$\mu : E \to [0, 1] \cap \mathbb{Q};\ \nu : E \to [0, 1] \cap \mathbb{Q}$$

and

$$\Phi_p^*((\mu(x), \nu(x))) \leq 1 ,$$

i.e.,

$$\varphi_p((\mu(x))^2 + q(\nu(x))^2) \leq 1 .$$

The hesitancy function $\pi_{d_{\Phi_p^*}} : E \to [0, 1]$ now is given by

$$\pi_{d_{\Phi_p^*}}(x) = 1 - \Phi_p^*((\mu(x), \nu(x)) ,$$

i.e., by

$$\pi_{d_{\Phi_p^*}}(x) = 1 - \sqrt{\varphi_p((\mu(x))^2 + q(\nu(x))^2)} .$$

Let p satisfy the conditions in (c). Then the normalized non-Archimedean norm $\Phi = \tilde{\Phi}_p$ generates $\mathbb{Q}^2$-ultrametric $d_\Phi = d_{\tilde{\Phi}_p}$ given for $(a_1, b_1), (a_2, b_2) \in \mathbb{Q}^2$ by the formula:

$$d_{\tilde{\Phi}_p}((a_1, b_1), (a_2, b_2)) \overset{\text{def}}{=} \tilde{\Phi}_p((a_1 - a_2, b_1 - b_2)) = \sqrt{\varphi_p((a_1 - a_2)^2 + (b_1 - b_2)^2)} .$$

The metric $d_{\tilde{\Phi}_p}$ generates the class ${}^{(\mathbb{Q}^2)}d_{\tilde{\Phi}_p}$-IFS($E$). For

$$\{\langle x, \mu(x), \nu(x)\rangle | x \in E\} \in {}^{(\mathbb{Q}^2)}d_{\tilde{\Phi}_p}\text{-IFS}(E)$$

we have

$$\mu : E \to [0, 1] \cap \mathbb{Q};\ \nu : E \to [0, 1] \cap \mathbb{Q}$$

and

$$\tilde{\Phi}_p((\mu(x), \nu(x))) \le 1 ,$$

i.e.,

$$\varphi_p((\mu(x))^2 + (\nu(x))^2) \le 1 .$$

The hesitancy function $\pi_{d_{\tilde{\Phi}_p}} : E \to [0, 1]$, now is given by

$$\pi_{d_{\tilde{\Phi}_p}}(x) = 1 - \tilde{\Phi}_p((\mu(x), \nu(x))) ,$$

i.e., by

$$\pi_{d_{\tilde{\Phi}_p}}(x) = 1 - \sqrt{\varphi_p((\mu(x))^2 + (\nu(x))^2)} .$$

Sufficient conditions for the existence of ${}^{(\mathbb{Q}^2)}d_{\Phi_p}$-IFS(E) in case (a) are given by Theorems 13–15.

Sufficient conditions for the existence of ${}^{(\mathbb{Q}^2)}d_{\Phi^*_p}$-IFS(E) in case (b) are given by Theorems 16–19.

Sufficient conditions for the existence of ${}^{(\mathbb{Q}^2)}d_{\tilde{\Phi}_p}$-IFS(E) in case (c) are given by Theorem 12.

References

1. Atanassov, K.: Intuitionistic Fuzzy Sets. VII ITKR's session (deposed in Central Sci. -Techn. Library of Bulg. Acad. of Sci. 1697/84) Sofia (1983) (in Bulgarian)
2. Atanassov, K.: A second type of intuitionistic fuzzy sets. BUSEFAL **56**, 66–70 (1993)
3. Atanassov, K.: Intuitionistic Fuzzy Sets. Springer Physica-Verlag, Heidelberg (1999)
4. Atanassov, K.: On Intuitionistic Fuzzy Sets Theory. Springer Physica-Verlag, Heidelberg (2012)
5. Bingham, N.H., Ostaszewski, A.J.: Normed groups: dichotomy and duality. LSE-CDAM Report, LSE-CDAM-2008-10rev
6. Bonsall, F., Duncan, J.: Numerical Ranges II. London Mathematical Society. Lecture Notes Series, vol. 10 (1973)
7. Bullen, P.S.: Handbook of Means and their Inequalities. Kluwer Academic Publishers, Dordrecht (2003)
8. Deza, M., Deza, E.: Encyclopedia of Distances. Springer, Heidelberg (2009)
9. Dougherty, G.: Pattern Recognition and Classification an Introduction. Springer, New York (2013)
10. Ireland, K., Rosen, M.: Classical Introduction to Modern Number Theory. Springer Physica-Verlag, New York (1990)
11. Koblitz, N.: *P*-adic Numbers, *p*-adic Analysis, and Zeta-Functions, 2nd edn. Springer, New York (1984)
12. Körner, M.-C.: Minisum Hyperspheres. Springer, Heidelberg (2011)
13. Krause, E.F.: Taxicab Geometry. Dover Publications, New York (1975)
14. Palaniapan, N., Srinivasan, R., Parvathi, R.: Some operations on intuitionistic fuzzy sets of root type. NIFS **12**(3), 20–29 (2006)
15. Parvathi, R., Vassilev, P., Atanassov, K.: A note on the bijective correspondence between intuitionistic fuzzy sets and intuitionistic fuzzy sets of *p*-th type. In: New Developments in Fuzzy Sets, Intuitionistic Fuzzy Sets, Generalized Nets and Related Topics. Volume I: Foundations, SRI PAS IBS PAN, Warsaw, pp. 143–147 (2012)
16. Pólya, G., Szegö, G.: Problems and Theorems in Analysis, vol. I. Springer, Berlin (1976)
17. Vassilev, P., Parvathi, R., Atanassov, K.: Note On Intuitionistic Fuzzy Sets of *p*-th Type. Issues in Intuitionistic Fuzzy Sets and Generalized Nets **6**, 43–50 (2008)
18. Vassilev, P.: A Metric Approach To Fuzzy Sets and Intuitionistic Fuzzy Sets. In: Proceedings of First International Workshop on IFSs, GNs, KE, pp. 31–38 (2006)
19. Vassilev, P.: Operators similar to operators defined over intuitionistic fuzzy sets. In: Proceedings of 16th International Conference on IFSs, Sofia, 910 Sept. 2012. Notes on Intuitionistic Fuzzy Sets, vol. 18, No. 4, 40–47 (2012)
20. Vassilev-Missana, M., Vassilev, P.: On a Way for Introducing Metrics in Cartesian Product of Metric Spaces. Notes on Number Theory and Discrete Mathematics **8**(4), 125–128 (2002)
21. Zadeh, L.: Fuzzy sets. Inf. Control **8**(3), 338–353 (1965)

Production Rule and Network Structure Models for Knowledge Extraction from Complex Processes Under Uncertainty

Boriana Vatchova and Alexander Gegov

Abstract This paper considers processes with many inputs and outputs from different application areas. Some parts of the inputs are measurable and others are not because of the presence of stochastic environmental factors. This is the reason why processes of this kind operate under uncertainty. As some factors cannot be measured and reflected into the process model, data mining methods cannot be applied. The proposed approach which can be applied in this case is based on artificial intelligence methods[1].

1 Introduction

Finding a relation between inputs and outputs of complex processes and building an adequate process model is the main control objective for these processes. The presence of uncertainty as a result of many factors, environmental behaviour and impossibility to measure all inputs makes difficult the modelling of these complex processes.

In this paper, the existing data mining methods [2–6] are supplemented by methods of random functions theory [7, 8] and multi-valued logic [8, 9]. The models of the processes are knowledge bases of production rules or multi-layer network structures which include probability of occurrence.

B. Vatchova (✉)
Institute of Information and Communication Technologies, Bulgarian Academy of Sciences, Sofia, Bulgaria
e-mail: boriana.vatchova@gmail.com

A. Gegov
School of Computing, University of Portsmouth, Portsmouth, UK
e-mail: alexander.gegov@port.ac.uk

V. Sgurev et al. (eds.), *Recent Contributions in Intelligent Systems*, Studies in Computational Intelligence 657,
DOI 10.1007/978-3-319-41438-6_20

2 Essence of the Method of MLPF

The processes which are investigated with the purpose to extract traditional knowledge depend on measurable factors like $\tilde{x}_i(\tau), i = 1 \div n$, which are called numerical values of the measurable factors. In this case, $\tau = 1, 2, 3 \ldots$ is the discrete time for each measurement and $W(t)$ is the set of immeasurable factors which is summarized as one *generalized input influence*.

The output $\tilde{y}_e(\tau), e = 1 \div m$, where m is the number of the processes, is derived as a result of inputs $\tilde{x}_i(\tau)$ and generalized factor $W(t)$ which has time delay τ_0. The output of the process $\tilde{y}(\tau)$ is a package of data sets with numerical values denoted by $\tilde{M}\{\tilde{x}_i(\tau), \tilde{y}_e(\tau)\}$, where $\tau = 1 \div N$ and N is the number of measurements for a package.

The non-measurable inputs $W(t)$ are supposed to change within a limited range and to be characterized by limited time modifications so that they can be interpreted as *pseudo-stationary stochastic processes.*

The numerical values for different inputs $\tilde{x}_i(\tau)$ and outputs $\tilde{y}_e(\tau)$ have different ranges because of the existence of different measurements.

The data set package $\tilde{M}\{\tilde{x}_i(\tau), \tilde{y}_e(\tau)\}$ is transformed from a package of numerical values $M\{x_i(\tau), y_e(\tau)\}$ to logical values $Lx_{ij}(\tau)$, $Ly_{eq}(\tau)$, where $j = 1 \div k_i$ is the number of x_i, k_i is the number of the logical values of x_i; $q = 1 \div k_e$ is the logical values of y_e, k_e is the number of the logical values of y_e. This transformation is calculated as a relation between numerical values and their max relative values for input $x_i(\tau)$ and output $y_e(\tau)$. The parameters k_i and k_e express the range of the proposed multi-valued logical system for each measurable factor x_i and for each output y_e [10]. The option for k_i and k_e depends on the range modification for inputs and outputs. The logical values $Lx_{ij}(\tau)$ and $Ly_{eq}(\tau)$ of each particular value τ define the *current data set* of logical values $LN(\tau)\{Lx_{ij}(\tau), Ly_{eq}(\tau)\}$. The combination of current sequence data sets for a particular sequence of time discrete moment τ is defined as a *package of logical values* $LM_s\langle LN(\tau)\{Lx_{ij}(\tau), Ly_{eq}(\tau)\}\rangle$, where s is the number of the package. The combination of current sequence data sets for a particular package $LM_s\langle LN(\tau)\{Lx_{ij}(\tau), Ly_{eq}(\tau)\}\rangle$ characterized by logical values for each input are defined as *grouping sequence data sets* $GLN_r[LM_s\langle LN_\tau\{Lx_{ij}(\tau), Ly_{eq}(\tau)\}\rangle]$ where $r = 1, 2, 3, \ldots$ is the number of the grouping sequence sets.

It is possible to include in one package of logical values $LM_s\langle LN(\tau)\{Lx_{ij}(\tau), Ly_{eq}(\tau)\}\rangle$ a number of *grouping sequence sets* $GLN_r[LM_s\langle LN_\tau\{Lx_{ij}(\tau), Ly_{eq}(\tau)\}\rangle]$.

The probability of occurrences $p\{GLN_r[LM_s\langle LN(\tau)\{Lx_{ij}(\tau), Ly_{eq}(\tau)\}\rangle]\}$ for each package $LM_s\langle LN(\tau)\{Lx_{ij}(\tau), Ly_{eq}(\tau)\}\rangle$ is calculated for every grouping sequence set.

The probability of occurrences of one grouping sequence set is calculated as a relation between the number of the current sequence sets included in a grouping sequence set and a current sequence set of a given package $LM_s\langle LN(\tau)\{Lx_{ij}(\tau), Ly_{eq}(\tau)\}\rangle$.

When the probability of occurrence of one grouping sequence set has a significant value then this grouping sequence set is perceived as *dominant grouping sequence set*. The current sequence sets which occur only once in a given package do not give reliable information for further investigations.

The dominant grouping sequence set GLN_r with its probability of occurrence of the package of real data sets ensures the existence of reliable relations between the logical values of inputs $Lx_{ij}(\tau)$, outputs $Ly_{eq}(\tau)$ and the probability of occurrences $p\{Ly_{eq}(\tau)\}$.

The relations between $Lx_{ij}(\tau)$, $Ly_{eq}(\tau)$ and GLN_r are presented in a table format which is similar to the form of the functions of multi-valued logic [11, 12]. These class functions are supplemented with a probability of occurrences as a result of an additional argument $W(t)$

$$Ly_{eq} = F_1\{GLN_r\} \qquad GLN_r(\tau) = F_2\{Lx_{ij}, W\} \tag{1}$$

$$p\{Ly_{eq}\} = P_1\{GLN_r\} \quad p\{GLN_r\} = P_2\{(Lx_{ij}, W)\} \tag{2}$$

where 1 expresses the *logical correspondence F*, 2 expresses the *probabilistic correspondence P*.

The two correspondences $F(F_1, F_2)$, $P(P_1, P_2)$ which are mutually related and they present new class functions in the multi-valued logic—*Multi-Valued Logic Probabilistic Functions* (MLPF) [10, 13].

The widely known functions of multi-valued logic express a correspondence between logical values of the inputs (arguments) and the outputs (function value) for deterministic subjects. The new class of functions MLPF corresponds to the general case for stochastic subjects with no apparent arguments $W(t)$ and non-stationary parameters.

The use of MLPF makes possible the description of processes from different application areas with multi inputs and outputs, nonlinearity, non-stationary and stochastic behaviour.

MLPF are presented mostly in a table format for a better visual form in comparison to the analytical form.

An example for MLPF is presented in the Table 1 for a subject with three measurable inputs Lx_1, Lx_2, Lx_3, two outputs Ly_1, Ly_2 and non-apparent factors $W(t)$.

Table 1 MLPF for three- degree logic

Lx_1		a_1	a_1	a_1	a_1	a_1	a_1	a_1	a_2	a_2	…..	…..	a_3
Lx_2		a_1	a_2	a_2	a_2	a_3	a_3	a_3	a_1	a_2	…..	…..	a_3
Lx_3		a_1	a_1	a_2	a_3	a_1	a_2	a_3	a_1	a_1	…..	…..	a_3
Ly_1	a_1	p_{1111}	p_{abcd}										
	a_2	p_{1112}		p_{1222}									
	a_3	p_{1113}											P_{3333}
Ly_2	a_1	p_{1111}							p_{abcd}				
	a_2	p_{1112}					p_{1322}						
	a_3	p_{1113}						p_{1333}					P_{3333}

Three degree multi-valued logic is used. The possible logical values of the inputs and the outputs are named as a_1, a_2, a_3 and the frequency of occurrences are named as P_{abcd}, where abc correspond to the logical values of the three inputs and d corresponds to the logical value of the output.

The logical values a_1, a_2, a_3 have a particular meaning, e.g. small, medium and large. Each given sequence set of logical values for the inputs Lx_1, Lx_2, Lx_3 (a_1, a_2 and a_3) corresponds to two possible logical values for Ly_1 and Ly_2 because of the non-measurable inputs $W(t)$. The sum of the frequencies of occurrence of the outputs Ly_1, Ly_2 ... for each sequence data set of logical values Lx_1, Lx_2 and Lx_3 is equal to 1.

3 Models of Multi-factor Processes Under Uncertainty Using MLPF

3.1 Production Rule Models

Using new data sets in real time creates packages of numerical values for inputs and outputs, which are updated values of MLPFs [11–13]. The model or the updated knowledge base is a combination of production rules with the following structure:

If ⟨logical values of measurable inputs⟩ Then ⟨logical values of the outputs *supplemented with aprobability of occurrences*⟩ or

If ⟨Lx_1, Lx_2, Lx_3⟩ Then ⟨Ly_1, Ly_2⟩.

3.2 Network Structure Models

The network structure includes three layers:

- The input layer has elements which correspond to the number of the measurable inputs and the number of the perceived logical values;
- The intermediate layer has elements which correspond to the number of the dominant input grouping sequence data sets;
- The output layer has elements which correspond to the number of the occurrence logical values for the output.

According to Fig. 1, if the process has more than one output then the network model is a composition of models.

The models for the other outputs Ly_2, Ly_3,… differ from the model for the output Ly_1 only by the links and the coefficients between the intermediate and the output layers. The logical values Lx_{1j}, Lx_{2j},…, Lx_{nj} are passed to the input layer. For each element of the intermediate layer there are signals from the elements of the input layer. The output of each element of the intermediate layer is passed to the input of

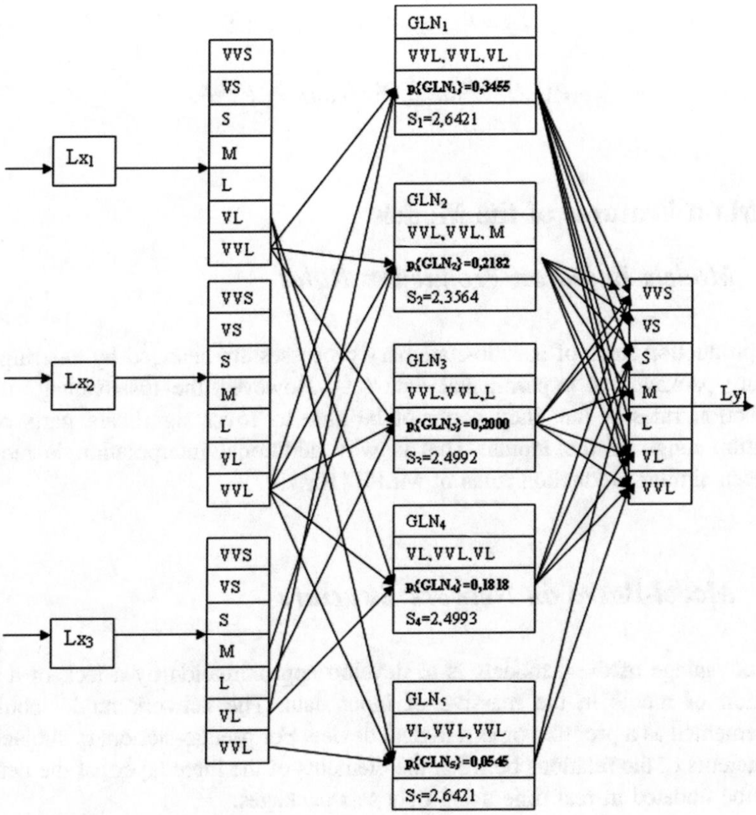

Fig. 1 Network model $Ly_1 = f(Lx_1, Lx_2, Lx_3)$ for a seven-degree logical system

the elements of the output layer. The links between these three layers are presented by coefficients which are elements of the matrixes.

R_{LxGLN_r} is a relation between the relative values corresponding to the logical values of the inputs $Lx_{1j}(\tau)$ of the input layer and the sum of the relative values included in the corresponding dominant grouping sequence set GLN_r of the intermediate layer.

$R^*_{LxGLN_r}$ is a relation between the frequency of occurrences of the sets of the input and the intermediate layers;

$R_{GLN_rLy_{eq}}$ is a relation between the relative values corresponding to the logical values of the elements of the intermediate and the output layers;

$R^*_{GLN_rLy_{eq}}$ is a relation between the frequency of occurrences of the elements of the sets of the intermediate and the output layers.

Using the network model, the logical values Ly with their probability of occurrence $p\{Ly\}$ for each sequence set and the logical input values LX with their probability of occurrence $p\{LX\}$ are derived as follows:

$$Ly = R_{GLN_r L_y} \times R_{LxGLN_r} \times LX \quad (3)$$

$$p\{Ly\} = R^*_{GLN_r L_y} \times R^*_{LxGLN_r} \times p\{LX\} \quad (4)$$

4 Main Features of the Models

4.1 Models Based on Production Rules

The production rules of pseudo-stationary processes are changed by entering new coming packages of experimental data sets. However, the disadvantage of the production rules is that they may not be able to cover significant parts of the variation range of the inputs. This is why additional interpolation is required between similar production rules of MLPF [10, 13].

4.2 Model-Based on Network Structure

The advantage of these models is to develop approximation by a lack of a combination of inputs in the massive of input data. The network model could be implemented as a program or as a logical device. For pseudo-stationary subjects the coefficients of the relations between the elements of the three layers of the network must be updated in real time using data sets packages.

5 Network Model Application for an Industrial Process

The process considered here is flotation of multi-component ore. A package of experimental input/output data sets with 56 measurements is available for this process. According to the range of modification and the desirable accuracy of the model presentation, a seven-degree logic system for presentation of inputs and outputs is perceived. The logical values are named as *VVS*, *VS*, *S*, *M*, *L*, *VL* and *VVL*, where *VVS* is 'very very small', *VS* is 'very small', *S* is 'small', *M* is 'medium', *L* is 'large', *VL* is 'very large' and *VVL* is 'very very large'. Table 2 presents the range of modifications and the mean values which correspond to the adopted logic values for the input, intermediate and output layers. The measurable inputs for Lx_1, Lx_2 and Lx_3 are grouped in five dominant grouping sequence sets: GLN_1, GLN_2, GLN_3, GLN_4, GLN_5. The probability of occurrence for dominant grouping sequence sets $p\{GLN_r\}$, the sum of the relative values of the inputs of the dominant grouping sequence sets and the relations between the input and the output layers are given in Fig. 1. The coefficients (relations) between the elements of the input and the intermediate layers are given in Tables 3, 4 and 5. The coefficients between the elements of the intermediate and the output layer are given in Table 6.

Using the network model, the logical values Ly_1 for each given sequence set of logical input values Lx_1, Lx_2, Lx_3 are derived. For example if there are logical input values Lx_1 = *VVL*, Lx_2 = *VVL* and Lx_3 = M then a grouping sequence set GLN_2 is activated. This grouping sequence set GLN_2 activates the logical output values Ly_1 as follows: *VVS* with a probability of occurrence $p\{VVS\} = 0.083$, *VS* with a probability of occurrence $p\{VS\} = 0.167$, *S* with a probability of occurrence

Table 2 Relative values and their corresponding logic values

	VVS	VS	S	M	L	VL	VVL
Min	0.0000	0.1429	0.2857	0.4285	0.5713	0.7141	0.8570
Max	0.1428	0.2856	0.4284	0.5712	0.7140	0.8569	0.9997
Mean value	0.0714	0.2142	0.3570	0.4998	0.6426	0.7855	0.9283

Table 3 Relations between mean values for Lx_1 and dominant grouping sequence sets

GLNr	GLN_1	GLN_2	GLN_3	2.4993	GLN_5
Lx_1	2.6421	2.3564	2.4992	GLN_4	2.6421
0.9283	2.8461				
0.9283		2.5384			
0.9283			2.6922		
0.7855				3.1817	
0.7855					3.3635

Table 4 Relations between mean relative values for Lx_2 and dominant grouping sequence sets

GLNr	GLN_1	GLN_2	GLN_3	GLN_4	GLN_5
Lx_2	2.6421	2.3564	2.4992	2.4993	2.6421
0.9283	2.8461				
0.9283		2.5384			
0.9283			2.6922		
0.9283				2.6923	
0.9283					2.8461

Table 5 Relations between mean relative values for Lx_3 and dominant grouping sequence sets

GLNr	GLN_1	GLN_2	GLN_3	GLN_4	GLN_5
Lx_3	2.6421	2.3564	2.4992	2.4993	2.6421
0.7855	3.3635				
0.4998		4.7146			
0.6426			3.8892		
0.7855				3.1817	
0.9283					2.8461

Table 6 Relations of frequency of occurrences between dominant grouping sequence sets GLNr and the output logic values Ly_1

$p\{GLN_r\}$	GLNr	VVS	VS	S	M	L	VL	VVL
0.3455	GLN_1	0.053	0.158	0.105	0.000	0.105	0.474	0.105
0.2182	GLN_2	0.083	0.167	0.333	0.167	0.000	0.167	0.083
0.2000	GLN_3	0.364	0.091	0.000	0.182	0.000	0.091	0.273
0.1818	GLN_4	0.200	0.000	0.200	0.000	0.000	0.500	0.100
0.0545	GLN_5	0.000	0.000	0.667	0.000	0.000	0.333	0.000

$p\{S\} = 0.333$, M with a probability of occurrence $p\{M\} = 0.167$, VL with a probability of occurrence $p\{VL\} = 0.167$ and VVL with a probability of occurrence $p\{VVL\} = 0.083$. For the investigated sequence input data set the biggest probability of occurrence for the output Ly_1 is $p\{S\} = 0.333$ with a logical value S (small). This low frequency of occurrence of the dominant logical value $Ly_1 = S$ implies that except the main three influenced inputs Lx_1, Lx_2, Lx_3,, other factors influence the output Ly_1 with a significant impact on the flotation process.

The possibility for other logical values to occur over the output Ly_1 is based on the existence of other immeasurable inputs such as a stage of oxide of the copper ore, an existence of minerals and other impurities. The network model that is based on the existence of 56 experimental data sets could be extended with additional sequence sets of logic input values, which do not appear in the existing data package [10, 11, 13]. The weight coefficients between the three layers for absent sequence data set are derived using the data from the tables of the relative values corresponding to the logical values of the elements of the input, intermediate and output layers. For example if $Lx_1 = L = 0.6426$; $Lx_2 = VVL = 0.9283$; $Lx_3 = VL = 0.7855$, then their sum is 2.3564, i.e. this non occur data sequence set is close to the inputs of the grouping sequence set GLN_2. The frequency of occurrence for Ly_1 according to Table 6 is: $VVS = 0.083$; $VS = 0.164$; $S = 0.333$, $M = 0.167$; $VL = 0.167$; $VVL = 0.083$. This example of a network structure includes the inputs of the relative values between 2.3564 and 2.6421. This limited range of inputs is due to the small number of experimental data sets and their small range deviations. The network model allows a set of combinations of logical inputs to be added. For example, if the range of modification of inputs changes with one logical value Lx_1, L, VL, VVL where $Lx_2 = VL$, VVL, $Lx_3 = S$, M, L, VL, VVL, then the number of possible inputs increases significantly. The network model allows an extension of the combination of logic values of the inputs without additional computations based on interpolation.

6 Derivation of Additional Knowledge Based on MLPF

The following knowledge attributes of the process are discovered:

6.1 System of Production Rules for Control

Each column of MLPF value is one production rule. New production rules [10, 12] are added using interpolation.

6.2 Static Characteristic of Multi-factor Processes

The static characteristic of the process is presented graphically. In Fig. 2, this is implemented using relative values of the inputs and outputs. The sum of the relative values of the inputs $Lx_i, i = 1 \div n$ of the grouping sequence sets $SGLN_r$ is presented on the abscise. The relative output values $Ly_e, e = 1 \div m$ for the outputs of the process are presented on the ordinate. The interval of inputs is determined by the character of the static characteristics of each input. These characteristics could be increasing, decreasing or extreme.

An example of static characteristics of the grouping sequence set GLN_r is given in Fig. 2.

where *LNr*—set of logical input values
Ly_{eq} (y_e)—logical and relative output values
SLNr—sum of relative values of inputs in the set *LNr*.

6.3 Significance of the Environmental Influence Over the Output of the Process

Relations are derived between the relative values of each input from each grouping sequence set and the relative values of the outputs.

Fig. 2 Static characteristics of multi-factor process

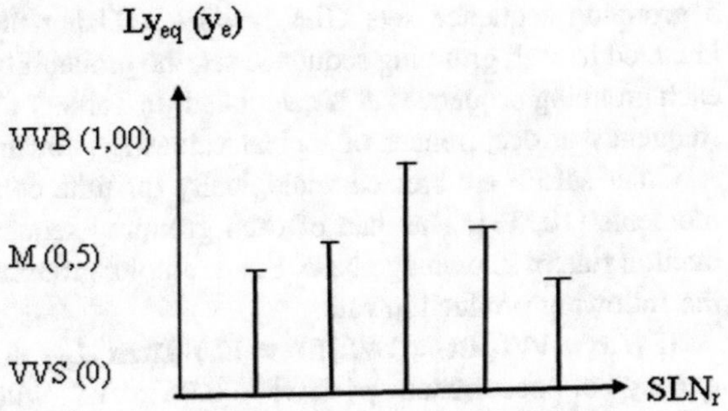

6.4 Influence of the Immeasurable Factors $W(t)$

Relative values of immeasurable factors $W(t)$ are derived in addition to the value of the sum of the relative values of the measurable inputs $X(t)$.

6.5 Assessment of the Time Delay Between the Inputs and the Outputs

The time delay is assessed using MLPF for different values of the time discrete moments τ between the experimental data for the inputs and the outputs [10].

6.6 A Reveal of the Existence of the New Period of Pseudo-Stationary

There are two ways to discover pseudo-stationarity

- Statistical methods for analysis of experimental data sets for each output: These methods are applied for a preliminary analysis of the experimental data sets with the purpose of choosing main parameters such as degree of the multi-valued logic, tsize of the data package, etc.
- Logical method based on time modifications of the relative values of a particular output with respect to the corresponding logic values of the inputs.

Example: The process with three inputs with logical values Lx_1, Lx_2, Lx_3 and three outputs Ly_1, Ly_2, Ly_3 is presented again. The experimental data sets are 56 and they are shown in Table 7. A seven-degree logic system is perceived where the inputs and the outputs have logical values: *VVS*, *VS*, *S*, *M*, *L*, *VL*, *VVL*, where *VVS* is 'very very small', *VS* is 'very small', *S* is 'small', *M* is 'medium', *L* is 'large', *VL* is 'very large', *VVL* is 'very very large' and the time delay is one discrete moment between the inputs and the outputs. The experimental input data sets are grouped in 5 grouping sequence sets $GLN_r, r = 1 \div 5$. Using the number of occurrences Z_r included in each grouping sequence set, the probability of occurrences $p\{GLN_r\}$ for each grouping sequence set is calculated. In Table 7 are also given the numbers and frequency of occurrences of logical values Ly_1 for the output y_1.

Other MLPF are created analogically for time delays $\tau = 2, 3, 4, 5$ time discrete moments [10, 13]. The data of each grouping sequence set of MLPF is one production rule of knowledge base. For example, grouping sequence set GLN_1 implies the following production rule:

If $\langle Lx_1 = VVL, Lx_2 = VVL, Lx_3 = VL \rangle$ **Then** Ly_1 has a value $\langle VVS \rangle$ with frequency of occurrence $p\{VVS\} = 0.053$; *VS* with frequency of occurrence $p\{VS\} = 0.158$; *S* with frequency of occurrence $p\{S\} = 0.105$; *M* with frequency of occurrence $p\{M\} = 0$; with frequency of occurrence $p\{L\} = 0.105$; *VL* with

Table 7 MLPF for a limited number of data sets Lx_1, Lx_2, Lx_3 and Ly_1

GLN_r		1	2	3	4	5
Z_r		19	12	11	10	3
p{GLN_r}		0.3455	0.2182	0.200	0.1818	0.0545
Lx_1		VVL	VVL	VVL	VL	VL
Lx_2		VVL	VVL	VVL	VVL	VVL
Lx_3		VL	M	L	VL	VVL
Ly_1	VVS	1	1	4	2	0
		0.053	0.083	0.364	0.2	0
	VS	3	2	1	0	0
		0.158	0.167	0.091	0	0
	S	2	4	0	2	2
		0.105	0.333	0	0.2	0.667
	M	0	2	2	0	0
		0	0.167	0.182	0	0
	L	2	0	0	0	0
		0.105	0	0	0	0
	VL	9	2	1	5	1
		0.474	0.167	0.091	0.5	0.333
	VVL	2	1	3	1	0
		0.105	0.083	0.273	0.1	0

frequency of occurrence $p\{VL\} = 0.474$; *VVL* with frequency of occurrence $p\{VVL\} = 0.105$. For each grouping sequence set, the significance of each input is determined from the frequency of occurrence of each logical value of the output y_1.

Table 7 implies that the input Lx_3 has maximal significance that is much higher than the significance of the inputs Lx_1 and Lx_2.

The degree to which the outputs are stochastic is assessed using the distribution of the frequency of occurrences of each output logic value.

The bigger the number of occurrences of logical values for a given output for different grouping sequence sets, the bigger the influence of the immeasurable factors.

The discovery of pseudo-stationarity is based on the derivation of the relations between the relative values of the inputs and the outputs. These relations should be similar to each other and within the series of discrete time moments in one pseudo-stationary time interval.

7 Conclusion

Two MLPF-based models for knowledge extraction from multi-factor, non-stationary, nonlinear complex processes are proposed. The model with updatable knowledge base is illustrated with real data sets for an industrial process from the mining industry. The difference between the two models is that the model with updatable knowledge base uses knowledge extraction in the form of production rule whereas the model with network structure uses a network whose elements can perform computational logical operations. The model with network structure is better for non-stationary processes than the model with updatable knowledge base because of its capability to interpolate new data.

The main strength of the proposed models is their suitability for a wide range of complex processes operating under uncertainty from different areas such as technology, the environment and others.

References

1. Lee, J. (ed.): Software Engineering with Computational Intelligence, Studies in Fuzziness and Soft Computing. Springer (2003)
2. Gray, J., Research, M., Han, J., Kamber, M.: Data Mining: Concepts and Techniques (The Morgan Kaufmann Series in Data Management Systems)", 2nd edn. Series Editors by Elsevier Inc. (2006)
3. Ruan, D., Chen, G., Kerre, E., West, G. (eds.): Intelligent Data Mining: Techniques and Applications (Studies in Computational Intelligence). Springer, Berlin, Heidelberg (2010)
4. Larose, D.: Data Mining Methods and Modles. A Wiley. New Jersey, Canada (2006)
5. Han, J., Kamber, M.: Data Mining Techniques. Morgan Kaufmann Publisher (2005)
6. Kandel, A., Last, M., Bunke, H.: Data Mining and Computational Intelligence. Physical-Verlag, Heidelberg (2001)
7. Kuznecov, V., Adelon-Velski, G.: Discrete mathematics for engineers. Moscow, Energoatomizdat (in Russian) (1998)
8. Lapa, V.: Mathematical bases of cybernetics. Kiev, Visha Shkola (1974) (in Russian)
9. Gotvald, S.: Multi-valued Logic. Introduction to Fuzzy Methods. Theory and Applications. Akademy–Ferlag (1989) (in German)
10. Vatchova, B.: Derivation and Assessment of Reliability of Knowledge for Multifactor Industrial Processes", PhD Thesis, 167 pages, Bulgarian Academy of Sciences, Sofia (2009) (in Bulgarian)
11. Gegov, E.A., Vatchova, B., Gegov, E.D.: Multi-valued Method for Knowledge Extraction and Updating in Real Time. IEEE'04, vol. 2, pp. 17-6–17-8. Varna, Bulgaria (2008)
12. Gegov, E., Vatchova, B.: Extraction of knowledge for complex objects from experimental data using functions of multi-valued logic. In: European Conference on Complex Systems '09, University of Warwick, Coventry, UK, 21–25 Sept 2009
13. Gegov, E.: Methods and Applications into Computer Intelligence and Information Technologies of Control Systems. Publisher "St. Ivan Rilsky", Sofia (2003) (in Bulgarian)

Printed in the United States
By Bookmasters

Printed in the United States
By Bookmasters